MTP International Review of Science

Biochemistry
Series One

Consultant Editors
**H. L. Kornberg, F.R.S. and
D. C. Phillips, F.R.S.**

Publisher's Note

The MTP International Review of Science is an important new venture in scientific publishing, which is presented by Butterworths in association with MTP Medical and Technical Publishing Co. Ltd. and University Park Press, Baltimore. The basic concept of the Review is to provide regular authoritative reviews of entire disciplines. Chemistry was taken first as the problems of literature survey are probably more acute in this subject than in any other. Physiology and Biochemistry followed naturally. As a matter of policy, the authorship of the MTP Review of Science is international and distinguished, the subject coverage is extensive, systematic and critical, and most important of all, it is intended that new issues of the Review will be published at regular intervals.

In the MTP Review of Chemistry (Series One), Inorganic, Physical and Organic Chemistry are comprehensively reviewed in 33 text volumes and 3 index volumes. Physiology (Series One) consists of 8 volumes and Biochemistry (Series One) 12 volumes, each volume individually indexed. Details follow. In general, the Chemistry (Series One) reviews cover the period 1967 to 1971, and Physiology and Biochemistry (Series One) reviews up to 1972. It is planned to start in 1974 the MTP International Review of Science (Series Two), consisting of a similar set of volumes covering developments in a two year period.

The MTP International Review of Science has been conceived within a carefully organised editorial framework. The overall plan was drawn up, and the volume editors appointed by seven consultant editors. In turn, each volume editor planned the coverage of his field and appointed authors to write on subjects which were within the area of their own research experience. No geographical restriction was imposed. Hence the 500 or so contributions to the MTP Review of Science come from many countries of the world and provide an authoritative account of progress.

Butterworth & Co. (Publishers) Ltd.

BIOCHEMISTRY SERIES ONE

Consultant Editors
H. L. Kornberg, F.R.S.
Department of Biochemistry University of Leicester and
D. C. Phillips, F.R.S., *Department of Zoology, University of Oxford*

Volume titles and Editors

1 **CHEMISTRY OF MACRO-MOLECULES**
Professor H. Gutfreund, *University of Bristol*

2 **BIOCHEMISTRY OF CELL WALLS AND MEMBRANES**
Dr. C. F. Fox, *University of California*

3 **ENERGY TRANSDUCING MECHANISMS**
Professor E. Racker, *Cornell University, New York*

4 **BIOCHEMISTRY OF LIPIDS**
Professor T. W. Goodwin, F.R.S., *University of Liverpool*

5 **BIOCHEMISTRY OF CARBO-HYDRATES**
Professor W. J. Whelan, *University of Miami*

6 **BIOCHEMISTRY OF NUCLEIC ACIDS**
Professor K. Burton, *University of Newcastle upon Tyne*

7 **SYNTHESIS OF AMINO ACIDS AND PROTEINS**
Professor H. R. V. Arnstein, *King's College, University of London*

8 **BIOCHEMISTRY OF HORMONES**
Dr. H. V. Rickenberg, *National Jewish Hospital & Research Center, Colorado*

9 **BIOCHEMISTRY OF CELL DIFFERENTIATION**
Dr. J. Paul, *The Beatson Institute for Cancer Research, Glasgow*

10 **DEFENCE AND RECOGNITION**
Professor R. R. Porter, F.R.S., *University of Oxford*

11 **PLANT BIOCHEMISTRY**
Dr. D. H. Northcote, F.R.S., *University of Cambridge*

12 **PHYSIOLOGICAL AND PHARMACO-LOGICAL BIOCHEMISTRY**
Dr. H. F. K. Blaschko, F.R.S., *University of Oxford*

PHYSIOLOGY SERIES ONE

Consultant Editors
A. C. Guyton,
Department of Physiology and Biophysics, University of Mississippi Medical Center and
D. Horrobin,
Department of Medical Physiology, University College of Nairobi

Volume titles and Editors

1 **CARDIOVASCULAR PHYSIOLOGY**
Professor A. C. Guyton and Dr. C. E. Jones, *University of Mississippi Medical Center*

2 **RESPIRATORY PHYSIOLOGY**
Professor J. G. Widdicombe, *St. George's Hospital, London*

3 **NEUROPHYSIOLOGY**
Professor C. C. Hunt, *Washington University School of Medicine, St. Louis*

4 **GASTROINTESTINAL PHYSIOLOGY**
Professor E. D. Jacobson and Dr. L. L. Shanbour, *University of Texas Medical School*

5 **ENDOCRINE PHYSIOLOGY**
Professor S. M. McCann, *University of Texas*

6 **KIDNEY AND URINARY TRACT PHYSIOLOGY**
Professor K. Thurau, *University of Munich*

7 **ENVIRONMENTAL PHYSIOLOGY**
Professor D. Robertshaw, *University of Nairobi*

8 **REPRODUCTIVE PHYSIOLOGY**
Professor R. O. Greep, *Harvard Medical School*

INORGANIC CHEMISTRY SERIES ONE

Consultant Editor
H. J. Eméleus, F.R.S.
Department of Chemistry
University of Cambridge

Volume titles and Editors

1. MAIN GROUP ELEMENTS—HYDROGEN AND GROUPS I–IV
Professor M. F. Lappert, University of Sussex

2. MAIN GROUP ELEMENTS—GROUPS V AND VI
Professor C. C. Addison, F.R.S. and Dr. D. B. Sowerby, University of Nottingham

3. MAIN GROUP ELEMENTS—GROUP VII AND NOBLE GASES
Professor Viktor Gutmann, Technical University of Vienna

4. ORGANOMETALLIC DERIVATIVES OF THE MAIN GROUP ELEMENTS
Dr. B. J. Aylett, Westfield College, University of London

5. TRANSITION METALS—PART 1
Professor D. W. A. Sharp, University of Glasgow

6. TRANSITION METALS—PART 2
Dr. M. J. Mays, University of Cambridge

7. LANTHANIDES AND ACTINIDES
Professor K. W. Bagnall, University of Manchester

8. RADIOCHEMISTRY
Dr. A. G. Maddock, University of Cambridge

9. REACTION MECHANISMS IN INORGANIC CHEMISTRY
Professor M. L. Tobe, University College, University of London

10. SOLID STATE CHEMISTRY
Dr. L. E. J. Roberts, Atomic Energy Research Establishment, Harwell

INDEX VOLUME

PHYSICAL CHEMISTRY SERIES ONE

Consultant Editor
A. D. Buckingham
Department of Chemistry
University of Cambridge

Volume titles and Editors

1. THEORETICAL CHEMISTRY
Professor W. Byers Brown, University of Manchester

2. MOLECULAR STRUCTURE AND PROPERTIES
Professor G. Allen, University of Manchester

3. SPECTROSCOPY
Dr. D. A. Ramsay, F.R.S.C., National Research Council of Canada

4. MAGNETIC RESONANCE
Professor C. A. McDowell, F.R.S.C., University of British Columbia

5. MASS SPECTROMETRY
Professor A. Maccoll, University College, University of London

6. ELECTROCHEMISTRY
Professor J. O'M Bockris, University of Pennsylvania

7. SURFACE CHEMISTRY AND COLLOIDS
Professor M. Kerker, Clarkson College of Technology, New York

8. MACROMOLECULAR SCIENCE
Professor C. E. H. Bawn, F.R.S., University of Liverpool

9. CHEMICAL KINETICS
Professor J. C. Polanyi, F.R.S., University of Toronto

10. THERMOCHEMISTRY AND THERMODYNAMICS
Dr. H. A. Skinner, University of Manchester

11. CHEMICAL CRYSTALLOGRAPHY
Professor J. Monteath Robertson, F.R.S., University of Glasgow

12. ANALYTICAL CHEMISTRY—PART 1
Professor T. S. West, Imperial College, University of London

13. ANALYTICAL CHEMISTRY—PART 2
Professor T. S. West, Imperial College, University of London

INDEX VOLUME

ORGANIC CHEMISTRY SERIES ONE

Consultant Editor
D. H. Hey, F.R.S.,
Department of Chemistry
King's College, University of London

Volume titles and Editors

1. STRUCTURE DETERMINATION IN ORGANIC CHEMISTRY
Professor W. D. Ollis, F.R.S., University of Sheffield

2. ALIPHATIC COMPOUNDS
Professor N. B. Chapman, Hull University

3. AROMATIC COMPOUNDS
Professor H. Zollinger, Swiss Federal Institute of Technology

4. HETEROCYCLIC COMPOUNDS
Dr. K. Schofield, University of Exeter

5. ALICYCLIC COMPOUNDS
Professor W. Parker, University of Stirling

6. AMINO ACIDS, PEPTIDES AND RELATED COMPOUNDS
Professor D. H. Hey, F.R.S., and Dr. D. I. John, King's College, University of London

7. CARBOHYDRATES
Professor G. O. Aspinall, Trent University, Ontario

8. STEROIDS
Dr. W. F. Johns, G. D. Searle & Co., Chicago

9. ALKALOIDS
Professor K. Wiesner, F.R.S., University of New Brunswick

10. FREE RADICAL REACTIONS
Professor W. A. Waters, F.R.S., University of Oxford

INDEX VOLUME

MTP International Review of Science

Biochemistry
Series One

Volume 4
Biochemistry of Lipids

Edited by **T. W. Goodwin, F.R.S.**
University of Liverpool

Butterworths · London
University Park Press · Baltimore

THE BUTTERWORTH GROUP

ENGLAND
Butterworth & Co (Publishers) Ltd
London: 88 Kingsway, WC2B 6AB

AUSTRALIA
Butterworths Pty Ltd
Sydney: 586 Pacific Highway 2067
Melbourne: 343 Little Collins Street, 3000
Brisbane: 240 Queen Street, 4000

NEW ZEALAND
Butterworths of New Zealand Ltd
Wellington: 26–28 Waring Taylor Street, 1

SOUTH AFRICA
Butterworth & Co (South Africa) (Pty) Ltd
Durban: 152–154 Gale Street

ISBN 0 408 70498 5

UNIVERSITY PARK PRESS

U.S.A. and CANADA
University Park Press
Chamber of Commerce Building
Baltimore, Maryland, 21202

> Library of Congress Cataloging in Publication Data
> Main entry under title.
> Biochemistry of lipids.
>
>   (Biochemistry, series one, v. 4)
>   1. Lipids. I. Goodwin, Trevor Walworth, ed.
> II. Series. [DNLM: 1. Lipids. QU85 B615 1974]
> QP1.B48 vol. 4 [QP751] 574.1′92′08s [574.1′9247]
>   ISBN 0–8391–1043–X   73–19646

First Published 1974 and © 1974
MTP MEDICAL AND TECHNICAL PUBLISHING CO LTD
St Leonard's House
St Leonardgate
Lancaster, Lancs
and
BUTTERWORTH & CO (PUBLISHERS) LTD

Typeset and printed in Great Britain by
REDWOOD BURN LIMITED
Trowbridge & Esher
and bound by R. J. Acford Ltd, Chichester, Sussex

Consultant Editors' Note

The MTP International Review of Science is designed to provide a comprehensive, critical and continuing survey of progress in research. Nowhere is such a survey needed as urgently as in those areas of knowledge that deal with the molecular aspects of biology. Both the volume of new information, and the pace at which it accrues, threaten to overwhelm the reader: it is becoming increasingly difficult for a practitioner of one branch of biochemistry to understand even the language used by specialists in another.

The present series of 12 volumes is intended to counteract this situation. It has been the aim of each Editor and the contributors to each volume not only to provide authoritative and up-to-date reviews but carefully to place these reviews into the context of existing knowledge, so that their significance to the overall advances in biochemical understanding can be understood also by advanced students and by non-specialist biochemists. It is particularly hoped that this series will benefit those colleagues to whom the whole range of scientific journals is not readily available. Inevitably, some of the information in these articles will already be out of date by the time these volumes appear: it is for that reason that further or revised volumes will be published as and when this is felt to be appropriate.

In order to give some kind of coherence to this series, we have viewed the description of biological processes in molecular terms as a progression from the properties of macromolecular cell components, through the functional interrelations of those components, to the manner in which cells, tissues and organisms respond biochemically to external changes. Although it is clear that many important topics have been ignored in a collection of articles chosen in this manner, we hope that the authority and distinction of the contributions will compensate for our shortcomings of thematic selection. We certainly welcome criticisms, and solicit suggestions for future reviews, from interested readers.

It is our pleasure to thank all who have collaborated to make this venture possible—the volume editors, the chapter authors, and the publishers.

Leicester   H. L. Kornberg

Oxford   D. C. Phillips

Preface

In the early years of the development of biochemistry, lipids were very much the Cinderella of cellular components in that they were frequently considered a nuisance and only contaminants of important components such as carbohydrates, amino acids and proteins. In the last decade or so the change in attitude has been impressive and biochemical studies on lipids have achieved a depth and sophistication undreamed of by early investigators. These developments have been due to a number of factors, chief of which are (a) highly developed techniques of identification, assay and structural determination, a *sine qua non* for all important biochemical advances; (b) advances in enzymology which allow non-aqueous substrates to be studied in detail and (c) continuously developing appreciation of the biological significance and function of lipids, for example as fat-soluble vitamins, hormones and structural components of membranes.

The chapters in this volume have been chosen to illustrate these various aspects of lipid biochemistry. Two specialised aspects of enzymology concern sterol and fatty acid biosynthesis. The techniques developed to study the final stages in sterol biosynthesis by insoluble enzymes are described in Chapter 1: the ways by which the problem of insoluble substrates reacting with particulate enzymes has been solved to a great extent, is described in detail. An equally impressive story is the elucidation (Chapter 3) of the structure of the multi-enzyme complex concerned with the formation of saturated fatty acids and the unravelling of the mechanism of control of this activity *in vivo*. Very closely related to this field is the problem of the biosynthesis of unsaturated fatty acids and this is considered in detail in Chapter 5. A number of diverse biosynthetic pathways has been revealed and their significance has been assessed critically as has the organisation of the desaturases involved and the stereochemistry of the reactions.

Perhaps one of the most unexpected but very significant concept of the function of lipids in general metabolism is discussed in Chapter 2. Here is described in detail the fascinating story of the discovery of the role of polyprenols (polyisoprenoids) as components of carbohydrate carriers in the formation of bacterial cell wall glycans and the ingenious extension of these studies of glycan formation in mammalian plant and bacterial systems.

In Chapter 4 we have a comprehensive review of the nature and metabolism of lipids in the nervous system and a discussion of their importance in the ageing process and in brain diseases. A similar comprehensive review of the nature, identification and biosynthesis of the fascinating prostaglandins is included in Chapter 6 which also deals in detail with pharmacological action and functions both real and alleged.

Finally in a chemically orientated chapter, the halogenated sulphatides are discussed. These compounds, which at the moment appear to be confined to the membrane of one algal genus *Ochromonas*, represent an extraordinary new class of compound which are sulphatides of chlorinated long chain alcohols: they can reasonably be visualised as being derived from oleic acid.

These seven chapters clearly reveal to some extent the tremendous field covered by lipid biochemistry and, written as they are by internationally known investigators in their chosen field, bring out the excitement and challenge of research in lipids biochemistry. I am most grateful to all the authors for the trouble they have taken in producing such attractive reviews.

Liverpool   T. W. Goodwin

Contents

Enzymes of sterol biosynthesis 1
J. L. Gaylor, *Cornell University*

Lipids in glycan biosynthesis 39
F. W. Hemming, *The University of Liverpool*

Biosynthesis of saturated fatty acids 99
P. R. Vagelos, *Washington University, St. Louis, Missouri*

The dynamic role of lipids in the nervous system 141
D. M. Bowen, A. N. Davidson and R. B. Ramsey, *The National Hospital, London*

The biosynthesis of unsaturated fatty acids 181
M. I. Gurr, *Unilever Limited, Sharnbrook, Bedford*

The prostaglandins 237
E. W. Horton, *University of Edinburgh*

The halogenated sulphatides 271
T. H. Haines, *City College of New York*

Index 287

1
Enzymes of Sterol Biosynthesis

J. L. GAYLOR
Cornell University

1.1	INTRODUCTION	2
	1.1.1 *Organisation of topics*	3
	1.1.2 *Past, present and future*	4
1.2	PATHWAY OF STEROL BIOSYNTHESIS	5
	1.2.1 *Sequence of reactions*	5
	1.2.1.1 Formation of mevalonic acid	6
	1.2.1.2 Transformations of mevalonic acid	7
	1.2.1.3 Squalene synthesis, squalene oxidation and cyclisations of squalene oxide	9
	1.2.1.4 Formation of product sterols from cyclised intermediates	9
	1.2.2 *Approaches used to investigate the sequence of enzymic steps*	13
	1.2.3 *Similarities and specificities of sterol biosynthetic enzymes*	15
	1.2.3.1 Qualitative presence or absence of enzymes	16
	1.2.3.2 Quantitative effect of the secondary process	16
	1.2.3.3 Similarities of enzymes in the primary process	19
1.3	METHODS OF INVESTIGATING PARTICLE-BOUND ENZYMES OF STEROL BIOSYNTHESIS	19
	1.3.1 *Investigation of particle-bound enzymes*	19
	1.3.1.1 Interruption of the multi-enzymic process	20
	(a) Drugs, metabolic inhibitors and other xenobiotics	20
	(b) Choice of incubation conditions	21
	(c) Action of enzyme on more than one sterol intermediate	21

1.3.2	Isolation of particle-bound enzymes	22
1.3.2.1	Direct extraction into buffer	22
1.3.2.2	Use of detergents	22
1.3.2.3	Partial proteolysis, lipolysis	23
1.3.3	Reconstitution of the multi-enzymic system	23
1.4	**CONTROL OF THE RATE OF STEROL BIOSYNTHESIS**	**24**
1.4.1	Control at the level of the component enzyme: β-hydroxy-β-methylglutaryl coenzyme A reductase	24
1.4.2	Control at the multi-enzymic level	26
1.4.3	Control at the genetic level	28
1.5	**RELATIONSHIP OF THE MULTI-ENZYMIC SYNTHESIS OF CHOLESTEROL TO OTHER MICROSOMAL PROCESSES**	**28**
1.5.1	Dynamic state of microsomal components	29
1.5.2	Microsomal mixed-function oxidases	29
1.5.3	Non-catalytic proteins of the microsomal multi-enzymic system	32

1.1 INTRODUCTION

The purpose of this article is to present a summary of several aspects of the enzymology of sterol biosynthesis. Knowledge of the principles of modern enzymology is assumed; but, to orient the reader, I have selected two important points that have been made by others about enzymology in general. First, to isolate an enzyme, you need a good source and a valid assay[*]. Secondly, 'Don't waste clean thoughts on dirty enzymes[†]'. Since the technology of purifying particle-bound enzymes, such as the microsomal enzymes of sterol biosynthesis, is only emerging at this time, each of us is guilty of trying to skirt the substantive messages, which are obvious in these statements. We rationalise that, for practical purposes, initial work may be carried out with partially purified enzymes; one assay may be suitable for study of an enzyme at all stages of purification; rat liver is a rich source of sterol biosynthetic enzymes, therefore an overwhelming fraction of work has been carried out with liver preparations, etc. Thus, in this summary we will examine some common practices for pitfalls, improvements, and, hopefully, durable unequivocal approaches. The bases of evidence will be examined, too, and illustrations will be used to the extent that unifying principles should emerge. Therefore, there really is a single purpose of this chapter; it is hoped that principles and points illustrated with specific examples will carry over into many areas of research and apply to investigation of enzymology with a variety of lipids and other natural products.

[*] Professor Marvin Johnson, Biochemistry Department, University of Wisconsin.
[†] Professor Efraim Racker, Section of Biochemistry and Molecular Biology, Cornell University.

Just as this work has application to analogous studies, we may draw upon efforts in related areas of experiment to develop our own rationale and methods. Indeed, the main point to be made in Section 1.2.2 (Approaches used to investigate the sequence of enzymic steps) was applied from Professor A. R. Battersby's eloquent analysis of approaches used to investigate alkaloid biosynthesis in a multi-enzymic system[1]. The reader is also directed to other related studies of biosynthesis of terpenoids and other acetogenins[2], such as in work on biosynthesis of gibberellins[3], monoterpenes[4], carotenoids[5,6], steroids[7], plant terpenoids[8], terpenoids of ecological significance[9] and alkaloids[10] (see also the compilation in Ref. 15 later).

1.1.1 Organisation of topics

Initially, the pathway of sterol biosynthesis will be described (Section 1.2). Details of the pathway will be used in illustrations, but points to be made will be illusory and not comprehensive. You will find that references will tend to be either to reviews, to recent reports or to articles that expand upon the points to be made rather than to document them. There are several *excellent* review articles on sterol biosynthesis that give details of experimental evidence, which is used to arrive at conclusions about the biosynthetic pathway*[11-21]. Furthermore, in the past two decades of work on the enzymology of sterol biosynthesis, since about 1950, the primary emphasis and the primary advancement has been in the understanding of the sequence of enzymic steps. Thus, review articles have tended to stress investigations of reactions in this sequence. Approaches used to study the pathways and the types of valid evidence have not been reviewed extensively, and these will be considered in Section 1.2.2. Particular stress will be placed on the enzymes that are bound to each other via subcellular membranes. Finally, consideration will be given to the striking similarities of the enzymes from various sources (Section 1.2.3).

Knowledge of pathways is absolutely basic, but it must be placed in a perspective of substantially larger goals. Although investigations of pathway bring the study to one logical conclusion at the point of solving the matter

---

* Several excellent review articles may be used for additional sources of reference materials. This review is not a 'review of reviews'; accordingly, the emphasis of the different review articles is given here for reference: three significant reviews appeared in 1965[11-13]; these articles tended to stress the chemistry of the compounds and the processes, but they are excellent summaries, albeit of differing lengths, of the work to that date. Subsequently, comprehensive reviews have been published[14,15], and these timely articles successfully brought information to a current status. Recent reviews have dealt with more restricted aspects of sterol biosynthesis: for example, Schroepfer et al.[16] reviewed the sequence of reactions in sterol transformations, and Fiecchi et al.[17] reviewed, particularly, the hydrogen exchange and elimination reactions of cholesterol biosynthesis. Summaries of the enzymology of sterol biosynthesis have appeared less frequently. Methods of investigation of enzymes of sterol biosynthesis were reviewed recently by Gaylor[18], use of inhibitors to study the multi-enzymic system of sterol biosynthesis was reviewed by Dempsey[19], and many features of the multi-enzymic nature of microsomal biosynthesis of sterols have been summarised[20]. Finally, methods of considerable value have been collected in a single volume of *Methods in Enzymology*[21]. Although this is not a review publication *per se*, it provides an excellent review in that details of methods are presented, and substantial descriptive narrative is incorporated into each method described.

of the sequence of enzymic transformations, several additional aspects of the enzymology of sterol biosynthesis emerged from these studies. Therefore, the next topic will be a more detailed résumé of methods used to study particle-bound enzymes of sterol transformations (Section 1.3.1). Considerable progress has been made in solving the technical problems of working with membranes and multi-enzymic systems[22] (see the entire volume of Ref. 22 for an analysis of multi-enzymic systems of significance to endocrinology, particularly steroid transformations). However, work with membranous systems is still quite new, and 'standard procedures' of isolation of particulate enzymes are not yet available (Section 1.3.2). Reconstitution studies are in initial stages of success (Section 1.3.3). Once again, the enzymologist who is studying sterol biosynthesis can benefit a great deal from efforts of others who have solved the technological problems with the use of other systems. For example, there is a direct analogy between membrane technology of micro-organisms[23], mitochondria[24] and microsomes.

Next, in Section 1.4, the control of the rate of sterol biosynthesis will be examined. The point is made that in a multi-enzymic system multiple loci of control may exist. Emerging knowledge of the role of sterols in membrane function and the membrane-bound enzymes of sterol biosynthesis allows us to start to relate these previously rather separate areas of study.

Finally, and in summary, the relationship of the multi-enzymic system of sterol biosynthesis will be related to other microsomal processes. There are very few, if any, enzymic processes of microsomes that fall exclusively in the domain of reactions of sterol biosynthesis. Each enzyme has its counterpart in other systems and processes, and each multi-enzymic system has its special relationship as it is integrated into other cellular processes and functions. For example, each of the mixed-function oxidases that exhibit marked stereospecificity in the attack of sterol substrates is part of an interwoven network of redox enzymes, which comprise the machinery for microsomal electron transport. Thus, in spite of the apparent specificity of each oxidase for its substrate, the terminal oxidase must interdigitate with electron carriers, which supply other microsomal oxidases. The mixed-function oxidases of microsomal sterol biosynthetic reactions may be viewed simply as multi-enzymic systems within the total multi-enzymic system.

1.1.2 Past, present and future

A separate discussion of the many ages and stages in the investigation of the enzymes of sterol biosynthesis hardly seems necessary. The unfolding drama of how we have arrived at the present stage with projections into the future is an exciting segment of modern science. Hopefully, the reader will recognise that there is a rich inheritance upon which we draw when we carry out experimental work. Furthermore, it should be equally obvious that much more work remains to be done in each area of investigation of enzymes of sterol biosynthesis. Clearly, the reader will recognise that an entire career could be devoted to almost any small part of this work. Therefore, in each sub-section of this review, future work will not necessarily be pointed out, but new investigations, approaches, and even unknown goals should become apparent as descriptions of current work are placed in perspective from the past to the future.

The point was made above that the technology of investigating the membranous enzymes of sterol biosynthesis has been evolved and adapted from investigations of other multi-enzymic systems. The reader is especially directed to methods that have been used to prepare sub-mitochondrial particles and soluble enzymes from mitochondria; these methods may be adapted easily for investigation of the microsomal enzymes of sterol biosynthesis. For example, most of the many procedures described in an entire volume of *Methods in Enzymology*[25] could be applied to related studies of microsomal enzymes. In addition to a direct analogy between these studies, investigation of mitochondrial electron transport and other mitochondrial enzymes has advanced somewhat more rapidly than have investigations of microsomal electron transport and other microsomal multi-enzymic systems. The reader is finally directed to an excellent comprehensive analysis of reconstitution of biological membranes, which was written recently by Razin[26].

In this introduction and in the remainder of this chapter, the reader should recognise that most of these investigations can be carried out under conditions that are either appropriate to most working situations or at least easily adapted. On the one hand, experiments may be performed with various organisms even in relatively modest field laboratory situations. For example, a refrigerated, preparative ultracentrifuge is not essential; microsomes may be prepared under appropriate conditions by precipitations at slower rotor speeds, such as those achieved in clinical centrifuges[27, 28]. Therefore, do not eliminate certain studies because equipment that others have used is not readily available; look for alternatives. On the other hand, investigations of membrane-bound enzymes and thorough study of membranes and fragments containing lipids and proteins have become highly sophisticated with the use of expensive modern equipment. Innovations in electron microscopy and x-ray analyses permit simultaneous detailed studies of membrane morphology and activities of membrane-bound enzymes[29].

1.2 PATHWAY OF STEROL BIOSYNTHESIS

1.2.1 Sequence of reactions

To put all readers in touch with the sequence of steps in sterol biosynthesis and to orient them for the further discussion of the enzymology of the process, this major section is introduced by a skeletal description of the sequence of transformations, structures and minimal belabourings of evidence, some of which will be considered later (see Sections 1.2.2 and 1.3). Thus, instead of cramming facts and reasoning into small spaces, as mentioned above please consider using comprehensive review articles as your next source of additional information, then turn to the original reports. The choice of minor subheadings (from Section 1.2.1.1 to 1.2.1.4) was made to distinguish: (a) reactions that are common to many biosynthetic pathways (e.g. Sections 1.2.1.1 and 1.2.1.2) from those that are related more specifically to sterol formation; and (b) enzymes that are found in the cytosol of cell-free preparations from those enzymes that are membrane-bound (e.g. Sections 1.2.1.3 and 1.2.1.4, and, in addition, β-hydroxy-β-methylglutaryl coenzyme A reductase).

1.2.1.1 Formation of mevalonic acid

Acetoacetyl coenzyme A is formed by self-condensation of acetyl coenzyme A. An equimolar amount of unesterified coenzyme A is released in the condensation reaction (Figure 1.1). The acetoacetyl coenzyme A thus formed

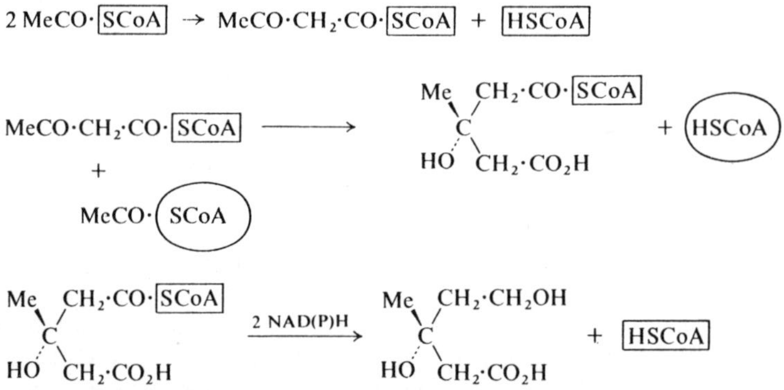

$$2 \text{ MeCO·SCoA} \rightarrow \text{MeCO·CH}_2\text{·CO·SCoA} + \text{HSCoA}$$

$$\text{MeCO·CH}_2\text{·CO·SCoA} + \text{MeCO·SCoA} \longrightarrow \underset{\text{HO}}{\overset{\text{Me}}{\diagdown}}\underset{\text{CH}_2\text{·CO}_2\text{H}}{\overset{\text{CH}_2\text{·CO·SCoA}}{\diagup}} + \text{HSCoA}$$

$$\underset{\text{HO}}{\overset{\text{Me}}{\diagdown}}\underset{\text{CH}_2\text{·CO}_2\text{H}}{\overset{\text{CH}_2\text{·CO·SCoA}}{\diagup}} \xrightarrow{2 \text{ NAD(P)H}} \underset{\text{HO}}{\overset{\text{Me}}{\diagdown}}\underset{\text{CH}_2\text{·CO}_2\text{H}}{\overset{\text{CH}_2\text{·CH}_2\text{OH}}{\diagup}} + \text{HSCoA}$$

Figure 1.1 Enzymic synthesis of mevalonic acid from acetyl coenzyme A. The identities of the two coenzyme A groups are indicated in the condensation that leads to hydroxymethylglutaryl coenzyme A formation. Thus, the coenzyme A molecule lost in condensation and reduction arises from acetyl coenzyme A and acetoacetyl coenzyme A, respectively. These relationships are essential to avoid confusion of the origin of each atom in the resulting mevalonic acid

condenses in the presence of β-hydroxy-β-methylglutaryl coenzyme A synthetase (EC 4.1.3.5, also called condensing enzyme) to yield β-hydroxy-β-methylglutaryl coenzyme A, as shown in Figure 1.1. After purification of the enzyme from yeast[30], Rudney and co-workers[31, 32] demonstrated unequivocally that the thioester bond of acetoacetyl coenzyme A remains unbroken during condensation and the unesterified coenzyme A released is derived from cleavage of the thioester bond of acetyl coenzyme A (see Figure 1.1). The same group of workers further demonstrated that acetoacetyl acyl carrier protein could be substituted for acetoacetyl coenzyme A, but the resulting rate of condensation dropped by *ca.* 85%. More recently, Middleton and Tubbs[33] reported an improved procedure for purification of the synthetase from yeast. Perhaps further work on the properties of this interesting enzyme will be carried out, since it is now more readily isolated.

Since the discovery of mevalonic acid as an acetate-replacing growth factor for *Lactobacillus acidophilus*[34], the compound has been shown to occupy a central position in the biosynthesis of sterols and other terpenoids[35].

STEROL BIOSYNTHESIS

This condition obtains because the enzymic reduction of β-hydroxy-β-methylglutaryl coenzyme A to mevalonic acid is essentially irreversible, and once formed, mevalonic acid is metabolised into terpenoids. Thus, this enzymic process is central to the regulation of sterol biosynthesis; properties of the enzyme will be discussed later in Section 1.4.1 when regulation of the biosynthetic rate is discussed.

The β-hydroxy-β-methylglutaryl coenzyme A reductase has been isolated and purified from both microbial sources[36, 37] and rat liver microsomes[38, 39] (Figure 1.1). The mammalian enzyme is tightly bound to the microsomal membranes, and special methods (see Section 1.3.1) were needed to liberate the enzyme in soluble form. The reduction occurs at the thioester linkage; reduction is thought to be stepwise and to proceed via an enzyme-bound intermediate.

1.2.1.2 Transformations of mevalonic acid

In the presence of ATP and appropriate phosphokinases, mevalonic acid is converted first to a 5-phosphate ester; then the phosphoester is converted into an ester of pyrophosphate (Figure 1.2)[35]. The pyrophosphorylated intermediate is phosphorylated again, only on the 3-hydroxyl group this time, to

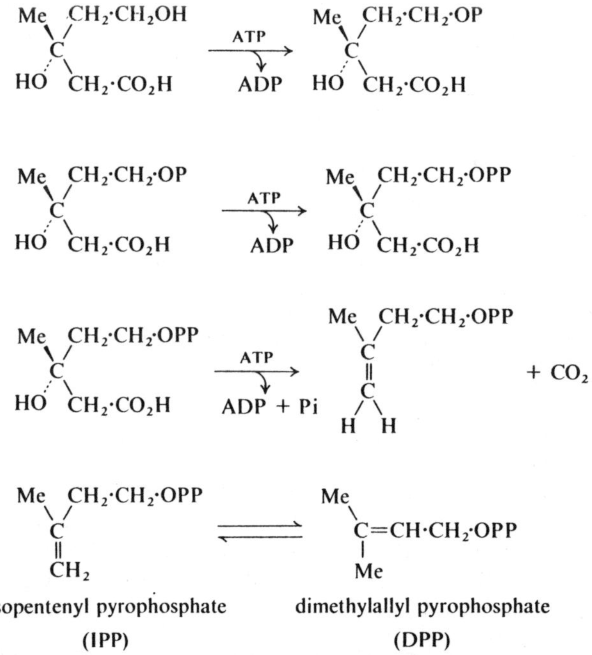

Figure 1.2 Formation of biosynthetic isoprene units, isopentenyl pyrophosphate and dimethylallyl pyrophosphate. Each phosphokinase and the isomerase has been isolated and purified. P = phosphate; PP == pyrophosphate esters

yield an unstable intermediate, which is converted to the pyrophosphate ester of isopentenol. The latter reaction involves the *trans*-elimination of the carboxyl and 3-hydroxyl (or phosphate) groups (Figure 1.2).

The resulting isopentenyl pyrophosphate may be isomerised to 3,3-dimethylallyl pyrophosphate in the presence of an isomerase[40]. The resulting pair of isomers constitute the biosynthetic isoprene units, which are utilised as building blocks for further transformation into terpenoids[11].

In the presence of condensing enzymes (prenyl transferases), geranyl pyrophosphate is formed by coupling of isopentenyl pyrophosphate to the

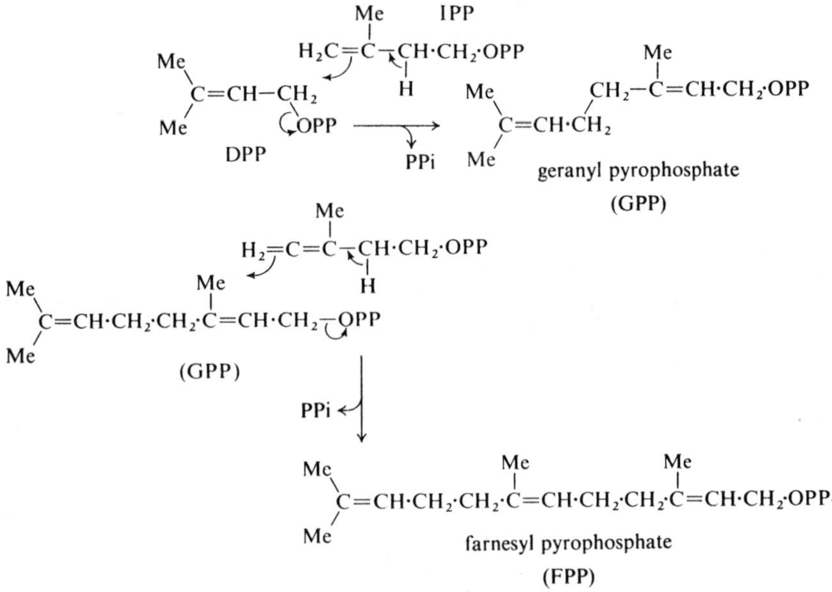

Figure 1.3 Condensation reactions of terpene biosynthesis. Facile equilibrium between isopentenyl pyrophosphate and dimethylallyl pyrophosphate exists. Condensation is between the exomethylene group of isopentenyl pyrophosphate and the growing chain of pyrophosphorylated intermediate. Each enzyme has been isolated, purified and characterised. PP_i = inorganic pyrophosphate

acceptor, dimethylallyl pyrophosphate (Figure 1.3). The geranyl pyrophosphate is further converted into farnesyl pyrophosphate by reaction with another isopentenyl pyrophosphate as shown in Figure 1.3. Particularly through the pioneering work of Cornforth, Popják and co-workers[35, 41], the stereochemistry of these processes has been elucidated, and it is on this basis that a considerable advance has been made into better understanding of the nature of enzymic reactions which occur much later in the process (see Section 1.2.2).

With the formation of farnesyl pyrophosphate, the process of sterol biosynthesis arrives at another branch-point. With the exception of β-hydroxy-β-methylglutaryl coenzyme A reductase, these enzymes are contained in the cytosol, or soluble portion of broken-cell preparations of liver. The enzymes that catalyse the remaining transformations are bound to microsomes, which are derived from the endoplasmic reticulum of the cell. All the enzymes required for the conversion of farnesyl pyrophosphate to cholesterol, with the possible exception of a soluble sterol carrier protein[42], are present in carefully prepared microsomes, and the enzymes are attached to the endoplasmic reticulum inside the cell[43].

1.2.1.3 Squalene synthesis, squalene oxidation and cyclisations of squalene oxide

The elucidation of the mechanisms of squalene synthesis and metabolism is clearly a remarkable scientific accomplishment. Because these reactions have been summarised in detail in review articles cited previously[7,9,10], and in articles not hitherto cited[44,45], these reactions will be only described here and purification of the enzymes will be discussed in detail later (Section 1.3.1).

Succinctly, the conversion of farnesyl pyrophosphate into squalene is catalysed by squalene synthetase, which is microsomal or bound to other membrane particles in cell-free preparations of both liver[46] and yeast[47]. The reaction (Figure 1.4) consists of a dimeric coupling, which occurs via an intermediate, presqualene pyrophosphate[46], which has been synthesised by Rilling and co-workers[48]. Further experimental evidence has been consistent with the proposal of Rilling that the catalytic process of squalene synthetase occurs in two discrete steps[49]. In further work from the same laboratory, the analogous formation of a prephytoene pyrophosphate in carotenoid biosynthesis has been demonstrated[50].

Squalene epoxidase catalyses the mixed-function oxidations of squalene into squalene oxide[44]. The mono-oxygenase requires a heat-labile protein of cytosol, a phospholipid and NADPH, in addition to oxygen (Figure 1.4)[51,52]. Although the soluble protein factor has been purified, the microsomal-bound epoxidase has not been obtained in soluble form.

Unlike squalene oxidase, which has not been purified, the squalene-2,3-oxide sterol cyclase has been solubilised from microsomal particles and partially purified[44,45]. The enzyme is isolated from the soluble fraction of yeast cell homogenates[53]. A similar squalene-2,3-oxide cycloartenol cyclase has been isolated from broken-cell preparations of *Ochromonas malhamensis* by deoxycholate solubilisation[54]. Cycloartenol appears to take the place of lanosterol as the first stable product of cyclisation in plant sterol biosynthesis. At the present time, there is neither unequivocal evidence for nor against concerted cyclisation and rearrangement[9].

1.2.1.4 Formation of product sterols from cyclised intermediates

The several reactions that must occur in this terminal part of sterol biosynthesis are summarised in Figure 1.5 for the steps involved in the conversion of lanosterol into cholesterol. Excellent summaries[10-16] of these reactions have been recorded in review articles and repetition of details is unnecessary here. Each of the three 'extra' methyl groups (in positions

4α, 4β and 14α) is oxidised to carbon dioxide as was first shown by Olson et al.[55]. The double bond in the side chain is reduced; the $\Delta^{8(9)}$-double bond is lost by isomerisation[56] and reduction; and a Δ^5-double bond is introduced[57]. The comprehensive summary of these reactions by Goad[15] is particularly noteworthy.

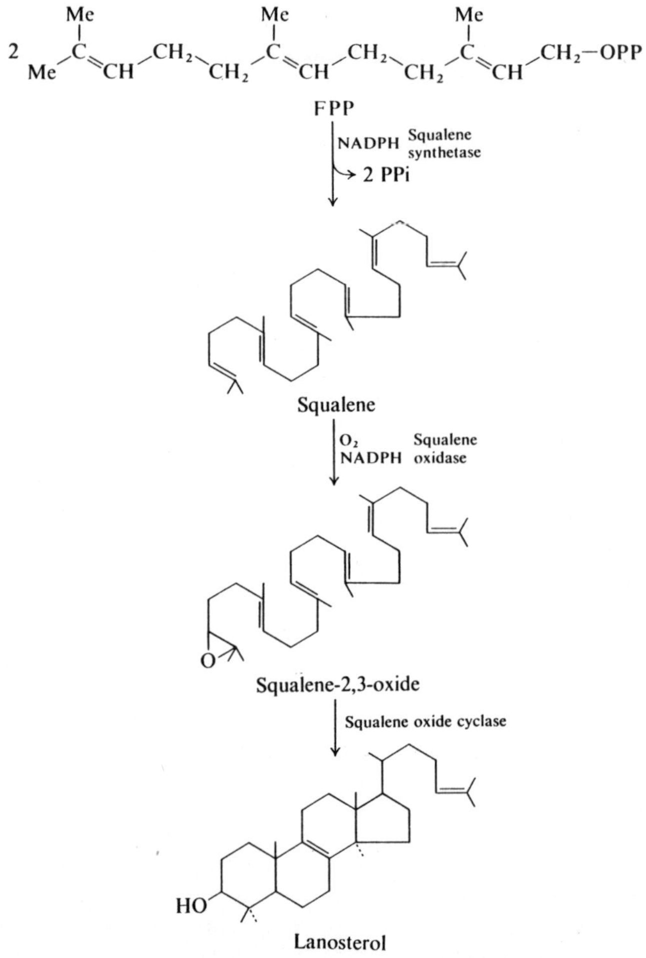

Figure 1.4 Formation and metabolism of squalene. Reductive condensation of two molecules of farnesyl pyrophosphate is catalysed by squalene synthetase. Oxidation of squalene is catalysed by squalene oxidase. Although a single arrow is indicated for the synthetase reaction, two steps have been delineated; in the oxidase, electron transport from NADPH to oxygen and substrate likely requires a multi-enzymic system. In contrast, the third reaction shown here as catalysed by the cyclase may be one of the most remarkable multivalent transformations in biochemistry, which has not, as yet, been dissected into individual, discrete steps

STEROL BIOSYNTHESIS

Much of the recent work in the author's laboratory has dealt with the process of enzymic demethylation of the 4-*gem*-dimethyl groups. Because frequent reference to this process will be made in later sections of this chapter, the reactions of oxidative demethylation are shown in Figure 1.6. Methyl sterol oxidase catalyses the mixed-function oxidative attack of the 4α-methyl group of 4,4-dimethyl-5α-cholest-7-en-3β-ol, a model substrate, to yield the 4α-carboxylic acid[58]. Oxidative attack is stereospecific[59]. The enzymic reaction is thought to be stepwise[60], from methyl→hydroxymethyl→formyl, and the same enzymic system appears to catalyse each step in the attack of both the 4-methyl and 4,4-dimethyl substrates. Thus, the mixed-function oxidative system is greatly simplified by employing a single complex oxidase in place of six separate multi-enzymic units. The resulting 4α-carboxylic acid is decarboxylated by an NAD^+-dependent oxidoreductase that yields the 3-ketosteroid

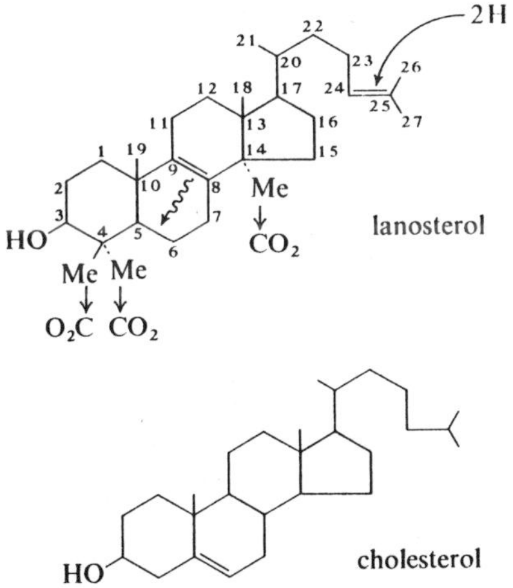

Figure 1.5 Conversion of lanosterol into cholesterol. This is drawn as a summary to illustrate the nature of the three types of conversion that occur in this transformation. Neither mechanism nor sequence of reactions should be implied from this drawing

product[61]. The 3-ketosteroid product must be reduced to a 3β-alcohol before sequential attack by methyl sterol oxidase[62]. As pointed out above, the resulting 4-monomethyl product is attacked by methyl sterol oxidase. The carboxylic acid and 3-ketosteroid intermediates of the removal of the second methyl group appear to be metabolised by the same enzymes that catalyse attack of the 4,4-disubstituted intermediates. The oxidase[63], decarboxylase[61] and reductase[18] have been isolated from liver microsomes and partially purified. Details of the enzyme isolation and purification will be described later (Section 1.3.1).

Thus, in a brief outline form, the enzymic transformations in sterol biosynthesis are summarised in Figures 1.1 to 1.6. In addition, transformations that impart distinguishing features will be considered later (Section 1.2.3) when the discussion will contrast biosynthesis of sterols in plants, animals and fungi. Because work on the soluble enzymes of the sterol biosynthetic pathway was carried out much earlier and adequate summaries of properties of the soluble enzymes are found in other reviews, the remainder of this chapter will deal with the insoluble, membrane-bound enzymes (Sections 1.2.1.3 and 1.2.1.4).

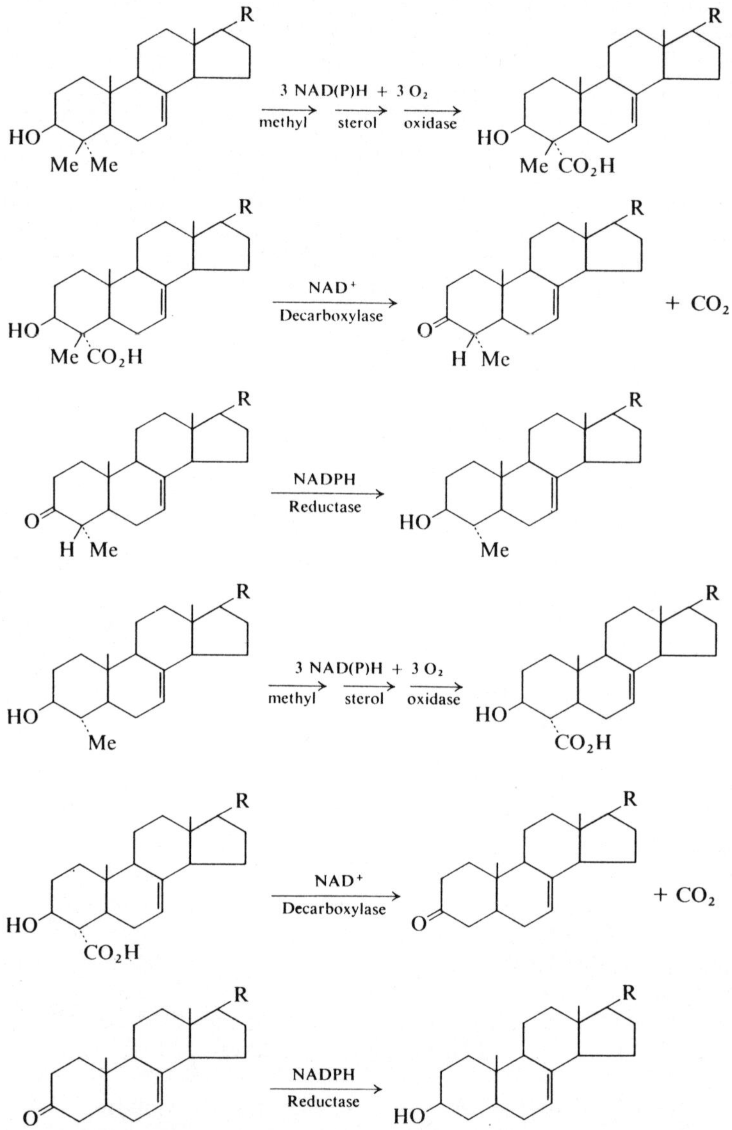

Figure 1.6 Multi-enzymic oxidative demethylation of the 4-*gem*-dimethyl group. The nature of the enzymic reaction is indicated in each case by demonstrated co-factor requirements. The trivial names have been selected accordingly. Stoichiometry is indicated where it is known

1.2.2 Approaches used to investigate the sequence of enzymic steps

The general approach to the investigation of a multi-enzymic sequence of reactions may be clearly distinguished from the actual methods that are used (see Section 1.3.1)[1]. There are two approaches that have been used in the investigation of the insoluble enzymes of sterol biosynthesis: the *organic* and the *enzymic* approaches[18]. Briefly, in the organic approach, proposed intermediates are synthesised and introduced into the multi-enzymic system; frequently, the intact organism is the source of the multi-enzymic system. The substance undergoes transformation to the end product, which is isolated. Frequently, the intermediate contains labelled atoms, and, by appropriate degradation methods, the origin of each atom in the product compound may be determined. In addition to the elucidation of the origin of each atom in the sought-for product, the organic approach is used first to establish the broad outlines of the sequence of reactions and to give the first indication about the nature of enzymic transformation. The organic approach can be carried to a very high degree of sophistication prior to adoption of the enzymic approach. In the investigation of the insoluble enzymes of sterol biosynthesis, the organic approach was used almost exclusively until recently[18, 20]. One remarkable collection of illustrations of the potentially high degree of sophistication of the organic approach was published recently by Witkop[64].

The organic approach is limited by the nature of the investigation to the extent that, in spite of the degree of sophistication, it cannot be used exclusively to select between alternate permissive and obligatory pathways; furthermore, the organic approach cannot be used exclusively to unravel unequivocal evidence about the nature of the individual catalytic events. Finally, the enzymic approach must be used to study fully the multi-enzymic nature of the process; for example, the investigation of control of the overall enzymic rate in the multi-enzymic system may be dependent upon the fact that the system is, in fact, multi-enzymic and membrane-bound.

Contrast of the organic and enzymic approaches is depicted in general terms in Figure 1.7. In the organic approach, compounds A, B, C, etc. are introduced into the multi-enzymic system, and conversions are catalysed by enzymes ab, bc, cd, etc. With accumulation of substantial amounts of evidence, such as subtle changes in structure of compounds, precursor–product studies with labelled compounds, etc. the outline of the biosynthetic pathway may be written. In the enzymic approach, single steps or only limited sequences of steps are allowed to occur; that is, although various intermediates may be introduced, as in the organic approach, the sequence of conversions is interrupted. If transformation D→E is to be studied, conversion E→F is prevented by one of several means. For example, when we were investigating the decarboxylase of methyl sterol demethylation (Figure 1.6), *sequential* conversion of the 3-ketosteroid product was prevented by preparing microsomes free of endogenous pyridine nucleotides and adding only NAD^+ to the microsomal suspension. *Concomitant* conversion of substrate D is indicated by the vertical arrow in Figure 1.7. That is, there must be prevention of side reactions that may be catalysed by other enzymes in the multi-enzymic system, which may also act on the same substrate.

Owing to low substrate specificities of many of the enzymes and the necessarily close juxtaposition of enzymes and intermediates in the multi-enzymic system, concomitant reactions are likely to occur. In the example of decarboxylase, substrate with a Δ^7-double bond was used to study the decarboxylase because, if the Δ^8-isomer had been used, $\Delta^8 \rightarrow \Delta^7$ isomerisation[56] would have occurred concomitantly since the isomerase does not require either co-factors or oxygen. In addition to elimination of essential co-factor and the appropriate choice of a substrate, unwanted sequential and concomitant reactions may be prevented by choice of anaerobic or aerobic conditions[65],

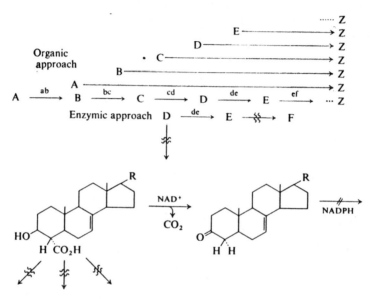

Figure 1.7 Contrast of the organic and enzymic approaches. Compounds are indicated as capital letters, and enzymes are indicated by blends of appropriate lower-case letters. Interrupted reactions are shown as cross-hatched arrows, ⟶

use of selective inhibitors[66], removal of the unwanted enzyme[67], etc. Examples of each of these techniques are given in the reference cited from work in this laboratory on enzymes of sterol biosynthesis. The point should not need further elaboration.

The enzymic approach is divided into five distinct stages (Table 1.1): (1) The existence of enzyme must be demonstrated. Preliminary evidence about the enzymic process is obtained in this investigation. (2) Next, it is necessary to investigate the enzyme bound to the multi-enzymic system to establish properties of the enzyme, which are characteristic of those features in its bound form. The enzyme must be isolated (3) free of other enzymes of the multi-enzymic system and purified extensively. Properties (4) of the isolated enzyme are investigated, and characteristics are compared with properties of bound enzyme studied in stage 2. Finally, it is essential to reconstitute (5) the multi-enzymic system[20]. Reconstitution experiments are needed to demonstrate the unaltered nature of the isolated enzyme. When properly carried out, the data obtained in these stages constitute the body of evidence in support of choices between proposed sequences of reactions. These stages will be illustrated in Section 1.3.1 below.

Table 1.1 Five stages of investigation in enzymic approach

Stage	Investigation
1	Existence of enzyme is demonstrated
2	Properties of the bound enzyme are studied
3	Enzyme is isolated from the multi-enzymic system
4	Properties of the isolated enzyme are studied
5	Segments of the multi-enzymic system are reconstituted

Schroepfer and co-workers[16] analysed similarly the evidence needed to support proposed sequences of reactions, which are based on the organic approach: '(a) Isolation from tissues in pure form and unequivocal establishment of structure; (b) Demonstration of the enzymic formation from a known precursor (acetate, mevalonate, squalene, etc.); (c) Demonstration of convertibility to cholesterol; (d) Demonstration of enzymic formation from postulated immediate precursor; (e) Demonstration of enzymic conversion to the next postulated intermediate'*.

In spite of the attempt to develop the concept of two approaches, it should be obvious that the approaches are interdependent. This relationship is illustrated by the similarities of some of the criteria for establishing the precursor–product relationship; these criteria were developed quite independently for the enzymic[20] and the organic[16] approaches. Furthermore, it should be obvious that the organic approach is used first. Hopefully, this analysis of approach and the criteria of evidence will be one of the durable contributions of this chapter.

1.2.3 Similarities and specificities of sterol biosynthetic enzymes

Before turning away from the topic of the study of sequence of reactions, some discussion of enzymic transformations that impart distinguishing features into the sterol products is needed. For example, plant and fungal sterols are differentiated by the presence of substituent groups on C-24. For some time, the differentiation reactions were shrouded in mystery. Simply, differentiation of sterol structure results from either the qualitative presence or absence of enzyme(s), quantitative amount of enzyme(s) or substrate specificity of the enzyme(s). Thus, there really is no mystery, but results from the enzymic approach have been needed and somewhat overdue to solve this part of sequence studies.

* Published with the kind permission of G. J. Schroepfer, Jr., Department of Biochemistry, Rice University. By courtesy of the Royal Society.

1.2.3.1 Qualitative presence or absence of enzymes

The effect of the action of more than one enzyme on a single substrate within a biosynthetic process is easily visualised if you think of the strong flux of *de novo* synthetic reactions as a vector, called the 'primary process', in one direction and the effect of the differentiating enzyme as another vector, at 90 degrees, called the 'secondary process' (Figure 1.8). The simple qualitative presence of differentiating enzymes ww' or uu'' *and* the generally low substrate specificity exhibited by the biosynthetic enzymes in the primary process, wx, wy, yz, etc., accommodate marked differentiation with the addition of only single enzymes. Thus, whole 'families' of analogous compounds are generated.

There is an excellent example in the literature on sterol biosynthesis: the differentiation of plant, animal, and yeast sterols, as shown in Figure 1.8a. For this analysis, consider reactions T→→Z as the primary process, which is the set of terminal reactions in the conversion of lanosterol (cycloartenol) to principal end products, i.e. ergosterol, cholesterol and sitosterol in yeast, animals and plants, respectively (Figure 1.8b). Thus, there is addition of the C_2 side-chain in plant sterol biosynthesis, and addition may occur before complete demethylation and nuclear transformation of the Δ^8-isomer into the Δ^5-isomer[6, 15]. In the yeast system, the methyltransferase exhibits pronounced selectivity for the demethylated intermediate, but side-chain methylation occurs prior to $\Delta^8 \to \Delta^7$ isomerisation and introduction of the Δ^5-double bond[68, 69]. Once again, the conditions required for simultaneous operation of primary and secondary processes are: (a) the presence of the differentiating enzyme(s); and (b) the low specificity of the enzymes of the primary process for the differentiated (i.e. W', X', Y'; U'', V'', W'', X'', Y'') intermediates.

1.2.3.2 Quantitative effect of the secondary process

If you refer to Figure 1.8 again, and simply consider the effect of quantitative differences that may exist between the primary and secondary processes, then you have the model of the quantitative effect. There are two obvious examples from animal sterol biosynthesis, which are reflected in (a) the relative abundance of Δ^{24}-sterols versus side-chain saturated sterols in some tissues, and (b) the abundance of Δ^7-isomers of biosynthetic intermediates in mammalian skin[70].

The unnecessary belabouring of differentiating between two 'pathways' of sterol biosynthesis (i.e. Δ^{24}-unsaturated versus side-chain saturated intermediates) was pointed out well in the review by Frantz and Schroepfer[14], Experimental evidence favouring one view over another may be obtained; however, the matter is reasonably simple when analysed as shown in Figure 1.8. The Δ^{24}-sterol reductase seems to be relatively non-specific in that it will reduce the side-chain of lanosterol, as well as partly demethylated intermediates, in addition to C_{27}-sterol products[71]. Thus, in the scheme shown in Figure 1.8, the reductase is present and active throughout the sequence of conversions but in different tissues, organisms, under different incubation conditions, etc. there may be a greater or lesser contribution of the reductase

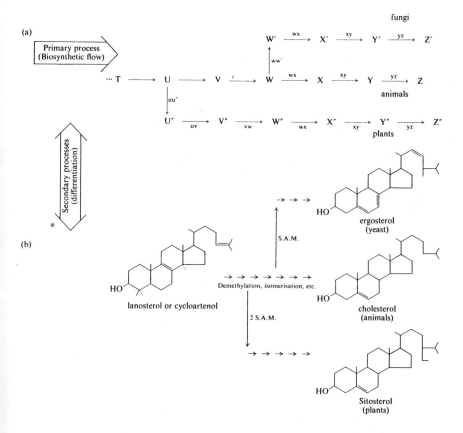

Figure 1.8 Illustration of the effect of differentiating enzymes of sterol biosynthesis. This primary process, left to right, is the flow of biosynthetic reactions that we described in Section 1.2.1. The effect of the secondary enzymes acting in concert (but at right angles) transforms intermediates from one type to another, which is characteristic of the organism in which the biosynthesis is occurring. This is illustrated in specific terms in the lower part of the figure b, in which side chain transformations as shown are the secondary process of nuclear transformations of sterols

(secondary process enzyme) when this rate is compared with the rate of the primary process.

The example from skin sterol biosynthesis is somewhat more complex because it appears to require two differentiating enzymes acting in concert; one is qualitative (as in Section 1.2.3.1), and the other is quantitative (as in this subsection). Mammalian skin, particularly rat skin, contains a number of sterols that possess a Δ^7-double bond in the nucleus. Thus, a $\Delta^{7,24}$-isomer of lanosterol has been identified[72] and evidence for the entire array of products of demethylation of the Δ^7-isomeric series has been obtained[70]. Since the mammalian $\Delta^8 \rightarrow \Delta^7$-isomerase is not operative with 14a-methyl-substituted sterols[56], Hornby and Boyd[73] reasoned that the $\Delta^{7,24}$-lanostadienol (Figure 1.9) could be formed directly by an alternative cyclisation of squalene oxide. Results from a relatively uncomplicated double-label experiment were consistent with the hypothesis that the isomeric C_{30}-sterol may be formed from squalene oxide in skin. Thus, Hornby and Boyd really demonstrated a

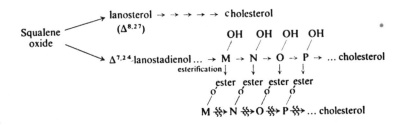

Figure 1.9 Illustration of the simultaneous effects of qualitative and quantitative differences in the enzymes of the secondary process of rat skin sterol biosynthesis. Once again, the primary process is shown, left to right, and the secondary transformations are shown as vertical arrows

qualitative difference (i.e. as in Section 1.2.3.1) of skin sterol biosynthesis, as shown as the cyclisation step in Figure 1.9. However, these observations only account for the qualitative presence of the Δ^7-isomers. Quantitatively, these compounds are very abundant in skin. As first shown by Wells and Baumann[74] and later verified by Frantz and co-workers[75], the Δ^7-isomeric compounds of skin are predominantly esterified. Thus, in a recent investigation of the relative rate of esterification of sterol intermediates and the rate of sterol demethylation, Brady and Gaylor[76] were able to show that the rate of esterification of Δ^7-sterol intermediates is six times faster in skin compared with liver. Furthermore, the esters were not metabolised to demethylated products. Thus, once formed, the esters of the partly demethylated intermediates accumulate. Operation in concert of the two secondary processes, one qualitative and one quantitative, may be all that is needed to account for the enormous quantity of methyl Δ^7-isomeric sterols in skin and the unusual accumulation of these intermediates. Therefore, only one different enzyme and relative abundance (activity) of another can lead to accumulation of compounds, which was once thought to be quite perplexing.

1.2.3.3 *Similarities of enzymes in the primary process*

There has been so much emphasis on the impact of the enzymes of the secondary process that the similarities of the enzymes of the primary process in a variety of organisms have been overlooked. *The most striking property of the sterol biosynthesis enzymes of different organisms, which catalyse reactions from squalene oxide to the principal end product, is their similarities. The next-most striking property is the generally low level of substrate specificity of these biosynthetic enzymes of the primary process.*

[If you have read this far, I am not going to distract you from the two statements above by burdening you with examples. There are several, in addition to those cited in Sections 1.2.3.1 and 1.2.3.3. Just think about it for a while, and see if you can challenge these sweeping statements, which seem appropriate in a chapter on 'Enzymes of Sterol Biosynthesis'.]

1.3 METHODS OF INVESTIGATING PARTICLE-BOUND ENZYMES OF STEROL BIOSYNTHESIS

In Section 1.2.2 you were told that 'Approaches' used to study the sequence of reactions in the biosynthetic process could be differentiated from the 'Methods' that are used. The purpose of this main section is to describe some methods that are in common use to the extent that you should be aware that the methods used in the two approaches are rather similar, and the methodology (apart from enzyme solubilisation and purification) is essentially identical. Thus, techniques for preparation, separation and identification of compounds, etc. are used in both approaches. On the other hand, the enzymic approach requires special techniques of enzymology, which have been adapted to the investigation of particle-bound enzymes. Thus, this section includes a discussion of methods used (a) to study the enzyme while particle-bound (Section 1.3.1), (b) to liberate the enzyme from particles (Section 1.3.2) and (c) to reconstitute multi-enzymic systems by combinations of purified enzymes and membrane components (Section 1.3.3).

1.3.1 Investigation of particle-bound enzymes

The first two steps in the enzymic approach require investigation of particle-bound enzymes to establish the presence of the enzyme and to investigate the principal characteristics of the particulate enzyme (Section 1.2.2). As indicated above and in Figure 1.7, the point was made that interruption of the enzymic process probably leads to the best evidence for existence of the enzyme and, simultaneously, conditions are established for the investigation of the properties of the enzyme that catalyses the reaction immediately preceding the inhibited enzyme. However, two other methods should be mentioned because they can lead to equally valid results, and, in some cases, the methods may be preferred (e.g. for simplicity). (a) Investigation of stereospecifically labelled (or doubly labelled) compounds, such as mevalonic acid, has been used successfully to show existence of enzyme (e.g. see work of

Hornby and Boyd above[73]) and to study preliminary charcteristics of the enzymic process. For examples of the latter statement, simply read chapters of reviews written by the people who have developed these techniques to high levels of sophistication (see Refs. 6, 15 and 17). (b) Study of transformations of model compounds by multi-enzymic systems may be enlightening. For example, much of the initial work on the nature of squalene oxidase and squalene oxide cyclase was carried out with various model compounds (see Section 1.2.1.3 and Ref. 9). Recently, Polito *et al.*[77] used combinations of study of model substrates and double isotope methods to investigate further the formation of squalene oxides and cyclisation products. The model substrate for methyl sterol oxidase, 4-hydroxymethylene-5α-cholest-7-en-3-one (Figure 1.10)* has also been used successfully to study the properties of

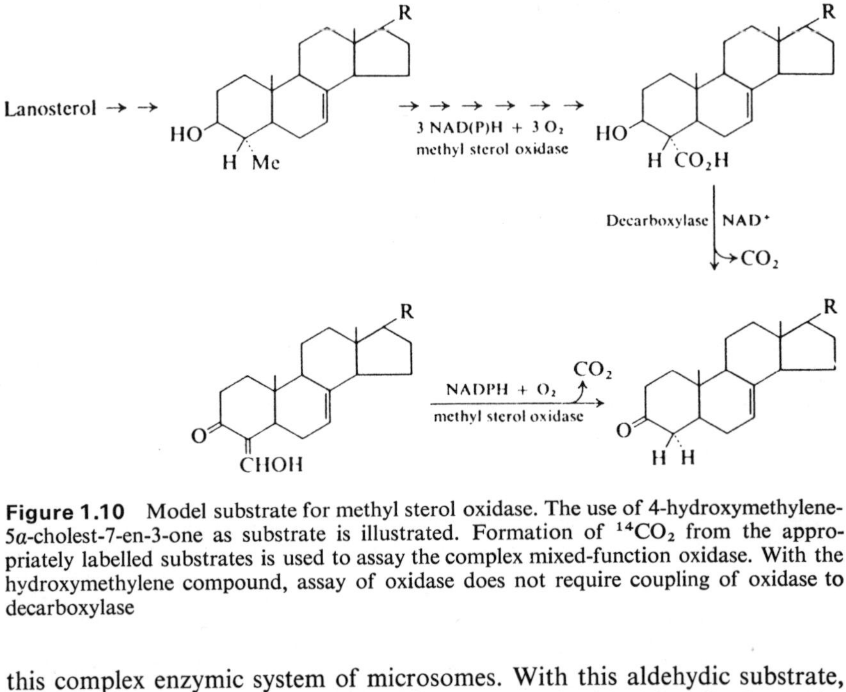

Figure 1.10 Model substrate for methyl sterol oxidase. The use of 4-hydroxymethylene-5α-cholest-7-en-3-one as substrate is illustrated. Formation of $^{14}CO_2$ from the appropriately labelled substrates is used to assay the complex mixed-function oxidase. With the hydroxymethylene compound, assay of oxidase does not require coupling of oxidase to decarboxylase

this complex enzymic system of microsomes. With this aldehydic substrate, mixed-function oxidation yields [^{14}C]-carbon dioxide directly from the substrate without the need to couple the oxidase to the decarboxylase of methyl sterol demethylation[18, 20, 78].

1.3.1.1 Interruption of the multi-enzymic process

(a) *Drugs, metabolic inhibitors and other xenobiotics*—Drugs are the largest single group of compounds that have been investigated for interruption of sterol biosynthesis. Since the earlier reports on inhibition of sterol

---

* Unpublished results, D. R. Brady and J. L. Gaylor.

Δ^{24}-reductase by triparanol[70, 71], many inhibitors have been found to be relatively specific. For example, bacitracin binds to pyrophosphate of isoprenols, such as farnesyl pyrophosphate, and the antibiotic thus may interfere with the coupling reactions[79]. Another drug, the oxalate salt of 2,2'''-[(1-methyl-4,4-diphenylbutylidene)bis-(p-phenyleneoxy)]bistriethylamin (SQ-10,591), also inhibits at the level of further metabolism of farnesyl pyrophosphate[80].

The most analytical use of drug inhibitors to solve problems related to sterol biosynthesis has been carried out in principally two laboratories. Dempsey and co-workers have used various drug inhibitors to unravel the steps in the terminal reactions of cholesterol biosynthesis[19, 81]. Dvornik and co-workers have used trans-1,4-bis-(2-chlorobenzylaminomethyl)cyclohexane dihydrochloride (AY-9944) as a selective inhibitor of the Δ^7-sterol reductase[82]. Use of the substituted sterol, cholestan-3β,5α,6β-triol, in two laboratories has demonstrated that this compound inhibits transformations, which occur in the A and B rings of sterols[83, 84].

Xenobiotics other than drugs have been used specifically. The fungicide Trimarinol appears to inhibit the terminal reaction of ergosterol biosynthesis in *Ustilago maydis*[85]. Finally, general metabolic inhibitors or regulatory substances, such as arsenite[86], adenosine-3',5'-cyclic phosphate[87] and β-mercaptoethylamine[88] have been shown to inhibit various reactions of sterol transformations.

(b) *Choice of incubation conditions*—Bloch and his co-workers have been able to study microsomal-bound squalene oxidase in the absence of contaminating cyclase, which catalyses the next sequential step (see Section 1.2.1.3), because these workers found that the cyclase could be inhibited by heating rat liver microsomal preparations to 50 °C for 5 min[51, 52]. Similarly, Rilling[46] successfully isolated presqualene pyrophosphate in rat liver microsomes by incubating farnesyl pyrophosphate and microsomes in the *absence* of NADPH, which is required for the conversion of the intermediate to squalene.

There are several examples of inhibition of oxygenases of sterol biosynthesis by elimination of oxygen[89] or addition of poisons, such as cyanide[63] or carbon monoxide[90]. Indeed, the NADPH-dependent 3-keto-steroid reductase of methyl sterol demethylation (Figure 1.6) may be investigated under anaerobic conditions, thus preventing oxidative attack of the 3β-hydroxy-4α-methyl sterol product that is formed by reduction of the ketone[18, 62].

(c) *Action of enzyme on more than one sterol intermediate*—The point has been made that the primary biosynthetic enzymes are strikingly insensitive to variations in the structure of sterol substrate; thus, they can accommodate compounds of subtle differences (Section 1.2.3.3). This means that the enzymes may act on more than one substrate. To establish evidence for one enzyme acting on more than one substrate, generally the enzyme is assayed with each substrate after various treatments, such as with proteolytic enzymes, heat, poisons, non-competitive inhibitors. The extent of inhibition produced by each treatment should be equal with each substrate. With this experimental approach, Miller was able to show that the 4α-methyl group of 4α-methyl-5α-cholest-7-en-3β-ol and of 4,4-dimethyl-5α-cholest-7-en-3β-ol is oxidised by the same methyl sterol oxidase of rat liver microsomes (Figure 1.6)[58,59]. Similarly, the 4α-carboxylic acids are decarboxylated by the same NAD-dependent enzyme[61]. Thus, methyl sterol demethylation at C-4 can be thought of as a cyclical process.

These examples have been given to illustrate methods that have been used to study microsomal-bound enzymes to establish the existence of the enzyme and to study properties of the enzyme prior to isolation and purification. Use of inhibitors may be effective in preventing sequential and concomitant attack (Figure 1.7; Section 1.2.2) of the substrate and product of the enzymic transformation under study. Obviously, at this point, the worker should know whether or not his sought-for enzyme catalyses more than one step in the biosynthetic process.

1.3.2 Isolation of particle-bound enzymes

Solubilisation methods used to obtain sterol biosynthetic enzymes from particles have been summarised in detail in another review[18]; in addition, reasons for the choice of various solibilisation techniques were given. It is not necessary to repeat the details of the procedures here, but the variety of methods available will be described, and citations to advances that have occurred in the past year will be added to the summary.

The principal goal at this stage of the enzymic approach is to isolate the sought-for enzyme free of other enzymes of the biosynthetic pathway. Furthermore, the enzyme must retain characteristics of the microsomal-bound enzyme (point 4, Section 1.2.2, enzymic approach).

1.3.2.1 Direct extraction into buffer

Direct extraction of enzyme from particles into dilute buffer should be tried first because this method eliminates the complicating effect of detergents or other solubilising agents. These latter substances must be removed to achieve criteria of solubility, and removal may be extremely difficult[91]. We have extracted an alcohol dehydrogenase from liver microsomes simply by washing the microsomes with dilute Tris-acetate buffer[67,92]. Krishna et al.[93] first successfully prepared soluble mammalian squalene synthetase by extraction of the enzyme from microsomes by dilute phosphate–bicarbonate buffer. Extraction efficacy may be improved by desiccating the microsomes first, either by lyophilisation or extraction with anhydrous solvent[18].

1.3.2.2 Use of detergents

Use of deoxycholate in solubilisation of yeast squalene synthetase[47], squalene oxide cyclases[45,51,53,54] and β-hydroxy-β-methylglutaryl coenzyme A reductase[39] has been mentioned in other sections of this review. Solubilisation by deoxycholate has been used for a number of microsomal-bound enzymes. However, bile salts inhibit the terminal reactions of sterol biosynthesis[94] and

STEROL BIOSYNTHESIS

the enzymes may be more labile in the presence of the detergent. However, methods are available for removal of deoxycholate by ion exchange chromatography;[47] also, glycerol and/or salts may be added to stabilise the isolated enzyme[44,95]. Furthermore, in a recent report on isolation of nucleoside diphosphatase from liver microsomes, Kuriyama[96] was able to obtain solubilisation of enzyme with a much lower concentration of deoxycholate when the treatment was carried out in an alkaline medium (pH 10.7). Because many of the difficulties have been overcome, deoxycholate and other ionic detergents probably occupy a central position of choice in initial selection of solubilising conditions.

The advantages of non-ionic detergents are several. The methods are mild and enzymes are relatively stable, even in the presence of detergent. Microsomal cytochromes b_5 and P-450 have been obtained by treatments with Triton X-100[97] and Lubrol[98], respectively. However, non-ionic detergents may be very difficult to remove; chromatography on Sephadex LH-20 yields separation of protein and detergents[91].

1.3.2.3 Partial proteolysis, lipolysis

Although partial proteolysis was one of the first methods used to obtain soluble enzymes from microsomes, the method has the severe disadvantage that alteration of the sought-for enzyme may occur during liberation from particles. Limiting use of the procedure to removal of unwanted protein prior to liberation of the sought-for enzyme with another procedure seems to be a more reasonable use of the digestion techniques[18]. However, partial proteolysis may be used as a last choice. The 5α-hydroxysterol dehydrase of yeast[99,100] resisted all other solubilisation techniques. Investigations with the highly purified enzyme have yielded considerable information about the mechanism of the enzymic process.

Lipolysis appears much less hazardous and more rewarding[18]. Procedures seem to be much more selective and analytical than those developed for partial proteolysis. Since the last review[18] use of one more phospholipase, D, has been reported[101], In addition, one note of caution should be added; fatty acids may alter electrophoretic movement and staining of membrane proteins after electrophoresis[102]. Phospholipase A is in common use and fatty acids are liberated by it.

Additional, recent study indicates that affinity labelling of steroid-binding sites of proteins may facilitate isolation of the sought-for enzyme[103].

1.3.3 Reconstitution of the multi-enzymic system

The final stage of the enzymic approach in work with the purified enzyme is to achieve reconstitution of the multi-enzymic system by combining the purified component enzymes and other constituents of the membranes (Section 1.2.2). Reconstitution must lead to aggregate properties, which are those exhibited by the original membrane-bound enzymes (Table 1.1, Stage

2). The criteria and procedure used in reconstitution experiments were described well in an article by Razin[26], which was cited earlier; this should be required reading for workers in the area.

Reconstitution may lie somewhere on the continuum between soluble enzymes acting freely in solution and an absolute requirement for the generation of a new membrane with the enzymes attached[23]. Furthermore, there are examples of functions of phospholipids as allosteric effectors[104] in addition to functions as membrane components. Although this work is in its infancy, regeneration of multi-enzymic systems promises to be an exciting area of future work. Clearly, the secrets of the next topic (control) may be revealed in the study of reconstitution.

1.4 CONTROL OF THE RATE OF STEROL BIOSYNTHESIS

Factors affecting the rate of sterol biosynthesis may be effective in altering the rate of enzymic formation of sterols in three ways: at the individual enzymic level; at the multi-enzymic sterol biosynthetic level; and at the genetic level. Each of these occurs with respect to other cellular processes. Examples of each of these contributing factors will be given.

1.4.1 Control at the level of the component enzyme: β-hydroxy-β-methylglutaryl coenzyme A reductase

Several endogenous and external factors may alter the catalytic rate of any enzyme in the multi-enzymic system. Most notable are those enzymic transformations that are either rate-limiting or at branch points in the pathway. The microsomal β-hydroxy-β-methylglutaryl coenzyme A reductase (HMG-CoA reductase) fulfils both of these criteria, and much of the study of control of sterol biosynthesis has been carried out with this enzyme.

Two independent factors control the rate of conversion of HMG-CoA to mevalonic acid. First, the enzymic activity is altered by effector substances or physiological factors, which result from administration of dietary cholesterol and Triton WR-1339, and which depress and stimulate HMG-CoA reductase, respectively[105, 106]. The effect of dietary cholesterol has been viewed as regulation via a negative feed-back mechanism because cholesterol is the end product of the biosynthetic conversion (equation (1.1) of Figure 1.11). Homeostasis of cholesterol metabolism would result. Second, with the important discovery of a diurnal rhythm of HMG-CoA reductase activity, which is reflected in rhythmic changes in acetate incorporation into hepatic cholesterol in mice[107] and rats[108] control at the level of enzyme amount was first demonstrated. Furthermore, this verified the possibility of cyclically altered amounts of HMG-CoA by carefully showing that the rhythmic changes in HMG-CoA reductase activity could be attenuated by treatment with either puromycin or cycloheximide. Although the factors that contribute to the rhythmic fluctuations in the amount of HMG–CoA reductase are not fully understood, several significant studies have been reported since the original publications of the phenomenon. In addition, control of the level of

HMG-CoA reductase is complicated even further because this is a microsomal enzyme and as one component in the multi-enzymic process, influence of enzymic activity and amount is not independent of control at the multi-enzymic level (see Section 1.4.2).

In the past few months, Slakey and his co-workers[109] verified the observed diurnal variation in HMG-CoA reductase; other enzymes of the biosynthetic pathway did not change similarly, and the rhythmic change in HMG-CoA reductase was shown to be quantitatively adequate to explain the changes in acetate incorporation into cholesterol. Although both fasting and re-feeding experiments[110] and alterations in light and dark periods[111] affected the cyclical changes in timing, amplitude, etc., the rhythmic changes were found to persist, but, quantitatively, the cycling of light and dark periods

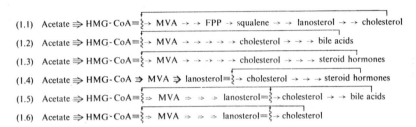

Figure 1.11 Illustration of end-product inhibition of the multi-enzymic biosynthesis of sterols. Arrows with many lines are used to illustrate larger fluxes of compounds with attenuation of synthetic rates, illustrated with the zigzag vertical line through the arrow. The result is a slower rate of conversion, which is shown as attenuation. Transformations in general tissues are illustrated

seems to be the most important factor[112]. Furthermore, workers in the same laboratory demonstrated that adrenalectomy did not affect the rhythmic changes in HMG-CoA reductase activity[113]. Although this work appears to be at variance with the results in the report of Hickman et al.[114], differences in experimental procedures may need to be resolved before the conclusions are reconcilable.

The rate of turnover of HMG–CoA reductase has been found to be very fast, and the half-life has been described as 'short' in the study by Slakey et al.[109] and a quantitative value has been given by Edwards and Gould[111]. These latter workers also made the important observation that the half-life of the enzyme, 4.2 h, is constant throughout the cycle, and that the rate of decay is unaffected by fasting. Thus, it appears that the cyclical change, albeit quite complicated in mechanism, may be the result primarily of cyclical changes in the rate of synthesis of HMG-CoA reductase.

Differentiation of direct feed-back effect of cholesterol on HMG-CoA reductase and an indirect effect on formation of HMG-CoA reductase is further complicated by the extended analogy (equation (1.2), Figure 1.11) that, in fact, bile salts are the end-products of sterol biosynthesis in liver, and, if analogy to end-product inhibition in microbial systems is valid, then bile salts should alter rates of HMG-CoA reductase. However, when studied *in vitro*, bile salts appear to be non-specific inhibitors at high non-physiological concentrations, for HMG-CoA reductions[115] and even the terminal lanosterol→cholesterol reactions[94] of cholesterol biosynthesis. However, in a very well controlled study, which took into account the cyclical changes in amount of HMG-CoA reductase, Hamprecht and co-workers[116] demonstrated that bile acids, *in vivo*, depressed the amount of HMG-CoA reductase activity in an experimental animal from which contributions of cholesterol were eliminated by cannulating the thoracic duct and thereby removing intestinally absorbed cholesterol from the animal. Thus, although the mechanism for this *in vitro* effect is not known, and the studies have not yet delineated whether or not there is only an effect of bile salts on amount of HMG-CoA reductase, I return to my first sentence of two paragraphs ago: There is an *interdependent* control of cholesterol biosynthesis at the level of HMG-CoA reductase. Control may be related to both the amount of enzyme synthesised and the fine-tuning imparted by enzymic effectors.

1.4.2 Control at the multi-enzymic level

The effects of end-product inhibitors on HMG-CoA reductase was discussed above. End-product inhibition obtains because this is a multi-enzymic system; thus, these two sub-sections are closely related. However, the purpose of the material in this part is to extend the analogy to other tissues, to other sites of inhibitory action within the sequence of reactions in addition to HMG-CoA reductase, and at least to make a case that additional work is needed very much. Finally, since control seems to be associated with the membrane-bound enzymes of sterol biosynthesis, and this article is to the subject of these enzymes, it seems appropriate to relate control of enzymic activity to the multi-enzymic nature of the system*.

Simple extension of equation (1.2) of Figure 1.11 to the steroid-secreting tissues generates equation (1.3), provided that the cholesterol is synthesised in the steroid-secreting tissues. Testicular tissue contains the multi-enzymic system of cholesterol biosynthesis, and endogenous cholesterol is used for steroid hormone biosynthesis. Thus, the effect of steroid hormone end products on cholesterol formation in testicular tissue was studied extensively. In 1965, we demonstrated that steroid hormones inhibit the conversion of lanosterol to cholesterol (equation (1.4), Figure 1.11) in testicular tissue; inhibitor steroid is non-competitive with respect to lanosterol and relatively specific[117]. When the work was initiated and steroid inhibition was first observed, we expected the steroid hormone to be a competitive inhibitor of lanosterol. The non-competitive nature of inhibition is consistent with the suggested regulatory role of steroid as an end-product effector substance. Specificity was further confirmed by demonstrating that inhibition varies *ca.* 100-fold with respect to structure of the steroid hormone[118]. Furthermore, inhibition is reversible, and we proposed that the steroid hormone endproduct functions as an allosteric effector in the control of cholesterol biosynthesis in this steroid-secreting tissue[119].

---

* Modern teenagers would probably call this a 'hang-up'. But, I cannot dissociate the tangled web of control of cholesterol biosynthesis from the observation that the controlling enzymes are bound to membranes, cholesterol and other hydrophobic substances (end products) are bound to membranes, etc.

As mentioned above with respect to non-competitive inhibition, to be a physiologically significant end-product inhibitor the steroid end product must be bound to a non-catalytic site(s) of the biosynthetic enzyme(s). This criterion was fulfilled recently[20]. Thus, with the simultaneous demonstration of stimulation of the testicular conversion of lanosterol to cholesterol by gonadotrophic hormones[120], we have proposed that the trophic hormone may participate in the control of the multi-enzymic system of cholesterol biosynthesis by affecting the amount of end-product steroid that is bound to the synthetic enzymes in the endoplasmic reticulum (Figure 1.12)[20]. The sterol substrates occupy a position in the endoplasmic reticular system that may be distinguished from the inhibitory steroid hormone binding site(s). Trophic hormone may alter the effective concentration of bound steroid inhibitor by

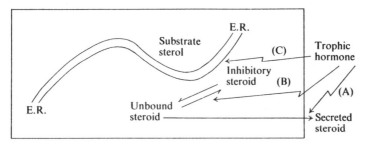

Figure 1.12 Proposed effect of trophic hormone on the multi-enzymic system of cholesterol biosynthesis. Enzymes of the system are bound to endoplasmic reticulum (E.R.). Membrane is indicated and binding of sterol substrate and steroid inhibitor to the multi-enzymic system of the E.R. is indicated. Sites of action of trophic hormone (either direct or via a chemical mediator) are indicated as A, B, and C

simply stimulating the secretion of steroid from the cell (A). Alternatively, the trophic hormone, via chemical mediator, may alter either the equilibrium between bound and unbound steroid (B) or the nature of the combination of the steroid inhibitor with the endoplasmic reticular system (C). The proposal must include the possibility that (B) and (C) in Figure 1.12 may result from alterations of steroid biosynthesis by trophic hormone, thus changing the population of steroids in the cell, which may, in turn, be less inhibitory towards cholesterol biosynthesis.

Assay of the effect of steroid hormones on the *individual* enzymes of the microsomal conversion of lanosterol to cholesterol has failed to show inhibition; because the collective multi-enzymic system is inhibited, understanding of control of cholesterol biosynthesis is going to even further require investigation of the nature of the associations of these enzymes. We are encouraged, however, that Ono and Imai[121] have reached similar conclusions and that work in each laboratory seems to be related to the dissolution of the end-product inhibitor in the microsomal membrane.

Since cholesterol is a major lipid component in membranes of higher animals, the relationship of cholesterol to the other lipid components of the membranes and to its own biosynthesis must be worked out prior to solving the problem of control of the rate of cholesterol biosynthesis in these membranes. Thus, cholesterol may serve as an end-product inhibitor *per se* (equation (1.6), Figure 1.11). Fortunately, the function of cholesterol in membranes is an active area of research[122, 123] and these studies may merge soon with others that are involved with study of cholesterol biosynthesis in non-steroid-secreting cells. For example, a Morris minimal-deviation hepatoma has a very rapid rate of cholesterol biosynthesis, but no significant changes in the amounts of individual component enzymes of microsomes from this tissue are observed*. There is no evidence that cholesterol in this tissue is further metabolised to other potential end-product inhibitors (i.e. equation (1.6), Figure 1.11).

Control of the multi-enzymic system at steps other than at the level of HMG–CoA reductase is not a new proposal nor limited to extrahepatic tissues. Slakey *et al.*[109] observed effects of fasting, etc. on the conversion of mevalonate to cholesterol in liver and they proposed that at least two sites must be involved in the hepatic system. Thus, bile salts may be mediators of feed-back inhibition to the other sites. Indeed, Moir *et al.*[124] demonstrated that removal of bile salts from the intestine by sequestering them to a dietary basic resin led to threefold stimulation of oxidative demethylation of methyl sterol intermediates. Therefore, the bile salts could alter more than one enzymic step in the sequence (equation (1.5), Figure 1.11).

1.4.3 Control at the genetic level

Finally, before closing this sub-section on control, the importance of genetic control and mutation on regulation of cholesterol biosynthesis should be emphasised. Two reports, which appeared within the last few months, further strengthen the view that considerably more work is needed[125, 126]. Furthermore, the finding of Kandutsch and Saucier[126] of the depressed HMG-CoA reductase levels in brains of three myelin-deficient mutants of mice is a recent finding of considerable importance.

1.5 RELATIONSHIP OF THE MULTI-ENZYMIC SYNTHESIS OF CHOLESTEROL TO OTHER MICROSOMAL PROCESSES

Since the report of Olson *et al.*[55] that microsomes contain all of the enzymes needed to catalyse the formation of cholesterol from squalene, there has been considerable interest in cholesterol biosynthesis as one process which occurs in microsomes (i.e. the endoplasmic reticulum of cells). The earlier emphasis on HMG-CoA reductase as a microsomal, rate-limiting enzyme (Section 1.4.1) and the frequent mention of other aspects of microsomal membranes in sterol biosynthesis have been used to make the point that detailed study of microsomes and the relationship of the multi-enzymic biosynthesis of cholesterol to other microsomal processes is quite essential.

---

* Unpublished observation of M. T. Williams and J. L. Gaylor.

1.5.1 Dynamic state of microsomal components

Chesterton reported that cholesterol biosynthesis occurs in the endoplasmic reticulum of the cell[43]. The microsomal fraction is derived from the endoplasmic reticulum by homogenisation and centrifugation of vesicles at *ca.* 105 000 × *g* for 1 h after removal of mitochondria from the homogenate by sedimentation at slower speeds, 10 000 × *g*. Several procedures have been reported for either further separation of microsomal vescicles[127] or alternatively preparations of relatively homogeneous derivatives of membranes from various parts of the cell[128]. Generally, enzymic or characteristic protein pigments are used to distinguish preparations that are obtained from various membranes[120, 129, 130].

One of the most remarkable features of microsomal enzymes is their dynamic state. The extremely short half-life of HMG-CoA reductase was mentioned above (Section 1.4.1). In addition, microsomal enzymes are induced by pretreatment of animals with certain drugs, carcinogens and other xenobiotics[131, 132]. The extent of induction varies with respect to genetic differences[133, 134]. Induction of changes in microsomal membranes may be clearly distinguished from changes in other membranes (e.g. members of the mitochondria)[135]. Phenobarbital, a commonly used agent to induce formation of microsomal enzymes, also stimulates phospholipid biosynthesis and phosphatidyl choline synthesis in rat liver microsomes is particularly elevated. Much of the recent work on the dynamic state of microsomal constituents has been carried out by Schimke. For example, Dehlinger and Schimke[137] recently showed that a variety of pretreatments of rats led to marked changes in amino acid incorporation into an electrophoretic component identified as cytochrome P-450 of liver microsomes. Since cytochrome P-450 is the terminal oxidase of many microsomal mixed-function oxidases, changes in cytochrome (P-450 may be reflected in many alterations of oxidase activities (see below). Thus the dynamic state of microsomal components will have an impact on the multi-enzymic biosynthesis of cholesterol in microsomes. Perhaps the dynamic state enables the control process to function optimally.

1.5.2 Microsomal mixed-function oxidases

Many reactions of cholesterol biosynthesis are catalysed by mixed-function oxidases. These are summarised in Figure 1.13. Squalene oxidase requires oxygen and reduced pyridine nucleotides (Section 1.2.1.3), which is characteristic of all external mixed-function oxidases[138]. The oxidase may not contain cytochrome P-450; it may be a functional flavoprotein oxidase[52]. Initial attack of each of the three 'extra' methyl groups of lanosterol is catalysed by a mixed-function oxidase[55]. The oxidation of the 4a-hydroxymethyl intermediate in demethylation of the 4-*gem*-dimethyl groups of C-4 is also catalysed by a mixed-function oxidase[58, 59]. Although Fried *et al.*[139] studied the metabolism of 32-oxygenated derivatives of Δ^7-lanostenol, unfortunately, the experiments were not carried out under conditions that would distinguish mixed-function oxidation from dehydrogenation. However, by analogy, oxidases 3 and 4 are indicated for these steps.

Oxidation of the 4-*gem*-dimethyl groups has been discussed (Section 1.2.1.4). Oxidative conversions continue until the carboxylic acid is formed for both mono-[58] and dimethyl-substituted[59] sterols (Figure 1.6). Each oxidase consumes NAD(P)H and oxygen.

$$R-H + O_2 + NAD(P)H \longrightarrow R-OH + H_2O + NAD(P)^+$$

(1) Squalene oxidase:

(2) Attack of the 14α-methyl group:

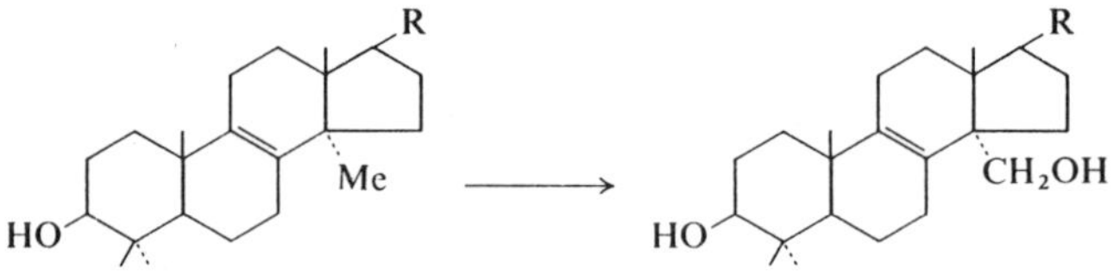

(3) 14α-hydroxymethyl ⟶ 14α-aldehyde
(4) 14α-aldehyde ⟶ 14α-carboxylic acid
(5) Attack of the 4α-methyl group:

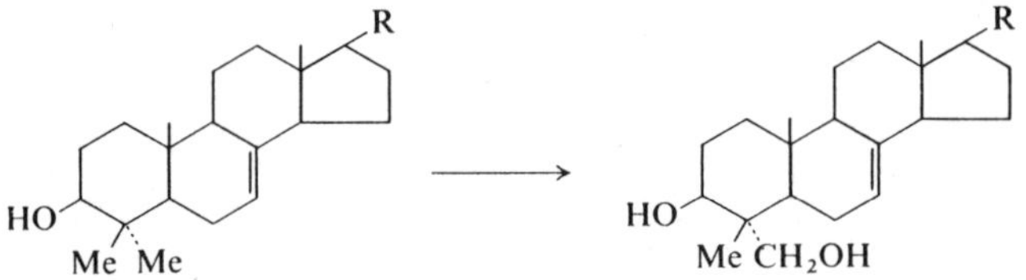

(6) 4α-hydroxymethyl ⟶ 4α-aldehyde
(7) 4α-aldehyde ⟶ 4α-carboxylic acid
(8)–(10) Repeat of (5)–(7) with 4-monomethyl substrate
(11) Δ⁵-dehydrogenase

Figure 1.13 Microsomal-mixed-function oxidases of sterol biosynthesis. A type-reaction is shown before specific illustrations. These are oversimplified because, in each case, additional enzymes, phospholipid, etc. may be needed to achieve electron transport from reduced pyridine nucleotide to the oxidase. Presumably, a complex of reduced oxidase, oxygen and substrate must be generated in the mixed-function oxidative process

The introduction of the Δ^5-double bond is catalysed by an oxygen-requiring enzyme of microsomes[57]. Although distinguishing properties of the oxidase have not been published, the enzyme is inhibited by cyanide[140], which is one property of the microsomal oxidases of C-4 demethylation[58, 59].

Some of the properties of the microsomal mixed-function oxidases of sterol biosynthesis have been studied. Particularly, the oxidase that attacks the 4α-methyl group has been shown to be different from the typical cytochrome P-450-dependent oxidases of liver microsomes.[63] For example, the oxidase is not strongly inhibited by carbon monoxide, but it is sensitive to inhibition by cyanide. The enzymic system has been separated from microsomes by treatment of the vesicles with deoxycholate. Resolution of microsomal oxygenases by chromatography of deoxycholate-solubilised particles yields a novel cyanide-binding haemoprotein of rat liver microsomes, which appears to be the terminal oxidase in methyl sterol demethylation[141]. The novel hemoprotein exhibits many spectral characteristics of cytochrome P-450, but it differs from the more abundant low-spin form of cytochrome P-450 principally by its higher affinity for cyanide[142]. Thus, many of the properties of cytochrome P-450 dependent oxidases, which act on drugs and xenobiotics, should be investigated in detail with sterol biosynthetic intermediates.

$$Me(CH_2)_7 \cdot CH_2 \cdot CH_2 \cdot (CH_2)_7 \cdot CO \cdot SCoA \quad \text{stearyl-CoA}$$

$$\underset{O_2}{NAD(P)H} \Bigg| \underset{\text{desaturase}}{\text{Acyl-CoA}}$$

$$\downarrow$$

$$Me(CH_2)_7 \cdot CH{:}CH \cdot (CH_2)_7 \cdot CO \cdot S \cdot CoA \quad \text{oleyl-CoA}$$

Figure 1.14 Another microsomal mixed-function oxidase, acyl-CoA desaturase, which is related to sterol oxidases

The microsomal mixed-function oxidases that attack methyl sterols and stearyl coenzyme A are similar (Figure 1.14).[63] Each oxidase appears to catalyse mixed-function oxidation of the appropriate substrate by transfer of electrons to the same cyanide-sensitive factor of liver microsomes[78]. Furthermore, control of the two processes appears to be independent, and each can be studied with the use of selective conditions of assay[78]. Work is in progress on isolation of these oxidases and reconstitution of the multi-enzymic mixed-function oxidations of each substrate. As pointed out by Raju and Reiser in a recent publication[143], many definite studies on the stearyl coenzyme A desaturase of liver microsomes are being delayed because a specific and well-characterised reconstitution system has not been obtained.

Recent relating of the microsomal ethanol-oxidising system of liver (Figure 1.13)[144] to the same novel haemoprotein raises some very interesting questions about the relationship of alcohol to sterol metabolism. Furthermore, ethanol feeding to rats simultaneously induces the appearance of the cyanide-binding form of cytochrome P-450 and elevated ethanol oxidation. Since much of the investigation of cholesterol metabolism is being carried out to answer questions of significance to cholesterol metabolism in humans, work on factors that may directly alter cholesterol biosynthetic rates through induction is of interest. These observations are indeed provocative, but further work must be done before the relationship should be thought of as more than a correlation.

1.5.3 Non-catalytic proteins of the microsomal multi-enzymic system

There have been reports of two types of non-catalytic proteins, which may participate in the microsomal system of sterol biosynthesis. These are the structural proteins of membranes and the carrier proteins of sterol biosynthetic intermediates.

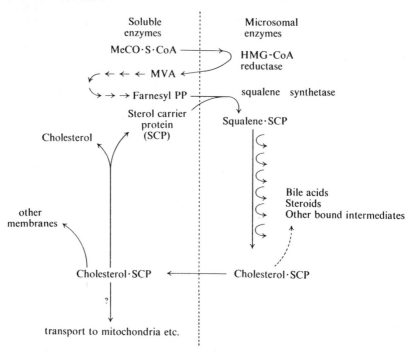

Figure 1.15 Function of a non-catalytic, sterol carrier protein (SCP) in sterol biosynthesis. Microsomal and soluble enzymes are depicted as being localised in two separate compartments; this is an oversimplification. However, the SCP is shown as facilitating the uptake of sterol substrate into the microsomal system and then remaining as a functionally distinct protein during the enzymic transformations, which occur in the microsomes (endoplasmic reticulum). SCP is also shown as part of the post-synthetic process of further transformation of cholesterol, incorporation into membranes, etc.

Structural proteins of mitochondria have been studied in detail[145]. Methodology and the possible roles of structural proteins in membrane structure, transport mechanisms and reconstitution studies have been summarised well by Kaplan and Criddle[146]. In addition, these authors included a section on microsomal membrane structural proteins in their review. From some of their own work on microsomal structural proteins, these workers make the point quite clearly that there is marked heterogeneity of these proteins. Thus, investigation of this class of non-catalytic proteins is under way but not of any marked consequence to date. Clearly, detailed studies are needed in which purification of the proteins is carried out in parallel with investigation of enzymic activities, analyses of component enzymes and lipids, and reconstitution.

Early reports from the laboratory of Dempsey and his co-workers[57, 147] contained evidence for a protein activator of microsomal sterol biosynthesis. The activator was obtained in the soluble fraction following isolation of microsomes by high-speed centrifugation. The activator has been purified *ca.* 300-fold, and very interesting activation and steroid substrate binding studies have been carried out[148-150].

Parallel studies in the laboratory of Scallen led to postulation of a soluble protein activator[151] and purification has resulted in the isolation of a protein, which has been called 'sterol-carrier protein'[152].

Considerable controversy exists at this writing over the structure of the carrier protein[153-155] and the principal workers in the field will simply have to continue their studies to the extent that thorough examination of the work, methods and conclusions in both laboratories are reconciled. However, there is essentially no controversy over the presumed role of the carrier protein (Figure 1.15). The function of the protein appears to be several-fold: (a) it has hydrophobic binding regions, which enables the protein to combine with sterol substrates; (b) it has suitably oriented hydrophobic regions, which contribute to the combination of the protein and the lipid matrix of the microsomal (i.e. endoplasmic reticular) membrane; (c) the protein has a pyridine nucleotide binding site[156] which allows close coupling of cofactor, substrate, and microsomal-bound enzyme. All of these properties, in addition to reversibility of combinations, ubiquity, etc., reinforce the notion that a non-catalytic protein could function as a substrate-binding polymer, which not only renders the water-insoluble substrate soluble, but imparts specificity, efficiency, etc. for the transformations that are catalysed by membrane-bound enzymes.

Recent work in other laboratories has not helped to resolve the controversy. For example, Calimbas[157] has reported the isolation and function of a sterol carrier protein in triterpene transformations in *Tetrahymena pyriformis*, and Rilling[158] has reported that a similar protein exists in high-speed supernatant fractions from yeast. On the other hand, Tai and Bloch[52] could not demonstrate a requirement for the carrier protein with partially purified oxidosqualene cyclase. Similar experiences in our laboratory with purified enzymes, even including those that have phospholipid attached to the purified protein, have not led to any clear-cut demonstration of a functional non-catalytic protein in these systems. But, these negative findings should not be viewed as evidence against sterol carrier protein. In these studies, there is an especially important parameter, which is not uniform from one laboratory to another, the assay. When someone says, 'we couldn't find "it" in our work', he means that, under the conditions of assay, with our experimental materials, etc., there was no demonstrable requirement.

Thus, this final word of caution brings me back to my initial comments in the introduction, and that is a good place to stop.

References

1. Battersby, A. R. (1967). *Pure Appl. Chem.*, **14**, 117
2. Richards, J. H. and Hendrickson, J. B. (1964). *The Biosynthesis of Steroids, Terpenes and Acetogenins* (New York: Benjamin)

3. Fall, R. R. and West, C. A. (1971). *J. Biol. Chem.*, **246**, 6913
4. Banthorpe, D. B., Charlwood, B. V. and Francis, M. J. O. (1972). *Chem. Rev.*, **72**, 115
5. Isler, O., Gutmann, H. and Solms, U., editors (1971). *Carotenoids*, Vol. 23 of Chemisch Reihe (Basel: Birkhäuser)
6. Goodwin, T. W. (1971). *Biochem. J.*, **123**, 293
7. Sih, C. J. and Whitlock, H. W., Jr. (1968). *Ann. Rev. Biochem.*, **37**, 682
8. Goodwin, T. W., editor (1971). *Proc. Phytochem. Soc. Symp., Aspects of Terpenoid Chemistry and Biochemistry* (New York: Academic Press)
9. Clayton, R. B. (1970). *Chemical Ecology*, 235 (E. Sondheimer and J. B. Simeone, editors) (New York: Academic Press)
10. Staunton, J. (1969). *Annu. Rep. Progr. Chem.*, **66**, 555
11. Bloch, K. (1965). *Science*, **150**, 19
12. Clayton, R. B. (1965). *Quart. Rev. Chem. Soc.*, **19**, 168
13. Olson, J. A. (1965). *Ergeb. Physiol. Biol. Chem. Exp. Pharmakol.*, **56**, 173
14. Frantz, I. D., Jr. and Schroepfer, G. J. (1967). *Annu. Rev. Biochem.*, **36**, 691
15. Goad, L. J. (1970). *Biochem. Soc. Symp. No. 29, Natural Substances Formed Biologically from Mevalonic Acid*, 45 (New York: Academic Press)
16. Schroepfer, G. J. Jr., Lutsky, B. N., Martin, J. A., Huntoon, S., Fourcans, B., Lee, W.-H. and Vermilion, J. (1972). *Proc. Roy. Soc. (London) B*, **180**, 125
17. Fiecchi, A., Kienle, M. G., Scala, A., Galli, G., Paoletti, E. G., Cattabeni, F. and Paoletti, R. (1972), *Proc. Roy. Soc. (London) B*, **180**, 147
18. Gaylor, J. L. (1972). *Advances in Lipid Research*, Vol. 10, 89 (R. Paoletti and D. A. Kritchevsky, editors) (New York: Academic Press)
19. Dempsey, M. E. (1968). *Ann. N.Y. Acad. Sci.*, **148**, 631
20. Gaylor, J. L. (1973). *Conf. on Multienzyme Systems in Endocrinology: Progress in Purification and Methods of Investigation, Ann. N.Y. Acad. Sci.*, in the press
21. Clayton, R. B. editor (1969). 'Steroids and terpenoids', *Methods in Enzymology*, Vol. 15 (New York: Academic Press)
22. Cooper, D. Y. and Salhanick, H. A. (1973). *Conf. on Multienzyme Systems in Endocrinology: Progress in Purification and Methods of Investigation, Ann. N.Y. Acad. Sci.*, in the press
23. Rothfield, L., Romeo, D. and Hinckley, A. (1972). *Federation Proc.*, **31**, 12
24. Kagawa, Y. and Racker, E. (1971). *J. Biol. Chem.*, **246**, 5477
25. Estabrook, R. W. and Pullman, M. E, editors (1967). 'Oxidation and phosphorylation', *Methods in Enzymology*, Vol. 16 (New York: Academic Press)
26. Razin, S. (1972). *Biochim. Biophys. Acta*, **265**, 241
27. Schenkman, J. B. and Cinti, D. L. (1972). *Life Sci.*, **11** (part II), 247
28. Kupfer, D. and Levin, E. (1972). *Biochem. Biophys. Res. Commun.*, **47**, 611
29. Culliton, B. J. (1972). *Science*, **175**, 1348
30. Stewart, P. R. and Rudney, H. (1966). *J. Biol. Chem.*, **241**, 1212
31. Stewart, P. R. and Rudney, H. (1966). *J. Biol. Chem.*, **241**, 1222
32. Rudney, H., Stewart, P. R., Majerus, P. W. and Vagelos, P. R. (1966). *J. Biol. Chem.*, **241**, 1226
33. Middleton, B. and Tubbs, P. K. (1972). *Biochem. J.*, **126**, 27
34. Skeggs, H. R., Wright, L. D., Cresson, E. L., MacRae, G. D. E., Hoffman, C. H., Wolf, D. E. and Folkers, K. (1956). *J. Bact.*, **72**, 519
35. Popják, G. (1970). *Biochem. Soc. Symp. No. 29, Natural Substances Formed Biologically from Melvalonic Acid*, 17 (New York: Academic Press)
36. Durr, I. F. and Rudney, H. (1960). *J. Biol. Chem.*, **235**, 2572
37. Bensch, W. R. and Rodwell, V. W. (1970). *J. Biol. Chem.*, **245**, 3755
38. Linn, T. C. (1967). *J. Biol. Chem.*, **242**, 984
39. Kawachi, T. and Rudney, H. (1970). *Biochemistry*, **9**, 1700
40. Lynen, F., Eggerer, H., Henning, U. and Kessel, I. (1958). *Angew. Chem.*, **70**, 739
41. Cornforth, J. W., Cornforth, R. H., Donninger, C. and Popják, G. (1966). *Proc. Roy. Soc. (London) B*, **163**, 492
42. Ritter, M. C. and Dempsey, M. E. (1971). *Biochem. Biophys. Res. Commun.*, **38**, 921
43. Chesterton, C. J. (1968). *J. Biol. Chem.*, **243**, 1147
44. Yamamoto, S. and Bloch, K. (1970). *Biochem. Soc. Symp. No. 29, Natural Substances Formed Biologically from Mevalonic Acid*, 35 (New York: Academic Press)
45. Dean, P. D. G. (1971). *Steroidologia*, **2**, 143

46. Rilling, H. C. (1970). *J. Lipid Res.*, **11**, 480
47. Schechter, I. and Bloch, K. (1971). *J. Biol. Chem.*, **246**, 7609
48. Altman, L. J., Kowerski, R. C. and Rilling, H. C. (1971). *J. Amer. Chem. Soc.*, **93**, 1782
49. Ogura, K., Koyama, T. and Seto, S. (1972). *J. Amer. Chem. Soc.*, **94**, 307
50. Altman, L. J., Ash, L., Kowerski, R. C., Epstein, W. W., Larsen, B. R., Rilling, H. C., Muscio, F. and Gregonis, D. E. (1972). *J. Amer. Chem. Soc.*, **94**, 3257
51. Yamamoto, S. and Bloch, K. (1970). *J. Biol. Chem.*, **245**, 1670
52. Tai, H.-H. and Bloch, K, (1972). *J. Biol. Chem.*, **247**, 3767
53. Schechter, I., Sweat, F. W. and Bloch, K. (1970). *Biochim. Biophys. Acta*, **220**, 463
54. Beastall, G. H., Rees, H. H. and Goodwin, T. W. (1971). *Fed. Europ. Biochem. Soc. Lett.*, **18**, 175
55. Olson, J. A. Jr., Lindberg, M. and Bloch, K. (1957). *J. Biol. Chem.*, **226**, 941
56. Gaylor, J. L., Delwiche, C. V. and Swindell, A. C. (1966). *Steroids*, **8**, 353
57. Dempsey, M. E. (1965). *J. Biol. Chem.*, **240**, 4176
58. Miller, W. L. and Gaylor, J. L. (1970). *J. Biol. Chem.*, **245**, 5369
59. Miller, W. L. and Gaylor, J. L. (1970). *J. Biol. Chem.*, **245**, 5375
60. Miller, W. L., Brady, D. R. and Gaylor, J. L. (1971). *J. Biol. Chem.*, **246**, 5147
61. Rahimtula, A. D. and Gaylor, J. L. (1972). *J. Biol. Chem.*, **247**, 9
62. Swindell, A. C. and Gaylor, J. L. (1968). *J. Biol. Chem.*, **243**, 5546
63. Gaylor, J. L. and Mason, H. S. (1968). *J. Biol. Chem.*, **243**, 4966
64. Witkop, B. (1971). *Experientia*, **27**, 1121
65. Moore, J. T., Jr. and Gaylor, J. L. (1969). *J. Biol. Chem.*, **244**, 6334
66. Tsai, S.-C. and Gaylor, J. L. (1966). *J. Biol. Chem.*, **241**, 4043
67. Moir, N. J., Miller, W. L. and Gaylor, J. L. (1968). *Biochem. Biophys. Res. Commun.*, **33**, 916
68. Lederer, E. (1969). *Quart, Rev. Chem. Soc.*, **23**, 453
69. Moore, J. T. Jr. and Gaylor, J. L. (1970). *J. Biol. Chem.*, **245**, 4684
70. Clayton, R. B., Nelson, A. N. and Frantz, I. D. Jr. (1963). *J. Lipid Res.*, **4**, 166
71. Gaylor, J. L. (1963). *Arch. Biochem. Biophys.*, **101**, 108
72. Gaylor, J. L. (1963). *J. Biol. Chem.*, **238**, 1649
73. Hornby, G. M. and Boyd, G, S. (1971). *Biochem. J.*, **124**, 831
74. Wells, W. W. and Baumann, C. A. (1954). *Arch. Biochem. Biophys.*, **53**, 471
75. Frantz, I. D., Jr., Dulit, E. and Davidson, A. G. (1957). *J. Biol. Chem.*, **226**, 139
76. Brady, D. R. and Gaylor, G. L. (1971). *J. Lipid Res.*, **12**, 270
77. Polito, A., Popják, G. and Parker, T. (1972). *J. Biol. Chem.*, **247**, 3464
78. Gaylor, J. L., Hsu, S. T., Delwiche, C. V., Comai, K. and Seifried, H. E. (1972). *Proc. 2nd Internat. Symp., Oxidases and Related Redox Systems*, (T. E. King, H. S. Mason and M. Morrison, editors) (Baltimore: University Park Press)
79. Stone, K. J. and Strominger, J. L. (1972). *Proc. Nat. Acad. Sci. (USA)*, **69**, 1287
80. Lerner, L. J., Harris, D. N., Yiacas, E., Hilf, R. and Michel, I. (1970). *Amer. J. Clin. Nutrition*, **23**, 1241
81. Dempsey, M. E. (1967). *Progr. Biochem. Pharmacol.*, **2**, 21
82. Dvornik, D., Kraml, M. and Bagli, J. F. (1966). *Biochemistry*, **5**, 1060
83. Witiak, D. T., Parker, R. A., Brann, D. R., Dempsey, M. E., Ritter, M. C., Connor, W. E. and Brahmankar, D. M. (1971). *J. Med. Chem.*, **14**, 216
84. Scallen, T. J., Dhar, A. K. and Loughran, E. D. (1971). *J. Biol. Chem.*, **246**, 3168
85. Ragsdale, N. N. and Sisler, H. D. (1972). *Biochem. Biophys. Res. Commun.*, **46**, 2048
86. Gaylor, J. L. (1963). *Arch. Biochem. Biophys.*, **101**, 409
87. Bloxham, D. P., Wilton, D. C. and Akhtar, M. (1971). *Biochem. J.*, **125**, 625
88. Hornby, G. M., Onajobi, F. D. and Boyd, G. S. (1970). *Biochem. Biophys. Res. Commun.*, **40**, 524
89. Miller, W. L., Kalafer, M. E., Gaylor, J. L. and Delwiche, C. V. (1967). *Biochemistry*, **6**, 2673
90. Gibbons, G. F. and Mitropoulos, K. A. (1972). *Biochem. J.*, **127**, 315
91. Gaylor, J. L. and Delwiche, C. V. (1969). *Anal. Biochem.*, **28**, 361
92. Bechtold, M. M., Delwiche, C. V., Comai, K. and Gaylor, J. L. (1972). *J. Biol. Chem.*, **247**, 7650
93. Krishna, G., Whitlock, H. W., Jr., Feldbruegge, D. H. and Porter, J. W. (1966). *Arch. Biochem. Biophys.*, **114**, 200

94. Miller, W. L. and Gaylor, J. L. (1967). *Biochim. Biophys. Acta*, **137**, 400
95. Tan, K. H. and Lovrien, R. (1972). *J. Biol. Chem.*, **247**, 3278
96. Kuriyama, Y. (1972). *J. Biol. Chem.*, **247**, 2979
97. Ito, A. and Sato, R. (1968). *J. Biol. Chem.*, **243**, 4922
98. Miyake, Y., Gaylor, J. L. and Mason, H. S. (1968). *J. Biol. Chem.*, **243**, 5788
99. Topham, R. W. and Gaylor, J. L. (1970). *J. Biol. Chem.*, **245**, 2319
100. Topham, R. W. and Gaylor, J. L. (1972). *Biochem. Biophys. Res. Commun.*, **47**, 180
101. Kapoor, C. L., Prasad, R. and Garg, N. K. (1972). *Biochem. J.*, **127**, 45
102. Fessenden-Raden, J. M. (1972). *Biochem. Biophys. Res. Commun.*, **46**, 1347
103. Sweet, F., Arias, F. and Warren, J. C. (1972). *J. Biol. Chem.*, **247**, 3424
104. Cunningham, C. C. and Hager, L. P. (1971). *J. Biol. Chem.*, **246**, 1583
105. Siperstein, M. D. and Fagan, V. M. (1966). *J. Biol. Chem.*, **241**, 602
106. White, L. W. and Rudney, H. (1970). *Biochemistry*, **9**, 2725
107. Kandutsch, A. A. and Saucier, S. E. (1969). *J. Biol. Chem.*, **244**, 2299
108. Back, P., Hamprecht, B. and Lynden, F. (1969). *Arch. Biochem. Biophys.*, **133**, 11
109. Slakey, L. L., Craig, M. C., Beytia, E., Briedis, A., Feldbruegge, D. H., Dugan, R. E., Qureshi, A. A., Subbarayan, C. and Porter, J. W. (1972), *J. Biol. Chem.*, **247**, 3014
110. Shapiro, D. J. and Rodwell, V. W. (1972). *Biochemistry*, **11**, 1042
111. Edwards, P. A. and Gould, R. G. (1972). *J. Biol. Chem.*, **247**, 1520
112. Huber, J. and Hamprecht, B. (1972). *Hoppe-Seyler's Z. Physiol. Chem.*, **353**, 307
113. Huber, J., Hamprecht, B., Müller, O.-A. and Guder, W. (1972). *Hoppe-Seyler's Z. Physiol. Chem,*, **353**, 313
114. Hickman, P. E., Horton, B. J. and Sabine, J. R. (1972). *J. Lipid Res.*, **13**, 17
115. Hamprecht, B., Nüssler, C., Waltinger, G. and Lynen, F. (1971). *Europ. J. Biochem.*, **18**, 10
116. Hamprecht, B., Roscher, R., Waltinger, G. and Nüssler, C. (1971). *Europ. J. Biochem.*, **18**, 15
117. Gaylor, J. L., Chang, Y., Nightingale, M. S., Recio, E. and Ying, B. P. (1965). *Biochemistry*, **4**, 1144
118. Lee, T. P. and Gaylor, J. L. (1968). *Steroids*, **11**, 699
119. Gaylor, J. L. and Lee, T. P. (1966). *Federation Proc.*, **25**, 221
120. Ying, B. P., Chang, Y. and Gaylor, J. L. (1965). *Biochim. Biophys. Acta*, **100**, 256
121. Ono, T. and Imai, Y. (1971). *J. Biochem. (Japan)*, **70**, 45
122. Engleman, D. M. and Rothman, J. E. (1972). *J. Biol. Chem.*, **247**, 3694
123. Hinz, H.-J. and Sturtevant, J. M. (1972). *J. Biol. Chem.*, **247**, 3697
124. Moir, N. J., Gaylor, J. L. and Yanni, J. B. (1970). *Arch. Biochem. Biophys.*, **141**, 465
125. Gaskin, F. and Clayton, R. B. (1972). *J. Lipid Res.*, **13**, 106
126. Kandutsch, A. A. and Saucier, S. E. (1972). *Biochim. Biophys. Acta*, **260**, 26
127. Dallner, G. and Ernster, L. (1968). *J. Histochem. Cytochem.*, **16**, 611
128. Fleischer, S., Fleischer, B., Azzi, A. and Chance, B. (1971). *Biochim. Biophys. Acta*, **225**, 194
129. Glaumann, H. (1971). *Structural and functional heterogeneity of the endoplasmic reticulum in the liver cell*, Dept. Pathol., Sabbatsberg Hosp., Karolinska Inst., Stockholm (Balder: Stockholm)
130. Brunner, G. and Bygrave, F. L. (1969). *Europ. J. Biochem.*, **8**, 530
131. Siekevitz, P. (1965). *Federation Proc.*, **24**, 1153
132. Ernster, L. and Orrenius, S. (1965). *Federation Proc.*, **24**, 1190
133. Nebert, D. W., Gielen, J. E. and Goujon, F. M. (1972). *Molec. Pharmacol.*, **8**, 651
134. Goujon, F. M., Nebert, D. W. and Gielen, J. E. (1972). *Molec. Pharmacol.*, **8**, 667
135. Kasper, C. B. (1972). *J. Biol. Chem.*, **246**, 577
136. Young, D. L., Powell, G. and McMillan, W. O. (1971). *J. Lipid. Res.*, **12**, 1
137. Dehlinger, P. J. and Schimke, R. T. (1972). *J. Biol. Chem.*, **247**, 1257
138. Mason, H. S. (1965). *Annu. Rev. Biochem.*, **34**, 595
139. Fried, J., Dudowitz, A. and Brown, J. W. (1968). *Biochem. Biophys. Res. Commun.*, **32**, 568
140. Dempsey, M. E. (1969). 'Steroids and terpenoids', *Methods in Enzymology*, Vol. 15, 501 (R. B. Clayton, editor) (New York: Academic Press)
141. Gaylor, J. L., Moir, N. J., Seifried, H. E. and Jefcoate, C. R. E. (1970). *J. Biol. Chem.*, **245**, 5511
142. Comai, K. and Gaylor, J. L. (1972). *J. Biol. Chem.*, 248 (in press)

143. Raju, P. K. and Reiser, R. (1972). *J. Biol. Chem.*, **247,** 3700
144. Joly, J.-G., Ishii, H. and Lieber, C. S. (1972). *Gastroenterology*, **62,** 174
145. Green, D. E., Haard, N. F., Lenaz, G. and Silman, H. I. (1968). *Proc. Nat. Acad. Sci. USA*, **60,** 277
146. Kaplan, D. M. and Criddle, R. S. (1971). *Physiological Rev.*, **51,** 249
147. Dempsey, M. E., Seaton, J. D., Schroepfer, G. J., Jr. and Trockman, R. W. (1964), *J. Biol. Chem.*, **239,** 1381
148. Ritter, M. C. and Dempsey, M. E. (1970). *Biochem. Biophys. Res. Commun.*, **38,** 921
149. Dempsey, M. E. (1971). *Chemistry and Brain Development*, 31 (R. Paoletti and A. N. Davidson, editors) (New York: Plenum Press)
150. Ritter, M. C. and Dempsey, M. E. (1971). *J. Biol. Chem.*, **246,** 1536
151. Scallen, T. J., Dean, W. J. and Schuster, M. W. (1968). *Biochem. Biophys. Res. Commun.*, **31,** 287
152. Scallen, T. J., Schuster, M. W. and Dhar, A. K. (1971). *J. Biol. Chem.*, **246,** 224
153. Scallen, T. J., Srikantaiah, M. V., Skrdlant, H. B. and Hansbury, E. (1972). *Federation Proc.*, **31,** 429
154. Ritter, M. C., Dempsey, M. E. and Frantz, I. D., Jr. (1972). *Federation Proc.*, **31,** 430
155. Dempsey, M. E., Ritter, M. C. and Lux, S. E. (1972). *Federation Proc.*, **31,** 430
156. Ritter, M. C. and Dempsey, M. E. (1971). *Federation Proc.*, **30,** 1159
157. Calimbas, T. (1972). *Federation Proc.*, **31,** 430
158. Rilling, H. C. (1972). *Biochem. Biophys. Res. Commun.*, **46,** 470

2
Lipids in Glycan Biosynthesis

F. W. HEMMING
The University of Liverpool

2.1	INTRODUCTION	40
2.2	POLYPRENOLS	41
	2.2.1 *Chemistry*	41
	2.2.1.1 *General*	41
	2.2.1.2 *Nomenclature*	42
	2.2.1.3 *Molecular size*	44
	2.2.1.4 *Stereochemistry*	44
	2.2.1.5 *Saturation*	45
	2.2.1.6 *Substituents*	46
	2.2.1.7 *Polyprenol phosphates and phosphate glycosides*	46
	(a) *General*	46
	(b) *Synthesis*	46
	(c) *Cleavage*	47
	2.2.1.8 *Preparation of radioactive polyprenols*	49
	2.2.2 *Distribution*	50
	2.2.2.1 *General*	50
	2.2.2.2 *Organisms and tissues*	51
	2.2.2.3 *Subcellular fractions*	51
	2.2.3 *Biosynthesis*	52
2.3	UNDECAPRENOL AND BACTERIAL WALL GLYCAN BIOSYNTHESIS	55
	2.3.1 *Introduction*	55
	2.3.2 *Polymannan*	56
	2.3.3. *Cellulose*	57
	2.3.4 *Peptidoglycan*	57
	2.3.4.1 *The synthetase cycle*	57
	2.3.4.2 *Rate-controlling factors*	60
	2.3.5 *'O'-antigen determinants*	61
	2.3.6 *Capsular exopolysaccharides*	64
	2.3.7 *Teichoic acids*	66

2.4	POLYPRENOLS AND MAMMALIAN GLYCAN BIOSYNTHESIS		70
	2.4.1 Introduction		70
	2.4.2 Dolichol phosphate derivatives		
		2.4.2.1 Glucosyl transfer	74
		2.4.2.2 Mannosyl transfer	78
		2.4.2.3 N-acetyl glucosaminyl transfer	81
	2.4.3 Retinol phosphate derivatives		82
		2.4.3.1 Introduction	82
		2.4.3.2 Mannosyl transfer	83
		2.4.3.3 Glucosyl transfer	84
		2.4.3.4 Galactosyl transfer	84
	2.4.4 Other lipid effects		84
2.5	POLYPRENOLS AND GLYCAN BIOSYNTHESIS IN YEASTS AND FUNGI		85
	2.5.1 Mannosyl transfer in yeast		85
	2.5.2 Mannosyl transfer in Aspergillus niger		86
2.6	POLYPRENOLS AND GLYCAN BIOSYNTHESIS IN GREEN PLANTS		87
	2.6.1 General		87
	2.6.2 Mannosyl transfer		88
	2.6.3 Glucosyl transfer		89
2.7	CONCLUDING DISCUSSION		89
	ACKNOWLEDGEMENTS		92
	NOTE ADDED IN PROOF		92

2.1 INTRODUCTION

The proper functioning of any living cell involves the control of phenomena at its surface and often beyond. This control is frequently mediated through polymers present in the vicinity of the cell surface and which are entirely carbohydrate (glycans) or contain a carbohydrate moiety attached to a peptide (peptidoglycans), a protein (proteoglycans) or a lipid portion (lipoglycans). It was in studies of the biosynthesis of this type of surface, or exterior, glycan in bacteria that it was first demonstrated that polyprenol phosphates play an essential role in glycosyl transfer.

At its simplest this process is summarised in general terms in Figure 2.1. The molecule of glycose (generalised sugar) is transferred from the nucleotide donor to a molecule of polyprenol monophosphate to form a molecule of polyprenol monophosphate glycose. Subsequently the glycose is transferred to the polymeric acceptor and polyprenol monophosphate becomes available to take part in the process again. The polyprenol phosphate glycoses are soluble in polar organic solvents and have frequently been termed lipid-linked intermediates. Both these intermediates and the transferases catalysing steps 1 and 2 (Figure 2.1) are membrane bound. In prokaryotic and possibly also in eukaryotic cells, the glycosyl donor is presented from one side of the membrane and the final polymeric product appears on the opposite side. It is clear that the polyprenol phosphates facilitate the trapping of the glycose molecules in the lipid-rich membrane and, together with the appropriate enzymes assist their transfer through the membrane to emerge in polymeric form on the other side. While the process of polymerisation is fairly well understood in most cases, vectorial aspects of the process remain to be clarified.

The initial elucidation of the structure and function of lipid-linked intermediates stimulated an enthusiastic burst of research activity in glycosyl transfer reactions, especially in bacterial systems and one reviewer, in 1969, was prompted to describe it as one of the most exciting developments of this decade[1]. What was an original process four years ago, now appears to be the normal procedure employed in the formation of bacterial surface glycans. It

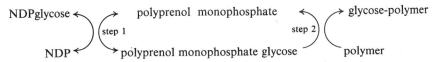

Figure 2.1 Generalised scheme showing the role of polyprenol monophosphate in the transfer of glycose from nucleotide donor (NDP) to glycan acceptor

also presents possible methods of controlling the processes. There is increasing evidence for an analogous scheme in some glycosyl transfers in mammalian and yeast systems. However, whether or not the scheme is general throughout the whole range of living organisms, especially in the plant kingdom, is still to be settled.

Although the concept of lipid-linked intermediates answers several questions, it raises many others. Many of these stem from the various modifications of the generalised scheme that are known to exist, from the great variety of polyprenols and glycans found in nature and from the possible relevance of the proposed pathways to the glycosylation of polymers in a controlled and biologically-significant way. It is with these problems in mind that the author has written this review of the evidence for and implications of the various examples of involvement of polyprenol phosphates in glycan biosynthesis. Beginning the review with a detailed discussion of the state of knowledge regarding the polyprenols is necessary in order to appreciate fully some of the discussion that follows. This approach also has the possible advantage that it has not been used before in several excellent reviews[1-8] dealing with lipid intermediates and it is a way into the subject with which the author is familiar.

2.2 POLYPRENOLS

2.2.1 Chemistry

2.2.1.1 General aspects

Naturally occurring polyprenols are primary monohydric alcohols with a carbon skeleton made up of several isoprene residues linked head to tail. The name is an abbreviated form of poly-isoprene-ol. In the generalised structure (1) the value of n is between 5 and 24 depending on the source of the polyprenol.

$$H{-}{\left[CH_2{\cdot}\underset{\underset{Me}{|}}{C}{:}CH{\cdot}CH_2\right]}_n{-}OH$$
(1)

$$\underset{R^1CH_2}{Me}{>}C{:}C{<}\underset{CH_2R^2}{H}$$
(2)

$$\underset{R^1CH_2}{Me}{>}C{:}C{<}\underset{H}{CH_2R^2}$$
(3)

Isoprene residues exist in the *cis* (2) or *trans* (3) form. In general the polyprenols found in nature belong to one of two groups: the all-*trans*- and the *cis,trans*-polyprenols. It appears that the biological function of each polyprenol is dependent upon its detailed stereochemistry. Usually the *cis,trans*-polyprenols are isolated as mixtures, or families of polyprenols, differing from each other only in the number of *cis*-residues. In such cases over 95% of the mixture can normally be accounted for by a range of compounds in which the value of n is within two of the median value.

One or more of the isoprene residues of some polyprenols is saturated by the addition of two extra hydrogen atoms. The presence of an exomethylene group as a further substituent has also been described.

2.2.1.2 Nomenclature

The structures of the polyprenols that have been isolated so far are summarised in Table 2.1. Details of their chemistry have been reviewed[9]. As more polyprenols have been isolated from an increasing number of sources it has become clear that they fall into only a small number of chemical types and that the trivial names give no indication of this. In an attempt to rationalise the situation and to use names that help identify the different chemical types Table 2.1 also includes abbreviated chemical names proposed by the author. The nomenclature adopted is based on that recommended by I.U.P.A.C.[10] for polyisoprenoid quinones. In particular the number of isoprene residues in each molecule is indicated by the postscript upper case Arabic numerals. Families of polyprenols are indicated by giving the range making up 95% of the mixture by weight. The position in the molecule of features worthy of detail is shown by designating the relevant isoprene residue by Roman numerals starting from the hydroxyl end of the chain and, in the case of a family of polyprenols, using the smallest polyprenol in the range indicated. For example, IV,V-di*trans*$_r$,poly*cis*-prenols-6→9 describes a family of prenols containing between six and nine isoprene residues, each molecule containing 2 *trans*-residues, the rest being *cis*. The presence in polyprenol-6 of the two *trans*-residues in positions 4 and 5 indicates that in each member of the family of polyprenols the *trans*-residues are immediately adjacent to the ω-isoprene residue. The evidence for the positions of the features is indicated in the following sections.

In many situations it is unnecessary to give the precise size of the molecules or the position of features. The names simplified by omitting the numbers are

Table 2.1 Summary of structures, names and sources of fully characterised polyprenols

	Trivial name	Source	Structure	Proposed abbreviated chemical name (see Section 2.2.1.2)
(a)	betulaprenols	silver birch wood[11,12]	ω-T-T-C-[C]-C-OH [1→4]	IV,V-ditrans-,polycis-prenols-6→9 (ditrans-,polycis-prenols)
(b)	bacterial prenols	bacteria[13–15]	ω-T-T-C-[C]-C-OH [5→7]	VIII,IX-ditrans-,polycis-prenols-10→12 (ditrans-,polycis-prenols)
(c)	dolichols	yeast[9,16] and P. cactorum[17] Mammals[9,16,18] Marine invertebrates[19]	ω-T-T-C-[C]-C-OH [9→13] ω-T-T-C-[C]-S-OH [12→16]	I-dihydro-,XII,XIII-ditrans-polycis-prenols-14→18 I-dihydro-,XV,XVI-ditrans-polycis-prenols-17→21 (dihydro-,ditrans-,polycis-prenols)
(d)	ficaprenols castaprenols, etc.	Green leaves[16,20–22]	ω-T-T-T-[C]-C-OH [4→8]	VI,VII,VIII-tritrans-,polycis-prenols-9→13 (tritrans-,polycis-prenols)
(e)	hexahydropolyprenols	Aspergillus[23] fumigatus	S-S-T-T-[C]-S-OH [14→18]	I,XVIII,XIX-hexahydro-,XVI,XVII-ditrans-polycis-prenols-19→23 (hexahydro-,ditrans-,polycis-prenols)
(f)	exo-methylene[24]-hexa-hydropolyprenols	Aspergillus[24] niger	S*-S-T-T-[C]-S-OH [14→18]	XIX exo-methylene-,I,XVIII,XIX-hexahydro-XVI,XVII-ditrans-,polycis-prenols-19→23 (methylene-,hexahydro-,ditrans-,polycis-prenols)
	solanesol	green leaves[25]	ω-T-T-T-[T]₄-T-OH	all-trans-polyprenol-9
	spadicol	spadix of Arum maculatum[26]	ω-T-T-T-[T]₅-T-OH	all-trans-polyprenol-10

ω = ω-isoprene residue, T = trans-isoprene residue C = cis isoprene residue S = saturated (dihydro) isoprene residue S* = exo methylene-dihydro isoprene residue.

included in brackets in Table 2.1. In general, they have the merit of containing much of the relevant information without being too unwieldy.

2.2.1.3 Molecular size

The size of the polyprenols is best determined by mass spectrometry. This technique will also confirm the polyisoprenoid nature of the compound and provide information regarding substituents. Reversed-phase partition thin-layer chromatography[16] gives a reliable indication of the number of isoprene residues present although extra care must be exercised if saturated residues or substituents are also present. Gas–liquid chromatography has been used for chain length measurements up to C_{65}[27]. A satisfactory n.m.r. spectrum will also give a reliable measure of molecular size[28] but this method requires several milligrams of sample whereas the other methods require considerably less.

2.2.1.4 Stereochemistry

The only satisfactory physico-chemical method for determining the total number of *cis*- and *trans*-isoprene residues in a polyprenol is n.m.r. spectrometry[28]. Using this technique it is also possible to confirm the presence or absence of an unsaturated ω-residue [R^1 = H in (2) and (3)] or a-residue [R^2 = OH in (2) and (3)] and, if present, to describe its stereochemistry. In the case of an ω-residue there are methyl groups both *cis* and *trans* to the olefinic proton. For this reason the ω-residue is usually ignored when commenting on the stereochemistry of polyprenols. In all of the di*trans*- and tri*trans*-poly*cis*-prenols that contain an unsaturated a-residue the configuration of the a-residue has been found to be *cis*.

Unfortunately, n.m.r. spectroscopy gives no direct information on the relative positions of the internal *cis*- and *trans*-residues. However, the fact that all polyprenols in the range di*trans*-,poly*cis*-prenols-6 to -9 for example contain a *cis*-a-residue, and two internal *trans*-residues suggests that they differ from each other only by the addition of *cis*-residues at the a-end of the molecule. This is strongly indicative of the two *trans*-residues being adjacent to the isoprene residue as indicated in Table 2.1. Biosynthetic studies (Section 2.2.3.2) support this conclusion and extend it to other poly*cis*-prenols.

The other group of naturally occurring compounds with polyisoprenoid moieties is made up of the ubiquinones plastoquinones and menaquinones, and their cyclic isomers[29, 30]. These compounds have all-*trans*-polyisoprenoid side chains and it is most likely that these are derived from the pyrophosphates of all-*trans* polyprenols including those mentioned in Table 2.1.

It has not been established if a particular stereochemical arrangement offers advantages in certain biological situations or if the differences simply reflect idiosyncracies of biosynthetic sequences that provide no functional advantage to the final product. Very little evidence is available regarding the possible effect of stereochemical differences on the physical properties or overall shape of the polyisoprenoid compounds. Nevertheless, since it is probable that their proper functioning is dependent upon the correct packing of the polyisoprenoid chain in the hydrophobic environment of membranes, it is necessary to consider what evidence is available.

On the basis of Langmuir trough experiments (J. Glazer, 1963, private communication) it was observed that at room temperature monolayers of both solanesol and dolichol (*ex* pig liver) gave cross-sectional areas per molecule much larger (three and four times, respectively) than expected for closely-packed vertically-orientated polyisoprenoid molecules. It was concluded that the monolayers possessed a 'liquid expanded' conformation and that the molecular cross-sectional area was related more closely to the length of the randomly-oscillating and rotating polyisoprenoid chains than to the stereo-chemistry of the constituent isoprenoid residues. This work also showed that the compressibility of the monolayers of solanesol was identical to that of monolayers of dolichol.

Although stereochemical differences appear not to be recognisable by Langmuir trough experiments they do give rise to changes discernible by reversed-phased partition chromatography. Thus substitution of one internal *cis*-residue for one internal *trans*-residue in undecaprenol results in a small but definite decrease in hydrophobicity of the molecule[14, 15].

A closer understanding of the functional significance of stereochemical differences, as well as chain length and substitution, will be dependent upon the extension of these studies to include other techniques and also to include the whole range of polyprenols and the effect of the presence of other membrane components on the parameters studied.

2.2.1.5 Saturation

Several polyprenols of eukaryotic organisms contain one, or occasionally more, fully hydrogenated isoprene residues. The α-isoprene residue of these partially saturated polyprenols is always saturated (see Table 2.1). One of the first polyprenols to be described contained a saturated α-residue. Because, when first discovered[31], the extreme length of the molecule was novel, it was given the trivial name dolichol (Greek *dolikos*: long). In fact several dolichols have now been described in which the common feature is not so much the extreme size but the presence of an α-saturated residue and of two internal *trans*-isoprene residues. The two cases of further saturation that have been reported[23, 24] show that in the *Aspergilli* polyprenols occur with an additional two saturated isoprene residues at the ω-end of the chain.

The presence of saturated isoprene residues in a polyprenol can be detected by spectroscopic, chromatographic and degradative procedures. Mass spectrometry[23] has been used to locate the presence of extra hydrogen atoms in terminal isoprene residues but it has not yet been demonstrated to be reliable in locating the position of internal saturated isoprene residues. N.M.R.[28] and i.r. spectrometry[21] can provide clear-cut evidence regarding the presence of a saturated or a normal α-isoprene residue. The presence of a saturated ω-isoprene residue is readily detected by n.m.r. spectroscopy[28]. Identification of the products of ozonolytic degradation of a polyprenol has been used to establish the presence of saturated α- and ω-isoprene residues and of a saturated isoprene residue immediately adjacent to the ω-residue[18, 23].

That saturation of isoprene residues renders the chain more hydrophobic is readily demonstrated by reversed-phase partition chromatography. Adsorption chromatography confirms the expectancy that saturation of the a-residue reduces slightly the polarity of the molecule. As yet there is little evidence regarding the biological consequence of the presence or absence of saturation in polyprenols but it is conceivable that it has an important influence on their binding to the hydrophobic areas of membranes.

2.2.1.6 Substituents

Apart from hydrogen the only substituent detected on the carbon chain of polyprenols has been an *exo*methylene group in the hexahydropolyprenols of *Aspergillus niger*[24]. Some polyisoprenoid quinones contain isoprene residues that carry oxygen functions[29] but these have not been detected in polyprenols.

The presence of the *exo*methylene group was demonstrated by mass, n.m.r. and i.r. spectroscopy.

2.2.1.7 Polyprenol phosphates and phosphate glycosides

(a) *General*—The *cis,trans*-polyprenols appear to function in biological systems as their phosphates. A biological role for both the monophosphates(4, 8) and diphosphates(5,9) has been demonstrated. The phosphorylated forms of the I-dihydropolyprenols, the alkyl phosphates(8,9) as well as those with the normal unsaturated a-residue, the allylic phosphates(4,5) are involved. When fulfilling their function these phosphates become glycosylated and give rise to allylic(6) and alkyl(10) monophosphate glycoses as well as allylic(7) and alkyl(11) diphosphate (poly) glycoses (see Figure 2.2.). The nomenclature, i.e. polyprenol monophosphate glycose and polyprenol diphosphate glycose adopted in this review is an extension of that used for dolichol derivatives by Leloir's group and based on analogy with the nucleoside diphosphate glycoses[32]. Consistency has been maintained by using the term diphosphate, rather than pyrophosphate in connection with polyprenol derivatives. Common usage has been acknowledged in retaining pyrophosphate in connection with short-chain prenol derivatives and for the inorganic form. The concentration of these polyprenol phosphates and derivatives in living organisms is low. In many eukaryotic organisms the unphosphorylated polyprenols are much more readily available. Several workers have avoided the technical problem of isolating in a pure state the natural phosphorylated derivatives by carrying out their chemical synthesis from the polyprenols.

(b) *Synthesis*—Two methods of phosphorylation have been used. The first method[33] is based on that of Cramer and Bohm[34] with modifications[32, 35] to suit the solubility properties of polyprenols and their derivatives. The polyprenol reacts with trichloroacetonitrile to form an intermediate imino ether which on treatment with bis-triethylamine phosphate yields the polyprenol phosphate. The second method has been developed by Warren[36] from that described by Khawja *et al.*[37]. This involves treatment of the polyprenol with o-phenylene phosphochloridate to yield the polyprenol o-hydroxyphenyl phosphate which is then cleaved oxidatively using lead tetra-acetate to form the polyprenol phosphate.

Both methods require anhydrous conditions; possibly under optimal conditions the second method gives the better yield. The second method also provides only the monophosphate, which may be important in some circumstances. The first method provides a mixture of monophosphates and diphosphates which may need to be separated by ion exchange chromatography. If the [^{32}P]phosphate is required, the first method is preferred due to the ease of making the bis-triethylamine salt of [^{32}P]phosphoric acid.

Warren[36] has made an important extension to these syntheses by forming in good yield ficaprenol diphosphate galactose. The method involves condensation of ficaprenol monophosphate with diphenylphosphochloridate to

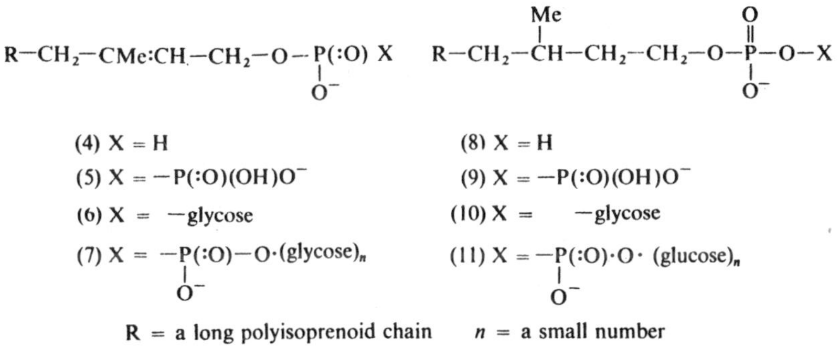

$$R-CH_2-CMe:CH-CH_2-O-P(:O)X \qquad R-CH_2-\underset{Me}{CH}-CH_2-CH_2-O-\underset{O^-}{\overset{O}{\underset{\|}{P}}}-O-X$$
$$\,\,\,O^-$$

(4) X = H   (8) X = H
(5) X = −P(:O)(OH)O$^-$ (9) X = −P(:O)(OH)O$^-$
(6) X = −glycose   (10) X = −glycose
(7) X = −P(:O)−O·(glycose)$_n$ (11) X = −P(:O)·O·(glucose)$_n$
     O$^-$ O$^-$

R = a long polyisoprenoid chain   n = a small number

Figure 2.2 Polyprenol phosphates and polyprenol phosphate glycosides

give the diphenyl diphosphate of ficaprenol. The diphenyl monophosphate group was then replaced by tetra acetyl galactose-1-phosphate and deacetylation yielded the required polyprenol disphosphate glycoside. The synthesis of other derivatives, including ficaprenol diphosphate N-acetyl glucosamine, ficaprenol diphosphate N-acetylmuramic acid and dolichol monophosphate mannose has also been achieved very recently (C. D. Warren and Y. Konami, 1972, private communication). The availability of this type of pure compound of unambiguous structure should facilitate several important studies related to the role of polyprenol phosphates.

(c) *Cleavage*—Relative rates of hydrolysis at low and high pH values have been critical aspects of the identification and purification of prenol phosphates and their derivatives. Allylic prenol phosphates are hydrolysed completely by exposure to pH 2 at 100 °C for 1–2 min whereas they are relatively stable at neutral and moderately alkaline pH. The extreme acid lability of allylic phosphates is a consequence of the double bond β- to the C–O bond. Alkyl monophosphates require extreme acid conditions before the phosphate can be removed. Not only is extreme acid lability characteristic of allylic phosphates but it liberates the isoprenoid portion in a form that is easier to characterise than is the parent phosphate. Unfortunately, this advantage is partially offset by the fact that the primary isoprenoid alcohol (13) is accompanied by the tertiary isomer(14) and frequently by hydrocarbon, presumably a result of dehydration (Figure 2.3). The formation of the tertiary isomer has been explained in the case of short chain polyprenol phosphates[38] by formation of a carbonium ion (12) involving C–O cleavage. Analogous reactions tend to switch increasingly to a bimolecular displacement mechanism as the polarity of the solvent decreases and hence lead to a higher ratio of primary to tertiary product[39]. This suggests a possible way of reducing the complexity of products. An alternative method is to employ a phosphatase for, with short chain polyprenol phosphates, alkaline phosphatase liberates the primary alcohol as a fairly clean product.

A survey of acid treatments employed to hydrolyse polyprenol phosphate glycoses shows that the order of lability of the bonds involved is allyl—

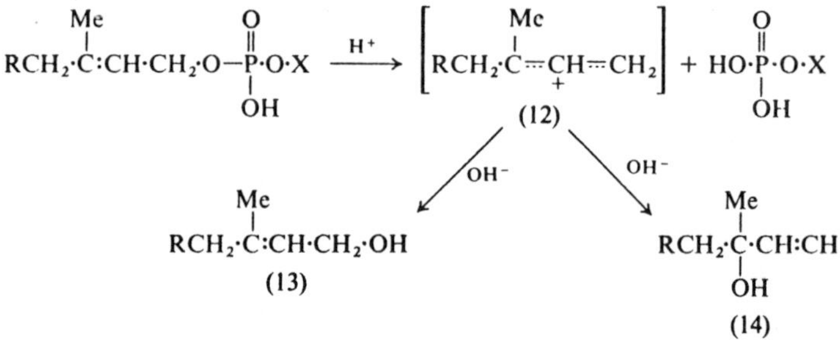

$$RCH_2 \cdot \underset{Me}{C{:}}CH \cdot CH_2 \cdot O - \underset{OH}{\overset{O}{P}} \cdot O \cdot X \xrightarrow{H^+} \left[RCH_2 \cdot \underset{Me}{\overset{+}{C}} {=\!=} CH {=\!=} CH_2 \right] + HO \cdot \underset{OH}{\overset{O}{P}} \cdot O \cdot X$$

$$(12)$$

$$\overset{OH^-}{\swarrow} \qquad \overset{OH^-}{\searrow}$$

$$RCH_2 \cdot \underset{Me}{C{:}}CH \cdot CH_2 \cdot OH \qquad\qquad RCH_2 \cdot \underset{OH}{\overset{Me}{C}} \cdot CH{:}CH_2$$

$$(13) \qquad\qquad\qquad (14)$$

Figure 2.3 Very mild acid hydrolysis of allylic prenol phosphate derivatives

phosphate > glycose-1—phosphate > phosphate—phosphate > alkyl—phosphate. Typical conditions used to hydrolyse the first three bonds have been pH 2, 100 °C, 1 min; pH 2, 100 °C, 10 min; and pH 1, 100 °C, 30 min, respectively. Investigation of the products after these different types of acid treatment has been used to partially characterise the type of compound under study. One group of workers[40] have used a relatively fast rate of hydrolysis of the glucose—1-phosphate bond to support the β rather than the α-configuration in their polyprenol monophosphate glucose.

All of these bonds are fairly stable at neutral and moderately alkaline pH. This enables the investigator to use mild alkaline treatment (pH 13, 37°, 15 min)[41] to deacylate most phospholipids and glycoglycerides, present in extracts from natural sources without damaging the polyprenol phosphate glycosides. Glycoglycerides are stable to the mild acid treatment described above. Thus stability to mild alkali and lability to mild acid treatment has been a useful criterion in identifying a glycolipid as a polyprenol phosphate glycose rather than a glycoglyceride.

More severe treatment with alkali can lead to cleavage of polyprenol phosphate glycoses although the rate of the process appears to be dependent upon the nature of the glycose and its anomeric configuration. For example, Wright[42] argued that since a sample of undecaprenol monophosphate glucose was stable to 1 M NaOH at 50 °C the glucose moiety must have been in the β-configuration. Had it been in the α-configuration (i.e. the hydroxyl on carbon atom 2 *cis* to that on carbon atom 1) the 1,2-cyclic phosphate of glucose would have been formed. Similar treatment of dolichol monophosphate glucose (0.1 M alkali 64 °C, 30 min) yields the 1,6-anhydroglucosan. By analogy with the rates of alkaline fission of arylglucosides[43] it has been suggested that the relatively fast rate of the cleavage indicates that the original configuration of the glucose residue may have been β.

Fission of the allyl-phosphate bond can be achieved by hydrogenolysis and this has the advantage over acid hydrolysis that the polyprenol moiety is converted to a single product, the saturated hydrocarbon. In the case of undecaprenol monophosphate glycoses the glycose 1-phosphates are liberated[42, 44]. This procedure would not be expected to cleave alkyl–phosphate bonds and so glycose 1-phosphate should not be liberated from dolichol monophosphate glycoses. This presents a convenient method of distinguishing between the two types of compound.

Treatment with hot phenol was one of the earliest methods used for separating the sugar from the lipid moiety in lipid intermediates[45]. It appears to cause cleavage mainly between the two phosphates of undecaprenol diphosphate glycose. The reaction has not been studied in any detail.

Phosphatases have been described that attack prenol monophosphates and pyrophosphates but not prenol phosphate glycoses. Intestinal and bacterial alkaline phosphatases readily catalyse the liberation of prenols from short-chain prenol diphosphates but they will not do so with polyprenol diphosphates[46] as substrates. More recently, Strominger's group[47] have described in detail a membrane-bound diphosphatase, probably widespread in bacteria, that catalyses the conversion of undecaprenol diphosphate to the monophosphate but not to the free prenol. The same laboratory has also isolated from *Staphylococcus aureus* a membrane-bound phosphatase that catalyses the release of undecaprenol from its monophosphate[48]. The fact that this enzyme was not detected in several other bacteria examined probably accounts for the presence of free undecaprenol in *S.aureus* but its absence from most other bacteria.

A prenol diphosphate phosphohydrolase was isolated from mammalian microsomes in 1966[48] and a little later a microsomal enzyme that would remove the pyrophosphate from presqualene diphosphate was described[50]. Neither of these enzymes was tested with polyprenol diphosphates or monophosphates as substrates. Since then, Seto's laboratory[51] has reported on an enzyme present in homogenates of rat liver and yeast which will remove phosphate from undecaprenol monophosphate but not from undecaprenol diphosphate. It would be interesting to know if this enzyme will utilise dolichol phosphates as a substrate. Not only would this be useful technically in characterising dolichol phosphate derivatives but it would also provide a possible mechanism for controlling glycosyl transfer reactions mediated by dolichol phosphates.

2.2.1.8 Preparation of radioactive polyprenols

Study of the metabolism of polyprenols and their phosphates is made easier if they can be obtained in radioactive form. All of the di*trans*-poly*cis*-prenols in Table 2.1 can be labelled with ^{14}C or ^{3}H by biosynthetic methods. When using whole organisms, the preferred precursor is radioactive mevalonate although this presents problems when using bacteria for the cell membrane of many appears to be impermeable to this compound. Yeast cells also will not metabolise exogenous mevalonate for the same reason. Lactobacilli are one of the few groups of bacteria that can incorporate exogenous mevalonate into undecaprenol[13, 14, 52-54]. Radioactive polyprenol phosphates have been formed by cell-free preparations of *Salmonella newington* using radioactive isopentenyl pyrophosphate and farnesyl pyrophosphate as precursors[55] (see Section 2.2.3) but this is not likely to be a convenient preparative approach. Probably the best way to obtain radioactive dolichols 14→16 is to administer radioactive mevalonate to *Phytophthora cactorum* for the mevalonate is not utilised for sterol biosynthesis in this organism and the dolichol can be obtained in quite good yield and of fairly high specific activity[56]. When using most eukaryotic organisms the problem of contamination of the polyprenol sample with radioactive sterols and the precursors can be avoided by using stereospecifically labelled [4S-4³H]mevalonate for this gives rise to

$$\text{ROH} \xrightarrow{\text{PBr}_3} \text{RBr} \xrightarrow{\text{EtO·C(:O)·CH}_2\text{·C(Me):O}} \text{RCH}_2\text{·C(Me):O} \xrightarrow{(^{14}\text{C·C})} \text{RCH}_2\text{·C(Me):}^{14}\text{CH·CH}_2\text{·OH}$$
(15)   (16)   (17)

Figure 2.4 Preparation of ¹⁴C-labelled polyprenols

³H only in *cis*-isoprene residues (see Section 2.2.3.2). This approach has been used quite successfully in regenerating rat liver (K. J. I. Thorne, 1972, private communication).

Although biosynthetic methods have the possible advantage of giving rise to radioactive derivatives in the natural site in the cell, the specific activity of the product is frequently too low to be useful in following changes in these compounds which function in catalytic amounts. A convenient method of obtaining ³H-allylic prenols of high specific activity is to oxidise the prenol to prenal and to then catalyse with potassium hydroxide the exchange of ³H with [³H]water. Reduction of the [³H]prenal with sodium borohydride yields [³H]prenol carrying up to six atoms of ³H per molecule on carbon atoms β, δ and δ' (CH₃ group) to the hydroxyl[57].

The chemical synthesis of [¹⁴C]prenols is more complicated but is best achieved by adding a ¹⁴C-labelled two-carbon moiety (usually [2-¹⁴C]bromoacetate) to the ketone (16) which is usually formed from the prenol(15) one isoprene residue shorter than the desired product(17) (Figure 2.4)[58, 59]. The final product(17) is a mixture of *a-cis-* and *a-trans-*isomers which can be separated chromatographically.

2.2.2 Distribution

2.2.2.1 General

Consideration of the distribution pattern of polyprenols must take account of the fact that these have been detected only in tissues where sufficient has accumulated following hydrolysis of prenol phosphates. The amounts of prenol phosphates required for glycosyl transfer appear to be vanishingly small and they can go undetected by normal analytical methods. Thus a high level of polyprenol in a tissue does not necessarily mean a high level of the functional form of polyprenol. It may simply reflect a high phosphatase activity and (or) a low rate of metabolism of the liberated polyprenol. Likewise a lack of detectable quantities of prenol should not necessarily be extrapolated to infer an abnormally low level of prenol phosphates. Nevertheless, the occurrence of free polyprenol is positive evidence of the formation of polyprenol phosphates, the product of polyprenol synthetase systems, and of the presence of phosphatase activity and, as such, is worthy of discussion.

2.2.2.2 Organism and tissues

Most of the information regarding the distribution of polyprenols throughout organisms is summarised in Table 2.1. It can be seen that the di*trans*-, poly*cis*-prenols have been isolated from bacteria and from non-photosynthetic plant tissue. The I-dihydro-,di*trans*-poly*cis* prenols are widely distributed throughout the animal kingdom and in yeasts and at least one fungus. Photosynthetic plant tissue yields tri*trans*-,poly*cis*-prenols, accompanied often by all *trans*-polyprenols. Saturated versions of this group of prenols, the hexahydro-di*trans*-poly*cis*-prenols, occur in some fungi. The concentration of polyprenols in plant tissues and possibly in others, increases with increasing age.

Only a few bacteria have been found that contain free polyprenol; most of that present is usually phosphorylated. In *S.aureus*, presumably harvested at approaching stationary phase, over 90% of the undecaprenol (0.03 µmol g^{-1} weight of cells) is free[15]. In *L.plantarum* approximately 40% of the polyprenol is free during log phase but this rises to *ca.* 75% in stationary phase (Thorne, 1972, private communication). In *A.niger ca.* 3% of the total polyprenols (0.2 µmol g^{-1} wet weight of cells) is present in phosphorylated form at the stationary phase[24].

Of pig tissues liver (70→140 µg/g wet weight) is the richest source of dolichols with intermediate levels (in order of decreasing concentration 40→10 µg/g wet weight) in spleen, kidney, pancreas, small intestine, brain and lung, low levels (5→1 µg/g wet weight) in spinal cord and fat and negligible quantities in blood, bone marrow, cartilage, heart, skeletal muscle, skin and tendon[60]. Of the dolichol present in the liver, *ca* 60% is esterified to fatty acids. Corresponding figures for spleen and kidney are 25% and <5%. The significance of these figures remains uncertain.

2.2.2.3 Subcellular fractions

A major part of the *cis,trans*-polyprenols present in mammalian tissues, in the *Aspergilli* and in silver birch wood is esterified to long-chain fatty acids. This renders the compounds so hydrophobic that it is doubtful if they can have any role other than possibly as a structural component. In both pig liver and *A.fumigatus* they are concentrated in a nuclear fraction of the cell whereas the unesterified polyprenol is found mainly in a mitochondrial fraction[62,63].

Very recently it has been reported that in rat liver the concentration of dolichol monophosphate (per mg of phospholipid) is highest in the nuclear, Golgi and rough endoplasmic reticulum fractions and lowest in plasma membranes and mitchondria with an intermediate level in smooth endoplasmic reticulum. The concentration of dolichol monophosphate in nuclear fractions raises the possibility of further functions for the compound[64].

In plant leaves only unesterified polyprenols have been detected although detailed searches for polyprenol phosphates have not been reported. Much of the increase in concentration of polyprenols in leaves with increasing age occurs in fatty droplets (osmiophilic globules) inside the chloroplast[65] where it is presumably devoid of metabolic influence. Of the remaining polyprenol some is associated with the chloroplast lamella and some with the cell wall[66]. It would be interesting to determine if polyprenol phosphates are also present in these particulate fractions.

In bacteria, undecaprenol and its phosphate derivatives are membrane bound. *L. plantarum* contains these compounds in both mesosomal and plasma membranes (Thorne 1972, private communication).

2.2.3 Biosynthesis

That polyprenols are formed by the standard terpenoid route from mevalonate has been demonstrated in several cases[9]. An appreciation of some of the stereochemical aspects of the pathway is necessary in order to understand the basis of some of the structures in Table 2.1. These features appear in Figure 2.5, which summarises the biosynthesis of the diphosphate of V-VId*itrans*-,-poly*cis*-prenol-7 (betulaprenol-7)(23). Of the stereochemical properties of mevalonic acid (18) it should be noted that carbon atom 4 carries two hydrogen atoms, one of which is pro-R (R in (18)) the other being pro-S (S in (18)). This configuration is retained during the formation of isopentenyl pyrophosphate (19) but in the isomerisation to dimethyl allyl pyrophosphate (20) the pro-S hydrogen is lost. Also during the further polymerisation steps to form geraniol pyrophosphate (21) and farnesol pyrophosphate (22) by addition of *trans* isoprene residues the pro-S hydrogen is lost. On the other hand, the further addition of *cis*-residues involves loss of the pro-R hydrogen atoms.

The elimination of hydrogens as described can be demonstrated by following the metabolism of [2-^{14}C, 4R-^{3}H]mevalonate and [2-^{14}C, 4S-^{3}H]mevalonate. Determination of the ratio of ^{3}H/^{14}C in the final product and comparison with the original mevalonate reveals which ^{3}H has been retained and which lost. It has been observed that in the biogenesis of *trans*-residues 4R-^{3}H is retained and 4S-^{3}H is lost while in the biogenesis of *cis* residues 4S-^{3}H is retained and 4R-^{3}H is lost. The presence of 4R-^{3}H and absence of 4S-^{3}H in a residue will thus characterise it as a biogenetically *trans*-residue even if its final stereochemistry is either ambiguous, as in the case of ω-residues, or lost, as in the case of saturated residues. Biogenetically *cis*-residues incorporate 4S-^{3}H but not 4R-^{3}H.

Betulaprenol-7 has been shown to contain three biogenetically *trans*- and four biogenetically *cis*-isoprene residues[67]. Since ω-isoprene residues appear always to be biogenetically *trans*, there is a precise correlation between the number of biogenetically *trans*- and the number of internal chemically *trans*- plus ω-residues and between the number of biogenetically *cis*- and the number of chemically *cis*-residues. This precise correlation has been demonstrated also with several other *cis,trans*-polyprenols (including bacterial undecaprenols[14] and mammalian dolichols[68]) and suggests that the stereochemistry of each isoprene residue is decided at the polymerisation step. The result does not allow *cis–trans* isomerism of any consequence after this step.

The location of the *trans*-residues at the ω-end of the molecule of betulaprenol-7 (Figure 2.5) is based partly on the observation that both di*trans*-farnesyl pyrophosphate and geranylnerol (ω-T-T-C-OH) would serve as precursors of betulaprenols but that geranylgeraniol (ω-T-T-T-OH) would not[67]. This result has been confirmed in *A. fumigatus* where, after incorporation of doubly-labelled mevalonates and ozonolytic degradation, the ω- and ψ-saturated residues of the hexahydropolyprenols were recovered and shown to be biogenetically *trans*. The saturated α-residue was biogenetically *cis*[69].

The results of these biosynthetic studies coupled with the chemical evidence discussed in Section 2.2.1.4 provide strong evidence in favour of the structures in Table 2.1 with the *trans*-residues permanently at the ω-end of the chains. Had they been arranged in a random fashion there would have been 36 stereochemical isomers of bacterial undecaprenol and 171 of dolichol-19. To reduce both of these numbers to unity is a major simplification.

What factors dictate the different stereochemistry of *cis,trans*-polyprenol synthetases, and whether or not these are open to manipulation are intriguing and unanswered questions. The control of the chain length specificity of these synthetases is equally obscure.

It must be assumed that the final products of a synthetase is the polyprenol diphosphate. It is likely that, in most bacteria, the functional form, the monophosphate, is produced by the action of a phosphatase (see Section 2.2.1.7(c)). In *S. aureus* a second phosphatase can remove the second phosphate group to leave the free polyprenol which may be converted back to the monophosphate under the influence of an ATP-dependent prenol phosphokinase. This latter enzyme is membrane-bound and is extremely hydrophobic, requiring the presence of phospholipids for activity and being soluble in butanol[70]. Control of the rates of these last two enzymes has been suggested as a possible way of altering the concentration of prenol phosphate and hence of the rate of bacterial wall biosynthesis[70].

The presence of a prenol phosphokinase has not been demonstrated in other bacteria or in eukaryotic organisms although some evidence for the slow pyrophosphorylation of short-chain prenols by plants exists[67,71]. The *S. aureus* enzyme exhibits some specificity for stereochemistry and possibly for chain length and (or) against α-saturation for on a molar basis pig liver dolichols were only 10% as effective a substrate as ficaprenols or betulaprenols. Solanesol, farnesol and hexahydropolyprenols were not phosphorylated[70].

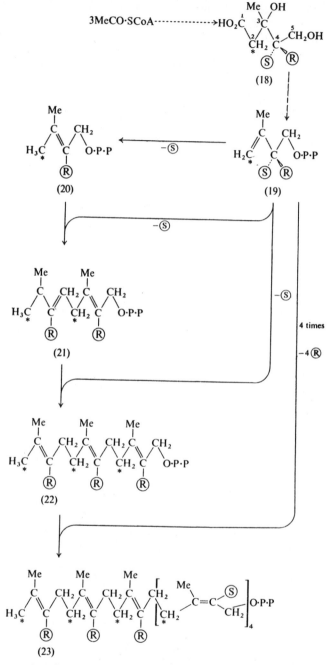

Figure 2.5 The biosynthesis of V,VI-di*trans*,poly*cis*-prenol→
diphosphate. S = hydrogen of the pro-*S* position of mevalonate,
R = hydrogen of the pro-*R* position, P = phosphate, C = carbon
derived from position 2 of mevalonate

Other outstanding problems in the biosynthesis of dolichols are the stage at which saturation of the α-residue occurs (before or after dephosphorylation?) and whether or not the monophosphate is formed directly from the pyrophosphate or by phosphorylation of the free dolichol. The liver polyprenol monophosphate phosphatase discussed in Section 2.2.1.7c has not been tested against dolichol phosphates.

The addition of isopentenyl pyrophosphate to farnesol pyrophosphate to form polyprenol phosphates, functional in 'O'-antigen biosynthesis in *S.newington* (Section 2.3.5) has been reported to be catalysed by a membrane fraction of this same organism in contrast to the formation of all-*trans* polyprenol phosphates which was catalysed by a soluble fraction[55]. In *L.casei*[72] undecaprenol appears to be formed in both mesosomal and plasma membranes. There was no evidence for the transfer of prenol from the former to the latter membranes. The subcellular distribution of the polyprenol synthetase systems in eukaryotic organisms has not been reported although recently Rudney's group[73] observed the presence in mitochondrial inner membranes of rat liver the capacity to form all-*trans*-nonaprenol pyrophosphate from isopentenyl pyrophosphate.

2.3 UNDECAPRENOL AND BACTERIAL WALL GLYCAN BIOSYNTHESIS

2.3.1 Introduction

The initial impetus to the study of lipid-linked intermediates in glycan biosynthesis has come from studies with bacteria. It is generally regarded that the demonstration of an alcohol-soluble 'intermediate' in cellulose biosynthesis by particulate fractions of *Acetobacter xylinum*[74] some 13 years ago was the first report of this type of process. However, it was only 7 years ago that the concept of lipid-linked intermediates was put on a firm basis first in the case of peptidoglycan in *S.aureus* and *M.lysodeikticus*[75] and then in the system forming the O-antigen determinants of *S.typhimurium*[76]. Almost concurrent reports 2 years later in the literature[78,79] established that in both of these glycan synthetase systems the lipid-linked intermediates were polyprenol derivatives. Since then lipid-linked intermediates have been recognised in synthetase systems forming at least six different types of bacterial wall glycans. In most of these cases the intermediate has been characterised by direct chemical analysis as an undecaprenol phosphate derivative. In some cases, the stereochemistry of the polyprenol has been established as being di*trans*, poly*cis*.

In the discussion that follows, of the schemes proposed for the involvement of undecaprenol in the biosynthesis of the different glycans, the order shows a gradual increase in complexity. Aspects not relevant to the involvement of lipid intermediates and therefore not included here can be found in one of the several recent reviews[1-8]. Also in an attempt to avoid repetition, details of a scheme will be discussed only if they illustrate a feature not discussed in earlier sections, or have particular importance.

2.3.2 Polymannan

Most of the work on the biosynthesis of bacterial polymannan has been done by Lennarz's group[79-81], using a particulate preparation of *M.lysodeikticus*. The polymer is made up of mannose residues linked by 1-2, 1-3 and 1-6 bonds with only a little branching. This *in vitro* biosynthetic work does in fact refer to chain lengthening rather than *de novo* synthesis of the polymer for the majority of the newly added mannose residues were found at the non-reducing termini.

The involvement of polyprenol monophosphate mannose in the transfer of mannose from GDP mannose to polymannan is summarised in Figure 2.6. When the particulate preparation, containing endogenous polymannan and lipid, was incubated with GDP [^{14}C]mannose a [^{14}C]mannolipid (24) was formed in addition to [^{14}C]polymannan and GDP. [^{14}C]Guanidine GDP mannose gave rise to [^{14}C]GDP. Mannose and phosphate were present in the purified lipid (24) in equimolar proportions and the release of mannose 1-phosphate upon gentle acid treatment and stability to mild alkali treatment suggested an allylic monophosphate mannose structure. Mass and n.m.r. spectrometry showed the lipid moiety to be di*trans*-poly*cis* prenol-11 accompanied by small quantities of the corresponding prenol-10 (4%) and prenol-12(6%). Incubation with undecaprenol [^{32}P]monophosphate led to the formation of [^{32}P]-lipid-linked intermediate (24). The transferase for reaction (a)

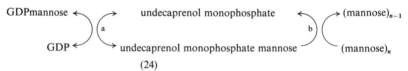

(24)

Figure 2.6 Undecaprenol monophosphate and mannosyl transfer in *M.lysodeikticus*. GDP = guanosine diphosphate (From Lennarz *et al.*[81], by courtesy of National Academy of Science, USA)

of acetone powder preparations was shown to require the presence of phosphatidyl glycerol, or other surfactants. This enzyme would also use the monophosphate of ficaprenols (tri*trans*-,poly*cis*-prenols) as a substrate. This requirement for phospholipids is a feature of several membrane bound glycosyl transferases. The lack of tight specificity to precise acceptor structure, appears to be characteristic of transferases using polyprenol phosphates as acceptors.

The kinetics of the transfer of ^{14}C from GDP[^{14}C] mannose to lipid and polymannan favoured the intermediate nature of the mannolipid. Incubation with undecaprenol monophosphate [^{14}C]mannose in the absence of GDP, resulted in the transfer of 50% of the ^{14}C to polymer (reaction (b)). The obligatory nature of this intermediate was demonstrated by the inhibitory effects of EDTA and Triton X-100. The former compound inhibited step (a) which requires Mg^{2+} but not step (b) whereas the latter inhibited step (b) but not step (a). Both, separately, inhibited the overall transfer of mannose from GDP mannose to polymannan.

Undecaprenol monophosphate mannose appears to be involved only in polymannan biosynthesis and not in the biosynthesis of dimannosyl diglyceride or of the oligomannosides attached to *myo*inositol, which also occur in *M.lysodeikticus*. This has been confirmed to hold also in *Mycobacterium smegmatis* and *Mycobacterium tuberculosis* (K. Takayama, 1972, private communication) where it has also been shown that the main polyprenol concerned is undecaprenol.

2.3.3 Cellulose

Acetobacter xylinum exports into its growth medium a cellulose which closely resembles that of higher plants. This is formed by transfer of glucose from UDP glucose to an endogenous primer by a particulate enzyme preparation. Progress on the involvement of a lipid-linked intermediate reported by Colvin[74] several years ago has been made only recently. Preliminary results[82] support the scheme in Figure 2.7.

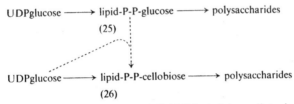

Figure 2.7 Possible role of lipid-linked intermediate in glycosyl transfer in *A.xylinum*. UDP-uridine diphosphate, P-phosphate (after Dankert *et al.*[82])

Incubations with UDP [^{14}C] glucose at 0° led to accumulation of ^{14}C in the lipids (25) and (26) but this passed on to polysaccharide at 30°C. Mild acid treatment of these glycolipids yielded pyrophosphate and glucose (or cellobiose) suggesting an allylic diphosphate glucose (or cellobiose) and not a monophosphate derivative as in polymannan biosynthesis. The production of cyclic sugar phosphates by exposure of the lipid-linked sugars to alkali suggested an α-glucosidic link to the phosphate. The lipid part of the intermediate has not yet been characterised. Investigation of its possible polyisoprenoid nature by biosynthetic means is hampered by the failure of this organism to take up mevalonate (F. W. Hemming, unpublished work, 1968).

2.3.4 Peptidoglycan

2.3.4.1 The synthetase cycle

At its simplest the peptidoglycan of *S.aureus* can be represented by the structure in Figure 2.8. It consists of polysaccharide chains made up of alternate *N*-acetyl muramic acid and *N*-acetyl glucosamine residues linked by β1-4 glycosidic bonds. These chains are cross-linked through the N-acetyl muramic acid residues by peptide chains. In the case of *S.aureus* the distal amino acid of the tetrapeptide chain on one residue is linked by a pentaglycine bridge to the penultimate amino acid of an identical tetrapeptide chain on an *N*-acetyl muramic acid residue of an adjacent polysaccharide chain. This 'tight' cross-linking produces a rigid structure. Most of the variations in peptidoglycan structure met in other bacteria involve differences in cross-linking and in substitution of the peptide chain.

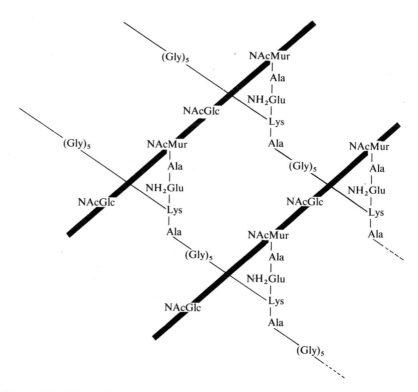

Figure 2.8 Schematised structure of peptidoglycan of *S.aureus*. N Ac Glc = *N*-acetyl glucosamine, N Ac Mru = N-acetylmuramic acid, Ala = alanine, Glu = glutamic acid, Lys = lysine, NH_2-Glu = glutamine, Gly-glycine (From Osborn[1], by courtesy of Annual Reviews)

Much of the biosynthetic work has been done by Strominger's group using particulate preparations of *S.aureus* and *M.lysodeikticus* and has been summarised recently[83]. The involvement of polyprenol phosphates in the chain lengthening of peptidoglycan of *S.aureus* is summarised in Figure 2.9. It is still uncertain if this scheme is also used for chain initiation. Figure 2.9, with minor modifications at steps (c) and (h), holds also in *M.lysodeikticus* and probably in at least ten other species, including gram +ve and gram −ve bacteria. After initial transfer of *N*-acetylmuramyl (pentapetide) 1-phosphate from nucleotide donor to undecaprenol monophosphate acceptor (step a) there is generation of a β1-4 glycosidic link (step (b)) to form a disaccharide (29) still attached to the undecaprenol phosphate carrier. Modifications of, and additions to, the peptide chain occur (step (c)) before the disaccharide is transferred to acceptor peptidoglycan (31) involving formation (step (d)) of another β1-4 glycosidic link and release of undecaprenol diphosphate. The other product (32) of this reaction is finally subjected to a transpeptidation (step e) resulting in a peptidoglycan (33) with a tetrapeptide chain rather than a pentapeptide chain as in the original donor (27). Undecaprenol diphosphate

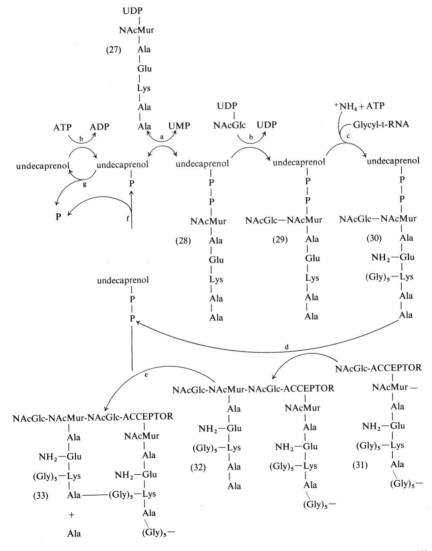

Figure 2.9 The peptidoglycan synthetase system of *S.aureus* (From Strominger et al.[83] by courtesy of Academic Press)

released at step (d) is dephosphorylated to the monophosphate (step (f)) before it enters the cycle of events again.

All the enzyme activities required are present in the particulate preparation used. The intermediates in the scheme have been isolated and studied. Weak acid hydrolysis (pH 4) of the disaccharide intermediate (29) yielded the corresponding disaccharide diphosphate which on further acid hydrolysis (1 M acetic acid) gave disaccharide and inorganic pyrophosphate. Using the [^{32}P]monosaccharide intermediate (28) it was shown that step (a) could be reversed in the presence of UMP but not UDP. The reversibility of this step (K = 0.25) had been demonstrated previously[84]. This observation suggested that step (a) preceded step (b) and this was confirmed by demonstrating that up to 75% of the radioactive disaccharide of the purified disaccharide intermediate could be transferred to polymer (33) whereas transfer of the radioactive monosaccharide from purified monosaccharide intermediate (28) to polymer (33) required the presence of UDP N-acetyl glucosamine. In these experiments the presence of nucleotide disphosphate sugars did not dilute the incorporation of radioactivity into polymer(33). Glycine-deficient peptidoglycan was not effective as an acceptor of glycine from glycyl-tRNA whereas the lipid-linked intermediates were, and of these the disaccharide intermediate (29) was the most efficient. Amidation has also been demonstrated to require a disaccharide-lipid-linked intermediate as acceptor but the precise sequence at step (c) is not clear. The structure of the lipid moiety of the intermediates was established as undecaprenol by mass spectrometry after acid hydrolysis. The stereochemistry of the prenol is based upon (i) the characterisation of the unesterified polyprenol of *S.aureus* as di*trans*-,poly*cis*-prenol-11, (ii) the demonstration that the monophosphate formed by enzymic phosphorylation of this prenol will act as an acceptor for step (a) (e.g. (9)) and (iii) the fact that undecaprenol phosphate isolated from the peptidoglycan synthetase system will function efficiently in the polymannan synthetase system as acceptor in step (a) (Figure 2.6). Although points (ii) and (iii) are weakened by the fact that ficaprenol phosphate will substitute quite well as acceptor in these steps, it seems almost certain that the intermediate contains di*trans*-poly*cis*-prenol-11.

2.3.4.2 Rate controlling factors

The phosphatases catalysing steps (f) and (g) and the phosphokinase catalysing step (h) have been described in Sections 2.2.1.7 (c) and 2.2.3, respectively. It has been suggested that variation in the relative rates of steps (f) and (g) offers a way of controlling the amount of undecaprenol monophosphate and hence the rate of the cycle. In this connection it was observed that the increased production of acidic fermentation products in *S.aureus* as it completes the log phase of growth might well favour the phosphatase (step)(g) with an optimum pH near 5, rather than the phosphokinase (step (h)) with an optimum pH between 7 and 8[83].

Reynolds[85] has shown that in *Bacillus megaterium* a shortage of undecaprenol monophosphate develops rapidly in particulate preparations that are incubated at 37C° and to a less extent at 30°C. It is argued that step

(f) is blocked at this temperature either because the enzyme is thermosensitive or because of disorganisation of the membrane. At present there is no evidence that this step is generally more thermosensitive than other steps in other bacteria but it is a possibility that should not be ignored.

The involvement of polyprenol diphosphate in the peptidoglycan biosynthetic cycle explains the antibiotic activity of bacitracin. This compound inhibits the biosynthesis of peptidoglycan and possibly of other wall polymers, and thus renders the bacterium much more vulnerable to physical and osmotic damage. The inhibition appears to be achieved by the bacitracin binding to undecaprenol diphosphate. In this way step (f) is blocked and the synthetic cycle is broken. The chelation of bacitracin to the diphosphate group is reinforced by hydrophobic bonding to the polyisoprenoid chain and steric factors consequent on this result in the binding being irreversible[83, 86]. The antibiotic has also been shown recently to inhibit prenylation of ubiquinone precursors in cell-free preparations of rat liver and *Rhodospirillum rubrum* by binding with all *trans* polyprenol diphosphates[87]. Polyisoprenoid quinones are important in bacterial electron transport and it may be that inhibition of their synthesis could play a part in the antibiotic activity of bacitracin. Even short-chain polyprenol pyrophosphates are bound sufficiently strongly by bacitracin for this compound to inhibit cholesterol biosynthesis in rat liver preparations[89].

At least three other antibiotics inhibit the peptidoglycan biosynthetic cycle. These are ristocetin, vancomycin and enduracidin which appear to block step (d).

2.3.5 'O'-antigen determinants

It is possible to classify the *Enterobacteriaceae*, especially the *Salmonellae*, by observing their antigenicity to different sera. The determinants of 'O'-antigenicity reside in a polysaccharide portion of the complex lipopolysaccharides found in the wall and outer membrane. The polysaccharide is made up of repeating tri- or tetra-saccharide units and is linked to a glucose residue of the lipopolysaccharide core (see Figure 2.10). In some sero-groups abequose

$$\begin{bmatrix} \alpha\text{-abequose} \\ |\ 1,3 \\ 4\text{-}\beta\text{-mannose-1,4-}\alpha\text{-rhamnose-1,3-}\alpha\text{-galactose-1-} \end{bmatrix}_n \quad \text{[polysaccharide-lipid]}$$

Antigenic determinant   Lipopolysaccharide core

Figure 2.10 The antigenic lipopolysaccharide of *S.typhitmurium*
(From Osborn,[1] by courtesy of Annual Reviews)

is not present and in others it is replaced by glucose. Anomeric differences occur and another variation is the presence of α-glucose linked 1–6 to the galactose residue. Figure 2.11 summarises the work mainly of Robbins and Wright[88-92] on the formation of the 'O'-antigen determinant of *S.newington*. The cycle is an extension of the basic theme discussed under peptidoglycan biosynthesis. A diphosphate bridge links undecaprenol to a monosaccharide

(34) and this is maintained during subsequent steps (b,c) leading to the formation of the lipid-linked trisaccharide repeating unit (36). In *S.typhimurium* this stage is extended by the addition of abequose (donated by CDP abequose) to form a tetrasaccharide repeating unit[93]. The polymerisation at the lipid-linked level is taken further by the transfer (step (d)) of preformed polymerised trisaccharide unit from another molecule of undecaprenol diphosphate (37) to the terminal mannose residue of the newly formed trisaccharide (36). For *de novo* synthesis steps (a), (b) and (c) will be carried out n times and steps (d) $(n-1)$ times. Figure 2.11 does in fact illustrate the

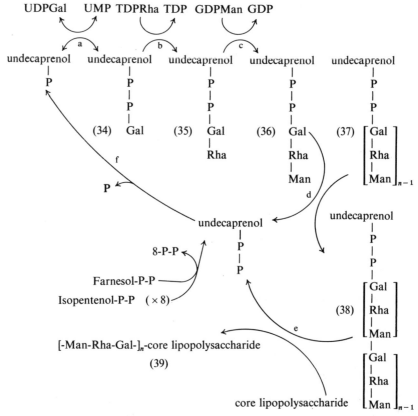

Figure 2.11 The synthetase system for the 'O' antigen determinant of *S.newington* (final cycle of process). Gal = galactose, Rha = rhamnose, Man = mannose, TDP = thymidine diphosphate (after Robbins, Wright *et al.*[89], by courtesy of National Academy of Science, USA)

final cycle in this process. When the chain of repeating trisaccharide units achieves the necessary length, the polysaccharide chain is then transferred (step (e)) from the undecaprenol diphosphate derivative (38) to the core lipopolysaccharide to form the complete antigenic lipopolysaccharide (39)[91, 92]. The undecaprenol diphosphate released by steps (d) and (e) is then converted to the monophosphate (step (f)) before re-entering the synthetase cycle.

A further interesting modification of the scheme concerns the introduction of glucose residues into certain antigenic determinants. For example, the determinant of *S.anatum* is normally as shown in Figure 2.11. However, after the bacterium has been rendered lysogenic by the temperate bacteriophages ε^{15} and ε^{34} the 'O'-antigen determinant has a β-galactose residue carrying an α-glucose. The presence of β-galactosyl residues is determined by the ε^{15} phage-specified polymerase (reaction (d)) in these organisms. The transfer of glucose is from UDP glucose via an undecaprenol monophosphate glucose intermediate in a manner analogous to that seen in Section 2.3.2[42,94]. Whether the final acceptor of glucose from this lipid intermediate is attached to lipopolysaccharide core (39) or to undecaprenol diphosphate oligosaccharide (38) has not been established but evidence favours the latter.

The enzyme activities for all the steps in the circle have been demonstrated in cell-free particulate preparations. Step (a) is readily reversible ($K = 0.5$) whereas the other steps are not. The intermediates (34), (35) and (36) have been isolated and purified. Their characterisation as undecaprenol diphosphate derivatives is quite secure and based on similar evidence to that for the peptidoglycan intermediates. In particular, mild acid hydrolysis yielded pyrophosphate and a lipid which was characterised as undecaprenol by mass spectrometry. The stereochemistry of the prenol has not been established although it probably contains two, or more, internal *trans* isoprene residues at the ω-end of the chain for a polyprenol phosphate formed from isopentenyl pyrophosphate and farnesyl pyrophosphate by particulate preparations of *S.newington* (Section 2.2.3) was functional in the 'O'-antigen determinant synthetase cycle[55]. Although ficaprenol monophosphate will also do this, it is most likely that the endogenous undecaprenol monophosphate concerned is of the di*trans*-poly*cis*-type. It is interesting that synthetic ficaprenol diphosphate galactose (Section 2.2.1.7(b)) was recently tested in the 'O'-antigen synthetase cycle (Figure 2.11) of *S.anatum*. It was active in the formation of polymer (39) but was less active than the corresponding intermediate (34) isolated from the micro-organism (A. Wright, Kanegasaki and Warren, 1972, private communication).

Bacitracin has been observed to inhibit 'O'-antigen determinant biosynthesis and confirms the diphosphate bridge in the lipid-linked intermediates. Further support is gained from the liberation of galactose 1-phosphate, or the corresponding oligosaccharide derivative, by treatment of the intermediates with hot phenol (see Section 2.2.1.7(c)). The sequence of addition of saccharides was shown by analysis of (34), (35) and (36), by the accumulation of (34), (35) or (36) when the appropriate nucleotide diphosphate glycose was absent from the medium and by demonstrating that exogenous (34) and (35) will act as acceptors of rhamnose and mannose respectively. Small amounts of the lipid-linked octasaccharide ((38), containing abequose, $n = 2$) have been isolated from *S.typhimurium* and enzyme activity for steps (d) (polymerase) and (e) have been observed. In experiments with *S.anatum*, Kanegasaki and Wright[92] have shown that exogenous [^{14}C]galactose-labelled (36) is incorporated into lipid-linked polymer ((38), in which $n = 20-33$) by cell-free particulate preparations. The value of n was based on the ratio of [^{14}C] galactitol to [^{14}C]galactose following mild acid hydrolysis and borohydride reduction of the product (38). The mechanism of step (d) was established by the results of pulse-chase experiments in which radioactive galactose incorporated first at the reducing end of the chain was displaced along the chain during the chase. This is the only account of glycan chain elongation from the 'activated' reducing end of the chain. It has been suggested that this mechanism offers advantages in the formation of membrane-bound polymers by virtue of keeping the point of addition of new polymer close to the active site of the membrane bound enzyme catalysing the addition[90,91].

Results of work in Wright's laboratory[92] suggest that the trisaccharide lipid-linked intermediate (36) is freely transferable within the membrane preparation from the mannosyl transferase, catalysing step (c), to the polymerase, catalysing step (d) and is not tightly bound in a complex containing enzyme activities to catalyse steps (a), (b), (c), (d) and (f). (Figure 2.11). In this experiment the cycle was blocked at step (f) by the presence of bacitracin. It was argued that if the prenol–phosphate intermediate was not freely transferable, addition of exogenous [^{14}C] disaccharide intermediate (35) and GDP mannose would result in the transfer of no more than one radioactive trisaccharide unit to form (38). If it was freely transferable, then there could be transfer of several trisaccharide units. Polymeric (38) with an average value of 17 for n was produced.

The evidence for an intermediate containing a monophosphate bridge for the addition of glucose residues in the lysogenic form of *S.anatum* is based partly upon reversal of the reaction by UDP and not by UMP. It is supported by failure to incorporate ^{32}P from [β-^{32}P]UDP glucose and the release of glucose and phosphate in equimolar amounts by mild acid treatment of the purified intermediate. Mass spectrometry of the lipid also released by acid showed it to be undecaprenol. The alkali stability of the intermediate (1 M NaOH, 50 °C) suggested an α-glucosyl linkage and this was confirmed by the liberation of α-glucose-1-phosphate during hydrogenolysis[42,94]. A similar system for glucose introduction into 'O'-antigen determinants of other sero groups of *S.typhimurium* and *S.enteriditis* has been described in detail by Nikaidos group [40,95].

The formation of polyprenol monophosphate glucose in particulate preparations of *Escherichia coli* and *Shigella flexneri* has also been demonstrated[96,97]. These two organisms will use ficaprenol monophosphate as an acceptor but not dolichol monophosphate or phytol monophosphate. A similar specificity was exhibited for galactosyl phosphate transfer, presumably via reaction (a) of Figure 2.11. As expected, mutants of *Sh. flexneri* that lack glucose in their antigen determinants appear to be defective in glucosyl transferase rather than short of polyprenol phosphate, for the poor uptake of glucose from UDP glucose by particulate fractions of these mutants failed to respond to addition of ficaprenol monophosphate.

2.3.6 Capsular exopolysaccharides

Several bacteria produce a capsule of polysaccharide overlaying the cell wall. Most of these polysaccharides consist of simple repeating oligosaccharide units (40, 41). Their biochemistry has been reviewed by Sutherland[98]. Most of

$$\begin{bmatrix} \text{-3-Gal-1-3-Man-1-3-Gal-1-} \\ 2 \\ 1 \\ \text{GlcA} \end{bmatrix}_n \qquad \begin{bmatrix} \text{-3-Gal-1-3-Gal-1-3-Glc-} \\ \\ \text{GlcA} \end{bmatrix}_n$$

(40) (41)

the biosynthetic studies have been carried out with strains of *Klebsiella aerogenes*.

Heath's laboratory has used a strain of this organism which produces (40) as its capsular exopolysaccharide[99]. As can be seen from Figure 2.12 the synthetase cycle for this polymer is closely analogous to that for 'O'-antigen determinants. The evidence for the two schemes is also very similar. One novel approach centred round the rapid reversibility of step (a) coupled with the essentially irreversible nature of step (b). After incubation with UDP [^{14}C]

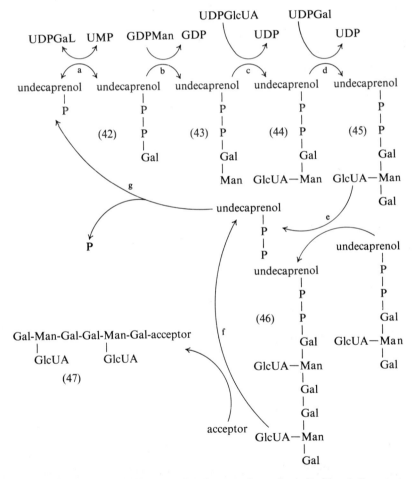

Figure 2.12 The synthetase system for the capsular polysaccharide of *K. aerogenes* (type DD45). Glc UA = glucuronic acid (after Heath *et al.*[99] by courtesy of American Society of Biological Chemists)

galactose the monosaccharide intermediate (42) became radioactive but, on the addition of a large excess of unlabelled UDP galactose, ^{14}C returned from (42) to the UDP galactose due to equilibration through step (a). However, if the initial incubation was with UDP [^{14}C]galactose in the presence of unlabelled GDP mannose, the radioactivity reached the disaccharide intermediate (43) and on addition of excess unlabelled UDP galactose there was no return of ^{14}C from lipid-linked intermediate to UDP galactose, due to the irreversibility of step (b).

Lipid-linked intermediates accumulated in incubations at 12°C but not at 37°C. A similar phenomenon was observed in the 'O'-antigen determinant cycle in *S. typhimurium*. All of the intermediates were isolated including a small amount of the lipid-bound octasaccharide (46). It is not certain if polymerisation goes further than this while bound to the lipid. Neither is it known if the tetrasaccharide is added at the reducing or non-reducing end of the growing polysaccharide chain. The lipid portion was identified as undecaprenol by mass spectrometry but there is no direct evidence regarding its stereochemistry. The undecaprenol monophosphate functions efficiently in the polymannan synthesising system (Section 2.3.2). The polymer product (47) was identified as capsular polysaccharide by its susceptibility to a specific depolymerase.

An analogous scheme has been proposed from Sutherland's laboratory for the biosynthesis of the tetrasaccharide units (41) of *K.aerogenes* type 8 although the lipid carrier was not characterised. The cycle was inhibited by bacitracin as expected. Some mutants of *Klebsiella* are resistant to bacitracin. Possibly these produce excess polyprenol monophosphate (Sutherland, 1972, private communication).

2.3.7 Teichoic acids

An important group of polymers in the cytoplasmic membranes and walls of gram +ve bacteria are the teichoic acids. At their simplest, these are made up of a chain of glycerol molecules linked 1–3 through a monophosphate bridge and carry a glucose or alanine residue on each glycerol residue (48). Most membrane teichoic acids are of this type. A second type of teichoic acid contains an analogous skeleton of polyribitol phosphate (linked 1–5) chains. Other variations include chains of poly-*N*-acetyl glucosamine phosphate (linked 1–6) as occurs in *Staphylococcus lactis* 2102. Mixtures of these also occur as in *S. lactis* I3 which contains the structure shown in (49). *Bacillus*

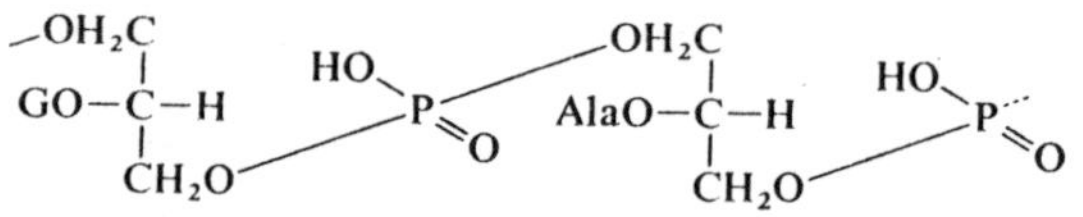

(48) G = glycosyl, Ala = Alanyl

[-3-glycerol-1-P-4-NAcGlc-1-P-]$_n$ [-3-glycerol-1-P-6-glucose-1-]$_n$
              (49)   (50)

licheniformis 9945 contains a different type of mixture since only one phosphate occurs in the repeating unit of glycerol phosphate and glucose (50). The biochemistry of teichoic acids has been reviewed by Baddiley[101].

Baddiley's group have shown[102] that in particulate preparations of *S.lactis* 2102 teichoic acid biosynthesis involves the scheme in Figure 2.13. The evidence is clearly in favour of an intermediate containing N-acetyl glucosamine linked to lipid through a diphosphate bridge. The effect of acid favours an allylic phosphate but the lipid has not been characterised further chemically. It should be noted that this scheme differs from those considered in previous sections in that a glycose phosphate is transferred (step (b) Figure 2.13) from the lipid-linked intermediate to the final acceptor (52). It follows that lipid monophosphate is released rather than the diphosphate, a situation consistent with the observed lack of inhibition of this process by bacitracin (see

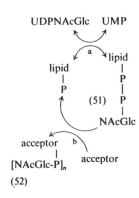

Figure 2.13 The synthetase system for teichoic acid in *S.lactis* 2102. (after Baddiley *et al.*[103], by courtesy of Macmillan Ltd)

Section 2.3.4.2). There is no evidence for polymerisation at the level of lipid-linked intermediate. A pulse-chase experiment followed by controlled acid hydrolysis of the polymer formed showed that chain lengthening occurred from the non-reducing end, i.e. from the glycosyl terminus rather than the phosphate terminus (compare Section 2.3.5).

In *S.lactis* I3 the scheme shown in Figure 2.14 appears to function although the particulate preparations used were not very effective in catalysing step (c). Analysis of the purified lipid-linked intermediate (53) and (54) confirmed the glycose and glycerol phosphate moieties and experiments with ^{32}P confirmed the presence of phosphate in both. The transfer of glycerolphosphate N-acetyl glucosamine phosphate and the liberation of lipid phosphate at step (c) was in keeping with a diphosphate bridge in (53) and (54) and the extreme acid lability of both intermediates suggested that an allyl phosphate was involved. The lipid moiety has not been fully characterised chemically but some interesting competition experiments[103] suggest that it is the same as that functioning in peptidoglycan biosyntheses, i.e. undecaprenol phosphate. It is suggested that the inhibition (by 22%) of teichoic acid (55) biosynthesis by the addition of UDP N-acetyl muramyl pentapeptide (equimolar to UDP N-acetylglucoasmine used in Figure 2.14) is due to competition for the lipid monophosphate. Consistent with this view was the demonstration that bacitracin and vancomycin failed to inhibit the cycle in Figure 2.14 unless UDP N-acetylmuramyl pentapeptide was also present. Inhibition was then 52% and 40%, respectively. This is seen as undecaprenol phosphate being trapped in the peptidoglycan synthetase cycle (Figure 2.9, Section 2.3.4). More recently, the biosynthesis of poly(glycerol phosphate) (57)[104] and of poly(glycerol phosphate glucose) (60)[44] in *B. licheniformis* has been shown to involve lipid-linked intermediates (Figure 2.15). The formation of (57) appears to involve a lipid-linked intermediate containing a diphosphate bridge (56) whereas those concerned in the formation of (60) each contain a monophosphate bridge (58, 59) between lipid and sugar moiety. The latter scheme is the first example of further substitution of a lipid monophosphate glycose prior to release of the lipid monophosphate.

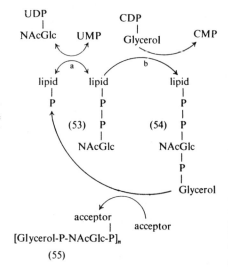

Figure 2.14 The synthetase system for teichoic acid in *S.lactis* 13. CDP = cytidine diphosphate (after Baddiley *et al.*[44], by courtesy of The Biochemical Journal)

Pulse labelling confirmed that (56) is a precursor of (57). The presence of the diphosphate bridge in (56) is supported by transfer of glycerol phosphate to (57) and production of lipid monophosphate in step (b). That the lipid phosphate was a monophosphate and of undecaprenol was indicated by competition experiments similar to those described in the formation of (55). Inhibition of the biosynthesis of glucose-containing teichoic acids by chloramphenicol, possibly at stage (c) Figure 2.15, has been reported[105].

As a result of detailed studies [104] of the competition between the peptidoglycan synthetase system (Figure 2.9) and the teichoic acid synthetase system (Figure 2.15) for undecaprenol phosphate in cell-free preparations of *B. licheniformis* it was concluded that once undecaprenol phosphate is being utilised in the peptidoglycan or teichoic acid synthetase system it is not available to other synthetase systems at the end of each cycle. It was suggested that each synthetase system is a multi-enzyme complex and that once the undecaprenol phosphate becomes associated with one of these it remains firmly bound until polymer synthesis is complete. Whether or not the undecaprenol phosphate can then become available for other synthetase systems is not clear from the experimental results. The evidence for this view is based on experiments in which the addition, together, of UDP *N*-acetylmuramy pentapeptide (0.1 µmol) and UDP N-acetylglucosamine (0.1 µmol) to an incubation containing CDP glycerol (0.2 µmol) and UDP glucose (0.1 µmol) inhibited the teichoic acid (57 + 60) biosynthesis over 1 h by 66%. In the reverse experiment peptidoglycan biosynthesis was inhibited by ca. 51.5%. The trapping of undecaprenol phosphate was proposed to explain the only small effect on inhibition of teichoic acid biosynthesis of (a) omission of UDP N-acetylglucosamine from the system (inhibition now 60%) and (b) addition of bacitracin to the system (inhibition now 76.5%). The concept is also supported by the increased inhibition of teichoic acid biosynthesis by preincubation with peptidoglycan precursor nucleotides in the absence

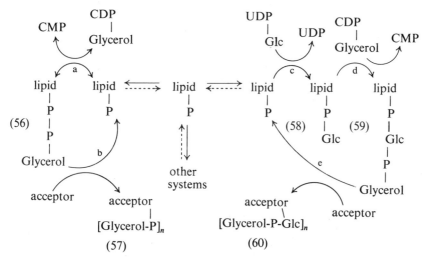

Figure 2.15 The synthetase systems for teichoic acids in *B. licheniformis* Glc = glucose (after Baddiley *et al.*[104] by courtesy of The Biochemical Journal)

(inhibition now 70.5%) and presence (inhibition now 83%) of bacitracin. All the inhibitory effects are clearly significant but some are quite small and the main difficulty in interpretation is the extent and nature of the 'trapping' of undecaprenol phosphate within a synthetase system. The results of work with 'O'-antigen determinant biosynthesis[92] (Section 2.3.5) suggest that within this synthetase complex the lipid-linked intermediate is freely exchangeable with exogenous intermediate.

The lipid-linked intermediate (58) (Figure 2.15) gave β-glucose 1-phosphate upon hydrogenolysis and glucose after mild acid treatment. Coupled with the fact that when (β-^{32}P]UDP glucose was used ^{32}P was not transferred to (58) and this provided strong evidence for an allylic monophosphate glucose. Production of the cyclic 1,2-monophosphate of glucose after extended alkali treatment supported a β-glucose 1-phosphate configuration. Identification of the allylic lipid as undecaprenol was again based on competition experiments.

It should be mentioned that one report[106] argues against the involvement of a lipid-linked intermediate in the transfer of glucose to teichoic acid. It was observed that during the glucosylation of teichoic acid by a solubilised transferase or by particulate preparations, at 37 °C, from *B.subtilis* only 0–2% of the ^{14}C provided as UDP [^{14}C]glucose could be recovered in the lipid fraction. Also the extraction of the solubilised enzyme with a mixture of chloroform and methanol had no effect on the incorporation of glucose into teichoic acid. This last result is difficult to understand especially in view of a recent observation that the solubilised teichoic acid synthetase of *B.licheniformis* (by a freezing and thawing technique) still uses lipid linked intermediates (N. Willmott, I. C. Hancock, and J. Baddiley, 1972, private communication). It is also worth noting that low recovery of ^{14}C in lipid in incubations at 37 °C has been observed also during the biosynthesis of cellulose, 'O'-antigen determinant and in capsular polysaccharides. However, in each case the incorporation increased considerably when incubations were carried out at lower temperatures. (See Sections 2.3.3, 2.3.5 and 2.3.6, respectively.)

2.4 POLYPRENOLS AND MAMMALIAN GLYCAN BIOSYNTHESIS

2.4.1 Introduction

It is clear from Section 2.3 that in bacterial systems polyprenol phosphates play an essential part in the transfer of sugars from nucleotide diphosphate sugars to a wide range of acceptors. These acceptors are usually membrane- or wall-bound and include homopolysaccharides, heteropolysaccharides, glycolipids and glycopeptides (peptidoglycans). Analogous groups of glycans are present in mammals. In view also of the presence of *cis,trans*-polyprenols (Section 2.2.2.1) investigations have been carried out concerning the possible formation of polyprenol phosphate glycosides, and their function, in mammalian tissues.

The analogous homopolysaccharide concerned is glycogen and it is relevant to note that glycogen synthetase is generally regarded as being bound to the glycogen particle which in turn is in intimate contact with membrane fractions. Also found in membrane fractions are the glycoproteins that are to be secreted from the cell or are to remain part of the membrane system of the cell. Several glycolipids connected with cell surface phenomena are also located in membrane structures.

Most interest in mammalian lipid-linked glucose carriers has centred round glycoprotein biosynthesis. A brief summary of the types of structures encountered may be useful in the ensuing discussion. More detailed reviews appear elsewhere[2, 107-109].

The glycoproteins are proteins that carry a carbohydrate polymer portion linked covalently. They are not characterised particularly by their amino acid content, except that they always contain asparagine or hydroxy-amino acids. They are usually differentiated on the basis of their carbohydrate composition. Most of those that have been isolated and studied in detail fall into one of three basic groups with the generalised structures (61), (62) and (63). However, throughout this discussion it should be appreciated that the writing of generalised structures in such a complex field is a major simplification of the situation which is justified when dealing with general points but may be misleading when discussing details.

$$\begin{bmatrix} \text{SA} \\ \text{or -Gal-NAcGlc} \\ \text{Fuc} \end{bmatrix}_{n_1} \quad \begin{bmatrix} \text{NAcGlc} \\ | \\ \text{Man} \end{bmatrix}_{n} \quad \begin{matrix} \text{H O} \\ | \; \| \\ \text{-NAcGlc-N-C-Asp} \\ | \\ \text{protein} \end{matrix}$$

(61)

$$\begin{matrix} & & \text{Thre} \\ \text{SA-NAcGal-O-} & \text{or} & \\ & & \text{Ser} \\ \text{NAcGal-Gal} & & | \\ | & & \text{protein} \\ \text{Fuc} & & \end{matrix} \qquad [\text{GlcUA-NAcGal}]_n\text{-GlcUA-Gal-Gal-Xyl-O-} \begin{matrix} \text{Thre} \\ \text{or} \\ \text{Ser} \\ | \\ \text{peptide} \end{matrix}$$

(62)　　　　　　　　　　　　　　　　(63)

The serum, or plasma, glycoproteins (61) contain branched oligosaccharide units in which n and n_1 varies from 1 to 5. The units make up to 40% of the weight of the molecule; there are up to several hundred of these per molecule, depending on the particular compound. Each glycan unit is linked through an N-glycosidic link between the N-acetylglucosamine residue at the reducing end of the unit and the amido group of an asparagine residue in the protein core. In general, mannose residues are located quite close to the core and galactose residues are more distal, whereas N-acetylglucosamine residues are found throughout the unit. Sialic acid or fucose residues are always situated at the non-reducing end of the chain. All plasma proteins, except albumin, appear to be glycoproteins of this type. An example is the oligosaccharide unit of α_1-acid glycoprotein (64) of human serum. The immunoglobulins often contain terminal fucose in place of sialic acid. The oligosaccharide units are synthesised by sequential addition of monosaccharides in the endoplasmic reticulum—Golgi system. The process is discussed in more detail later.

$$\begin{matrix}
\text{SA} & \text{SA} & \text{SA} & \text{SA} \\
\alpha \, | & \alpha \, | & \alpha \, | & \alpha \, | \\
\text{Gal} & \text{Gal} & \text{Gal} & \text{Gal} \\
\beta \, | & \beta \, | & \beta \, | & \beta \, | \\
\text{NAcGlc} & \text{NAcGlc} & \text{NAcGlc} & \text{NAcGlc} \\
& \beta \searrow \quad \swarrow \beta & & \beta \searrow \quad \swarrow \beta \\
& \text{Man} & & \text{Man} \\
& \alpha \searrow & & \alpha \, | \\
& & \text{NAcGlc} & \\
& & \beta \, | & \\
& & \text{Man} & \text{SA-NAcGal-O-serine} \\
& & ? \, \alpha \, | & | \\
(64) & & \text{NAcGlc} & \text{protein} \\
& & | & (65) \\
& & \ldots \text{Asp} \ldots \text{protein} &
\end{matrix}$$

The mucin type (62) contain several hundred small carbohydrate units, usually di- or penta-saccharides linked to threonine or serine residues of a protein chain through an O-β-glycosidic bond with N-acetyl galactosamine. In contrast to the serum type these glycoproteins do not contain mannose or N-acetylglucosamine but do contain N-acetyl galactosamine. If present, sialic acid residues are always terminal in both types. Most plasma-membrane glycoproteins are of this type. Ovine submaxillary mucin with the disaccharide unit shown in (65) is a simple and typical example of this type. Glycosylation of these glycoproteins generally occurs in the Golgi apparatus.

The third group, the mucopolysaccharides or proteoglycans are often not classified as glycoproteins because of the small proportion of protein (or peptide) in the molecule. A short polypeptide chain carries the long unbranched polysaccharide (63) in which n is several hundred. The presence of uronic acids and absence of sialic acid also distinguishes these glycan chains from those of serum and mucin glycoproteins. The structure (63) represents the glycan portion of chondroitin—a typical member of this group. Synthesis of the polysaccharide chain occurs in the endoplasmic reticulum—Golgi system. There is no evidence for the involvement of polyprenol phosphates in the process.

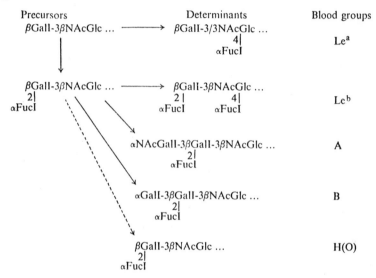

Figure 2.16 The synthesis of Lewis blood-group determinants. Fuc = fucose, Nac Gal = N-acetylgalactosamine, (From Ginsburg[111], by courtesy of Academic Press)

One of the most convincing demonstrations of the importance of the proper control of glycosylation to the biological properties of the final product is provided by work on the human blood-group substances. The determinants of blood group specificity are small oligosaccharides carried on lipids or proteins found on the surface layers of cells and in some body fluids. When attached to proteins these compounds are glycoproteins of the mucin type (Figure 2.16). It has been shown that the formation of the glycosyl transferases catalysing the introduction of these sugar residues is under tight genetic control[110, 111]. Since the relative rates of the glycosylations will also be relevant, any rate-controlling factor in this process will be of biological importance. However, no evidence has been published that involves polyprenol phosphates in the process.

Other cell surface phenomena appear to be mediated through oligosaccharide units attached to lipids. One of the major classes of glycolipids concerned is the gangliosides. The final stages in the biosynthesis of these is summarised in Figure 2.17[112]. Normally all five compounds are formed in cultured fibroblasts but if a virus, which will produce hepatoma in liver, is introduced, there is a block between (66) and (67). Normally cell-to-cell contact facilitates this step, (66)→(67), and this appears to be important in maintaining confluent growth probably through contact inhibition. When the sialyl transferase activity (for (66)→(67)) is missing, contact inhibition is lost and there is a consequent lack of growth restraint. Such cells are described as transformed.

It has been suggested by Roseman[113] that contact recognition and adhesion of one cell to another may involve the binding of surface-bound highly specific glycosyl transferases of one cell with substrate (an oligosaccharide chain) present in the surface of the other. The hypothesis also offers an explanation of the dissociation of cells. The presence of an appropriate nucleoside diphosphate sugar enables the transferase to glycosylate the oligosaccharide and

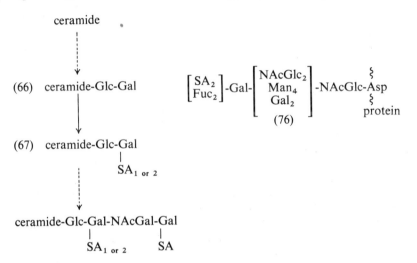

Figure 2.17 The transfer of sugars to ceramide. SA = sialic acid

the highly-specific transferase then no longer binds to the modified glycan chain. This serves again to illustrate the biological potential of factors that may affect the rates of glycosylation reactions. However, as yet, there has been no report of the involvement of polyprenol phosphates in phenomena of this type.

Usually not all the carbohydrate chains of glycoproteins and glycolipids are complete. This microheterogenity has been explained[109] on the basis that the monosaccharides are added sequentially and that as the chains grow, steric factors hinder the approach of all the glycosyl transferases. Another relevant factor may be the speed of passage through, or change of, the endoplasmic reticulum and Golgi apparatus. An understanding of the problems involved requires mention of current views on the relationship between the membrane systems of cells and the biosynthesis of these compounds. This will also complete the background to assist the possible roles of prenol phosphates to be seen in perspective.

Most work has been with liver cells and a view that is generally accepted is probably best presented by reference to a recent paper on[114] the biosynthesis of serum glycoproteins in rat liver. Figure 2.18 illustrates the process of development from rough to smooth endoplasmic reticulum through the Golgi apparatus to the plasma membrane[115,116]. The core protein is made on polysomes attached to the membrane of the rough endoplasmic reticulum (stage (a)). The first N-acetylglucosamine residue is probably added while the protein is still attached to the polysome and before the protein chain is complete (stage (b)). Mannose and further residues of N-acetylglucosamine are then added still at the rough endoplasmic reticulum stage, within the membrane, the protein no longer being attached to the polysomes (stage (c)). As the membrane loses its polysomes and becomes smooth endoplasmic reticulum, further molecules of N-acetylglucosamine (and possibly mannose) are added and the addition of galactose begins (stage (e)). However, most of the galactose and all of the sialic acid (or fucose) is added in the Golgi apparatus (stage (f)). At present, technical limitations make it extremely difficult to distinguish between a gradual change in properties of the membrane and the addition of sugar residues, as depicted in Figure 2.18, and a more clearcut stepwise progression in which the addition of different sugar residues to different positions along the oligosaccharide is associated only with certain morphological forms of the membrane. The scheme in Figure 2.18 accommodates most observations made using liver cells and probably holds also for the cells of several other tissues. One point of contention is probably the stage at which mannose residues are added. There is some evidence [118,119] that mannosyl transferase activity is associated with smooth endoplasmic reticulum which is at variance with the scheme. Other workers[114,129,132] (and F. W. Hemming, 1969, unpublished work) locate this activity in both smooth and rough endoplasmic reticulum, with a preponderance in the latter.

The scheme (Figure 2.18) is based in part on radioautographic evidence with whole cells but most of the evidence has been gained from work with cell-free particulate preparations. A close study of the glycosyl transferase activities of microsomal preparations has led to the conclusion that in some cases at least, polyprenol phosphates act as glycosyl acceptors. In addition some of the resultant glycosides have been shown to act as intermediates in the transfer of sugars to proteins. A detailed discussion of the different transferases concerned follows.

2.4.2 Dolichol phosphate derivatives

2.4.2.1 Glucosyl transfer

Much of the impetus to the establishment of dolichol phosphate derivatives as intermediates in mammalian glycosyl transferase systems has come from the studies of the metabolism of UDP glucose by microsomal preparations of rat liver carried out in Leloir's laboratory. These preparations which probably contain membranes from rough and smooth endoplasmic reticulum and from the Golgi apparatus catalyse the series of reactions summarised in Figure 2.19. The difficulty of obtaining sufficient of the lipid-linked mono- and oligo-saccharides (69)–(71) in a pure condition explains why the lipid moiety has not been characterised by unequivocal chemical analysis. The lipid-linked monosaccharide was studied first and it has been shown[32] that its properties are consistent with the acceptor (68) being dolichol monophosphate and with the intermediate (69) consisting of this joined to glucose through a β-glycosidic linkage.

When the particulate preparation was incubated with UDP [^{14}C]glucose, [^{14}C]glucose was incorporated into a lipid (step (a), Figure 2.19) which behaved as a single weakly-acidic compound in several chromatographic systems. The transfer of [^{14}C]glucose to lipid was increased by the addition of

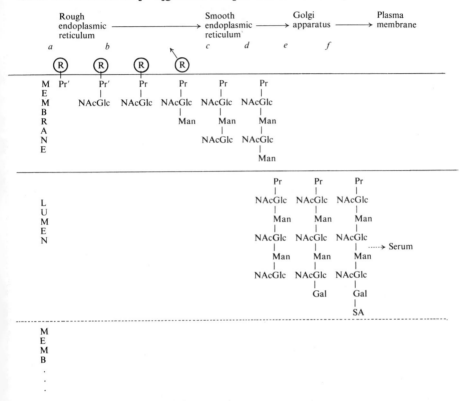

Figure 2.18 Schematic representation of glycosylation of protein to form serum glycoprotein during the development of the membranes of rough endoplasmic reticulum into plasma membranes in rat liver. R = ribosomes, Pr = protein, Pr' = incomplete protein. (From Redman et al.[114], by courtesy of The Rockefeller Institute)

dolichol monophosphate (prepared by the chemical phosphorylation of purified dolichol) or a lipid isolated from liver which was identical to dolichol monophosphate chromatographically and in its stability to dilute acid and alkali. The intermediate (69) can be deglucosylated rapidly by mild treatment with acid and by severe treatment with alkali (see Section 2.2.1.7(c)). The stability of the acceptor (68) to relatively severe acid treatment proved both the absence of a double bond β to the phosphate group and also that it was a monophosphate rather than a diphosphate (Section 2.1.7(c)). The β-configuration of the glucose residue in (69) was suggested by the rate of production of

1,6-anhydroglucose upon relatively severe alkali treatment (Section 2.2.1.7(c)). When [β-^{32}P]UDP glucose was incubated with the preparation ^{32}P was not recovered in the lipid[32, 120] confirming the monophosphate bridge. Slight reversal of step (a) by excess UDP has also been observed[97].

Under optimal conditions the amount of dolichol monophosphate glucose (69) formed is directly proportional to the quantity of dolichol monophosphate added. Using UDP [^{14}C]glucose of high specific activity this offers a sensitive method of assaying dolichol monophosphate. The method has been used to determine the sub-cellular distribution of dolichol monophosphate[65] and also to assay its release by acid hydrolysis of dolichol diphosphate derivatives (70, (71).

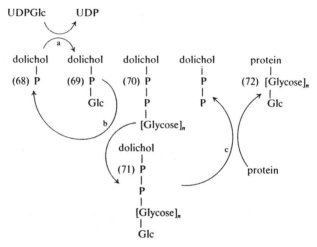

Figure 2.19 The role of dolichol phosphates in the transfer of glucose to protein in rat liver (From Leloir et al.[20-123], by courtesy of Academic Press, Elsevier and National Academy of Sciences, USA)

Initially, the nature of the acceptor of glucose from dolichol monophosphate glucose was not clear. There was evidence that the glucose eventually became associated with protein[32]. There was also evidence against involvement in the biosynthesis of collagen (a glycoprotein known to contain glucose) or of gangliosides (Figure 2.17)[120]. The former was tested with a skin enzyme preparation and the latter with chick embryo brain microsomes. In this second system both radioactive dolichol monophosphate glucose (acid-labile) and glucosyl ceramide (acid-stable) were formed from UDP [^{14}C]-glucose but there was no evidence for transfer of [^{14}C]glucose from one lipid to the other. The [^{14}C]glucose from dolichol monophosphate [^{14}C]glucose did, in fact, become associated with protein in a similar manner to that in liver.

Subsequent work[121, 122] has shown that part of the [^{14}C]glucose associated with protein as a result of incubations of liver microsomes with UDP [^{14}C]glucose or with dolichol monophosphate [^{14}C]glucose can be removed by solution in special mixtures of chloroform–methanol–water (optimal ratios 3:3:1 by volume) but not by the standard mixture of chloroform–methanol (2:1 by volume) usually employed to dissolve polyprenol phosphate glycosides. Mild acid treatment of the product with special solubility properties (71) yielded a lipid-soluble compound, presumably dolichol disphosphate which on further acid treatment was converted into a lipid-soluble material which will substitute for dolichol monophosphate in step (a). A water-soluble product of acid hydrolysis of (71) had the properties of an uncharged oligosaccharide containing 20 unidentified neutral monosaccharide units. Relatively severe alkaline treatment of (71) yielded a water-soluble [^{14}C]product of identical size carrying a negative charge that could be removed by alkaline phosphatase. This suggested a diphosphate bridge linked to C-1 of a monosaccharide residue in the oligosaccharide. The molecular weight of the original incubation product (71) determined by gel filtration of the deoxycholate inclusion compound was consistent with the structure (71) in which $n = 19$. Chromatography on DEAE cellulose was also consistent with a diphosphate bridge. In fact the compound was not degraded by treatment with hot phenol (see Section 2.1.3(c)) and it was suggested that a monophosphate bridge was still a possibility. However, the balance of evidence favours a diphosphate bridge.

Demonstration of the transfer of the oligosaccharide from (71) to endogenous protein (step (c), Figure 2.19) in microsomal preparations was dependent upon the presence of deoxycholate and of Mn^{2+}. Triton X-100 would not substitute for deoxycholate although this was the case in steps (a) and (b). The final product was identified as a protein partly by its solubility characteristics in solutions of phenol, sodium dodecyl sulphate (containing urea), trichroacetic acid (insoluble) and several organic solvents (insoluble) and also by its degradation by protease and alkali. The last two treatments yielded the oligosaccharide moiety, possibly also containing a few amino acid residues, thus confirming the transfer of the entire oligosaccharide in step (c)[123].

The nature of the glycoprotein produced remains unknown. Most of the glycoproteins discussed in Section 2.4.1 are formed by sequential addition of monosaccharides to the protein core rather than by addition of a preformed oligosaccharide and none of the oligosaccharide moieties contain glucose distal to the protein core. Evidence has recently been reported[124] for the formation of a glucoprotein as an intermediate in glycogen biosynthesis. However, it seems unlikely that the two areas of work are directly related because (a) glycogen synthetase can be recovered in a subfraction of the cell different from that used above[123], (b) step a (Figure 2.19) is much more sensitive to EDTA than is the glycogen synthetase system and (c) neither α- nor α- plus β-amylase degraded the oligosaccharide liberated from (71) by acid methanolysis.

The formation of a compound with the properties expected of dolichol monophosphate glucose by liver microsomal preparations has been confirmed in the laboratories of Chojnacki[96,97] and Hemming (unpublished work). The former group has shown that ficaprenol monophosphate is less effective (30%) than dolichol monophosphate as an acceptor of glucose. Phytol monophosphate had no activity in this respect. The same specificity was found with rat brain microsomes. The stimulating effect of dolichol monophosphate of glucosyl transfer to lipid was accentuated by the presence of diglyceride although diglyceride itself had no effect.

While dolichol monophosphate, or a compound able to mimic it in step (*a*) (Figure 2.19), appears to be concentrated mainly in the nuclear membrane, Golgi membrane and rough endoplasmic reticulum the enzyme catalysing step (*a*) is concentrated mainly in the outer membranes of mitochondria and in smooth endoplasmic reticulum; there is considerable activity also in the membranes of the nucleus, Golgi apparatus and rough endoplasmic reticulum[64]. On the other hand the enzymic activity for step (*b*) appears to be concentrated either in Golgi membranes and smooth endoplasmic reticulum or in rough and smooth endoplasmic reticulum depending on the method of assay[64]. Possibly a more detailed study of the products formed by each cell fraction will rationalise what is at present a very confusing picture.

This aspect of glucosyl transfer is complicated further by an earlier report[125] on the biosynthesis of mitochondrial glucoproteins showing that UDP glucose:glucoprotein glycosyl transferase activity was associated with the inner membranes of rat liver mitochondria. The activity of this glucosyl transferase was observed to be much higher than of those using GDP mannose, GDP fucose or UDP xylose. It was also demonstrated that UDP glucose:lipid glucosyl transferase activity was concentrated in the outer membrane of the mitochondria. In fact three lipids were formed, one of which appeared to be a glucosyl diglyceride. The identity of the other two has not been reported.

2.4.2.2 Mannosyl transfer

The first report of a possible role for a mannolipid as an intermediate in mannosylation of glycoprotein came from Eylar's laboratory[119]. Using a preparation of smooth endoplasmic reticulum from rabbit liver [^{14}C]mannose was transferred from GDP [^{14}C]mannose to acid-labile, alkali-stable lipid and to protein and on addition of unlabelled GDP mannose the radioactivity associated with lipid fell while that associated with protein continued to rise slowly. Comparison of results from liver and oviduct (both tissues active in secretion of glycoproteins containing mannose) with those from myeloma tumour (not active in secretion of glycoproteins) led these authors to suggest that possibly glycosyl lipids were intermediates in the formation only of secreted glycoproteins and not of membrane-bound glycoproteins. The partially purified mannolipid contained a mannose:phosphate ratio of between 1:1 and 2:1 and was chromatographically very similar to undecaprenol monophosphate mannose. A similar mannolipid is formed by rat brain microsomes[127].

More recently the transfer of mannose to mannolipid and glycoprotein by microsomal preparations of pig liver has been studied in detail. The scheme in Figure 2.20 summarises some of the results of this study. Incubation of the preparation with GDP [^{14}C]mannose can lead to transfer of up to 40% of the mannose to endogenous lipid and up to 3% to endogenous protein. In the presence of exogenous dolichol monophosphate both figures can be doubled. The mannolipid formed chromatographs as a single compound and has properties similar to those of dolichol monophosphate glucose (Section 2.4.2.1). The firm identification (P. Evans, and F. W. Hemming, 1972, unpublished observations) of dolichol in (74) formed in large scale incubations is based on its physical properties (n.m.r., i.r.) and on the incorporation into (74) of [^{3}H]dolichol monophosphate by microsomal preparations[128, 129]. Chemical analysis showed equimolar proportions of mannose and phosphate and the presence of a monophosphate bridge and absence of a β-double bond is supported by the results of mild treatment with acid and severe treatment with alkali and hydrogenation. Acid treatment gave rise to only mannose and no mannose-1-phosphate or inorganic phosphate. Hydrogenation[42, 44] (see Section 2.1.7(c)) did not liberate mannose or mannose 1-phosphate. In both physical and chemical properties the mannolipid (74) was identical (Evans and Hemming, unpublished observations, 1972) to synthetic dolichol monophosphate mannose provided by Dr. C. D. Warren (see Section 2.2.1.7(b)). It is also relevant that an endogenous lipid closely resembling dolichol phosphate mannose has been observed in whole liver and in liver cell cultures (Oliver and Hemming, unpublished observations).

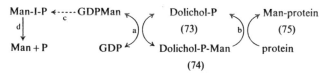

Figure 2.20 The role of dolichol monophosphate in the transfer of mannose to protein in pig liver (From Hemming *et al.*[128,129] by courtesy of Academic Press and Cambridge University Press)

The reversal of step (*a*) (Figure 2.20) by GDP, but not by GMP, confirmed the monophosphate bridge[128-130]. The rapid and ready reversibility of step (*a*) and, in these preparations, the slow rate of step (*b*), explains why the addition of excess unlabelled GDP mannose to a preparation containing dolichol phosphate [^{14}C]mannose leads to a rapid loss of most of the radioactivity from lipid (74) without a correspondingly large and rapid increase in radioactivity associated with protein. The fact that the pool size of GDP mannose is now greater than that of (73) and (74) and the rapid equalisation of specific activities results in the recovery of most of the radioactivity in the water-soluble GDP mannose. The activity of enzymes catalysing steps (*c*) and (*d*) in some preparations coupled with the rapid and ready reversibility of step (*a*) and the slow rate of step (*b*) explains why the ^{14}C transferred from GDP [^{14}C]mannose to (74) reaches an early peak and then drops off with only a slow and small transfer of ^{14}C to protein. The GDP mannose is slowly removed by conversion to mannose 1-phosphate and mannose (steps (*b*) and (*c*)) and step (*a*) goes into reverse resulting in the gradual appearance of most of the ^{14}C in the form of mannose and mannose 1-phosphate. Obviously in systems of this sort, a large fall in the level of the lipid-linked sugar during straightforward incubations and in pulse-chase experiments requires careful interpretation.

While step (a) (Figure 2.20) has been studied in detail, step (b) has, as yet, received less attention. Nevertheless, although it appears to proceed very slowly in microsomal preparations, there is good evidence for its occurrence. The transfer of 4% of the mannose of exogenous dolichol monophosphate mannose to protein has been reported[128, 129] and the transfer of mannose from GDP mannose to protein was increased by exogenous dolichol monophosphate. It was also observed that the presence of excess GDP at the beginning of an incubation inhibited transfer of mannose from GDP mannose to both dolichol monophosphate (73) and protein. However, when added after mannose had been transferred to both lipid and protein it caused a marked fall in mannose content of lipid (74) but did not affect the mannose content of protein[128, 129]. This is clearly in accord with Figure 2.20.

The nature of the glycoprotein (75) has not been investigated. The mannose remained associated with TCA-precipitable protein during gel chromatography and following extraction with detergents and organic solvents. No evidence has been found in this system for the formation of an intermediate analogous to compound 71 in Figure 2.19 which associates with protein (see Section 2.4.2.1). It is likely that the glycoprotein fraction (75) is a precursor, or mixture of precursors, of serum glycoproteins (61) (see Section 2.4.1).

Confirmation of reactions (a) and (b) in Figure 2.21 has come recently from Heath's group[131] (Heath, private communication, 1972) working on the biosynthesis of a glycoprotein by a mouse myeloma tumour. This tumour produces an immunoglobulin containing a glycan chain shown in (76). The presence of dolichol in the intermediate was shown by mass spectrometry and the reversal of step (a) by GDP, but not GMP, was confirmed. Additional evidence for step (b) was provided by the inhibitory effect of adding excess EDTA to a microsomal preparation that has formed (74) from GDP mannose. The EDTA completely inhibited the further activity of step (a) but allowed the transfer of mannose to protein to continue to an extent roughly proportional to the amount of (74) that had been formed prior to addition of EDTA. The presence of excess EDTA prior to build up of (74) completely inhibited transfer of mannose from GDP mannose to both (74) and protein. It was also observed that removal of lipid from a microsomal preparation by extraction with butanol inhibited the transfer of mannose from GDP mannose to lipid as protein but still allowed the transfer of mannose from exogenous dolichol phosphate mannose to protein.

In pig liver microsomes GDP mannose is a much more efficient donor of glycose to dolichol monophosphate than are other nucleotide diphosphate sugars. It has been observed that, using equimolar amounts of donor, the transfer of glycose to endogenous lipid when using UDP N-acetylglucosamine, UDP glucose, UDP galactose, ADP glucose and UDP glucuronate as donors was ca. 5%, 2%, 2%, <1% and <1%, respectively of the corresponding transfer of mannose from GDP mannose. (Evans and Hemming unpublished, 1972.) In rat liver microsomes exogenous dolichol monophosphate increases the transfer to lipid from only the first two of this list and from GDP mannose [120]. In both pig liver and mouse myeloma systems the presence of excess UDP N-acetylglucosamine did not affect the transfer of mannose to lipid or protein and in the myeloma system UDP glucose and

UDP galactose were without effect. However, the pig liver system was markedly inhibited by excess of either UDP glucose or, surprisingly, ADP glucose. The effect of UDP glucose can be interpreted as competition for the limited amount of dolichol phosphate acceptor. The complete lack of effect of excess UDP N-acetylglucosamine is surprising both in view of the possible formation of dolichol monophosphate or diphosphate N-acetyl-glucosamine and also in the light of the relative positions of N-acetylglucosamine and mannose in the serum glycoproteins (61).

The formation of mannolipid in rat liver microsomes is stimulated tenfold by the addition of optimum quantities of dolichol monophosphate[120]. This is approximately the same degree of stimulation reported for glucolipid formation. It is considerably more than obtained under similar conditions with pig liver microsomes, the basal rate of the latter being much higher. This may reflect a higher concentration of endogenous dolichol monophosphate in pig liver microsomes. Both mannolipid and mannoprotein are formed by mitochondrial membranes[125, 126]. In this organelle the inner membrane appears to catalyse the transfer of mannose from GDP mannose to protein and the outer membrane to catalyse its transfer to lipid. As with glucose (Section 2.4.2.1), the lipid formed was reported to contain three mannolipids, one of which corresponded to a mannosyl diglyceride, the other two remain unidentified.

The specificity of the enzyme catalysing step (a) (Figure. 2.20) to polyprenol phosphate has been investigated[128, 129]. It was shown that betulaprenol monophosphate caused as large an increase (fourfold) in the transfer of mannose to lipid as did dolichol monophosphate whereas solanesol monophosphate was only 65% as effective and ficaprenol monophosphate only 30% as effective. Farnesol monophosphate and cetyl monophosphate caused no stimulation. It thus appears that the enzyme catalysing step (a) has only loose specificity to the stereochemistry, chain length and saturation of the α-isoprene residue of the polyprenol concerned. Nevertheless, it may be significant that the most effective polyprenols tested were di*trans*,poly*cis*-prenols and that the polyprenol involved in bacterial glycosyl transfer is also of this type. The enzyme catalysing step (b) (Fig. 2.20) exhibits a much tighter specificity to dolichol as part of structure (74) for betulaprenol monophosphate and solanesol monophosphate were only 30% as effective as dolichol monophosphate in stimulating the transfer of mannose from GDP mannose to protein. The monophosphates of ficaprenol, farnesol and cetyl alcohol were ineffective in this respect.

2.4.2.3 N-acetyl glucosaminyl transfer

Molnar's group reported[130] the transfer of N-acetylglucosamine from UDP N-acetylglucosamine to an acid-labile, alkali stable lipid by rat and rabbit liver microsomes. At the same time the sugar was also transferred to protein and in a pulse-chase experiment the amount of lipid-linked sugar fell while that of protein-bound sugar rose. Both UMP and UDP inhibited the reaction, a fact consistent with the presence of a diphosphate bridge (assuming UDP→UMP + P). Subsequent work[132] has confirmed this for the ^{32}P and

^{14}C of [β-^{32}P]UDP [^{14}C]N-acetylglucosamine were transferred to the lipid in equimolar amounts. Leloir's group[120] has shown that exogenous dolichol monophosphate will act as an acceptor of N-acetylglucosamine. Thus the evidence is in favour of the scheme in Figure 2.21 although it is perhaps premature to consider it established. In particular the nature of the endogenous lipid acceptor and step (b) require further investigation.

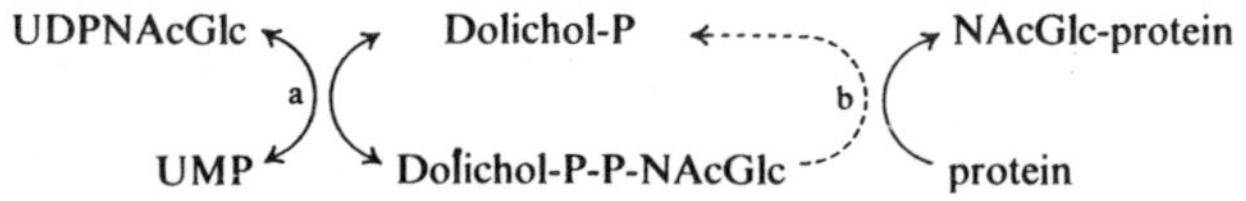

```
UDPNAcGlc ↘   ↗ Dolichol-P   ←------↘   ↗ NAcGlc-protein
          ) a (                   b )  (
     UMP ↙   ↘ Dolichol-P-P-NAcGlc ---↗   ↘ protein
```

Figure 2.21 The possible role of dolichol monophosphate in N-acetyl glucosaminyl transfer in mammalian liver (From Molnar et al.[132] by courtesy of Elsevier)

The work of Bosmann[133] shows that a solubilised N-acetylglucosaminyl transferase (from liver) functions quite well in the absence of exogenous lipids, although the effect of adding lipids was not tested. Apart from differences in specific activity no major differences could be found between the system using endogenous protein as acceptor and that using as exogenous acceptor the serum glycoprotein fetuin that had had the distal sialic acid, galactose and N-acetylglucosamine residues removed. This arises the possibility that the introduction of N-acetylglucosamine into endogenous protein normally studied is also a distal process. Possibly lipid-linked intermediates are concerned only with glycosyl introduction proximal to the protein core.

Although it has been reported that rat liver microsomes catalyse the transfer of N-acetylgalactosamine from UDP N-acetylgalactosamine to an endogenous lipid acceptor the failure of exogenous dolichol monophosphate to stimulate the transfer[120] suggests that the endogenous acceptor was not a polyprenol monophosphate.

2.4.3 Retinol phosphate derivatives

2.4.3.1 Introduction

An involvement of retinol in membrane biochemistry has been a recurring feature of the conclusions of several studies concerning the mode of action of this fat-soluble vitamin. The results of other experiments have pointed to a role in mucopolysaccharide biosynthesis. These two themes, of the many still receiving active investigation[134], have been extrapolated and merged in recent reports of an essential function as the lipid moiety of lipid linked sugars involved in the biosynthesis of glycoproteins. The first hint of the possibility of such a role was the observation[135] that in vitamin A-deficiency the transfer of N-acetylglucosamine to a specific fucose-containing glycoprotein of the intestinal mucosa was depressed. Whereas no further evidence has been published in favour of a role in N-acetylglucoasaminyl transfer (in fact there is evidence against such a function in rat liver microsomes), strong cases have

(77)

been made for a role in mannosyl, galactosyl and glycosyl transfer. It is relevant to note that retinol is a cyclised form of tetradehydro tetraprenol (77).

2.4.3.2 Mannosyl transfer

The transfer of mannose from GDP mannose to lipid in ATP-supplemented liver microsomal preparations from mildly retinol-deficient rats is half of that in identical preparations from normal rats[136]. Addition of retinol to the incubation medium restored this to normal. The addition of retinol in the absence of ATP was ineffective. On an equimolar basis exogenous dolichol was only half as effective as exogenous retinol but exogenous dolichol monophosphate was 16 times more effective. It was suggested that the low basal rate in retinol-deficient preparations was possibly in part due to endogenous dolichol monophosphate. The concurrent production of dolichol monophosphate mannose complicated the analysis of the retinol product for although separation of the two products could be achieved chromatographically analytical results suggested that some purified preparations contained a mixture of the two. Nevertheless, addition of [15-^{3}H]a-retinol to incubations containing GDP[^{14}C]mannose gave rise to [^{3}H, ^{14}C] mannolipid containing approximately equimolar quantities of retinol and mannose. A chromatographically-identical [^{3}H, ^{14}C]mannolipid was isolated from liver microsomes of a retinol-deficient rat that had been dosed with [^{3}H]a-retinol over 4 days and had received an injection of [^{14}C]mannose 20 min before sacrifice. A similar result followed incubation with GDP [^{14}C]mannose of the microsomes of another retinol-deficient rat dosed with [^{3}H]retinol in the same way. The presence of a phosphate group in the lipid was demonstrated by the incorporation of ^{32}P from [γ-^{32}P]ATP added to the incubation medium. That a monophosphate bridge was involved was suggested by the failure of ^{32}P to be transferred from [β-^{32}P]GDP mannose and by the reversal of the transfer of mannose by GDP. Mild acid treatment of the [^{3}H, ^{14}C]lipid released a [^{3}H]lipid identified as retinol and [^{14}C]mannose. Although the mannolipid was stable to mild alkali, severe alkali treatment yielded mannose 1-phosphate. Hydrogenation of the purified [^{3}H, ^{14}C]lipid yielded a [^{3}H]hydrocarbon and [^{14}C]mannose. While hydrogenolysis of an allylic phosphate mannose is expected it is surprising that mannose rather than mannose 1-phosphate was produced (see Section 2.2.1.7(c)). However, the presence of retinol in the lipid fraction appears well-established.

Thus there is a strong body of evidence to favour the formation of retinol monophosphate mannose in microsomes of mildly retinol-deficient rats by a reaction analogous to step (a) in Figure 2.20. The results also support the presence of a retinol phosphokinase in these preparations although the formation, or presence, of retinol phosphate has not been demonstrated directly. A preliminary report[138] indicated the transfer of mannose from this retinol-intermediate to protein. It has also been reported briefly[139] that retinoic acid is incorporated into a lipid that contains mannose and phosphate but detailed evidence for this has not yet been published. However, since it is generally accepted that retinoic acid cannot be converted to retinol it is unlikely that this lipid is the same as that described above.

2.4.3.3 Glucosyl transfer

During the course of investigations into the incorporation of exogenous retinol into a mannolipid by microsomes of retinol-deficient rats (see Section 2.4.3.2) De Luca et al.[137] observed the formation of a retinol-containing glucolipid when the incubation medium contained UDP glucose instead of GDP mannose. The glucolipid was shown to be acid-labile but no further evidence is available. The same group has also reported[139] an intermediary role for a glucosylated form of retinol in the glucosylation of collagen catalysed by human platelet membranes.

2.4.3.4 Galactosyl transfer

Two laboratories have investigated the possible transfer of galactose from UDP galactose to endogenous lipid or to exogenous dolichol phosphate in rat liver microsomal preparations and have found this to be negligible[96,97,120]. In pig liver microsomes the transfer to endogenous lipid is also very small (see Section 2.4.2.2). Bosmann's group report[125,126] that although the inner membranes of mitochondria will transfer galactose to endogenous protein, the transfer to lipid by either inner or outer membranes is insignificant. However, there is one report[130] of transfer of galactose to form an acid-labile lipid-soluble material by rat liver microsomes but further details have not yet been published.

The most positive report of transfer of galactose to a lipid-linked 'intermediate' is of preliminary nature from Peterson's laboratory[140]. The conclusion here was for the production in microsomal preparations from a mouse mastocytoma of a retinol phosphate galactose. The results of hydrogenolysis favoured an allylic monophosphate galactose, yielding galactose monophosphate, but there was a puzzling stability to acid treatment (0.1 M HCl, 30 °C, 45 min, see Section 2.2.1.7(c)). The transfer of galactose from this compound to material insoluble in lipid and aqueous solvents was also reported. A related [^{14}C]lipid was also formed from GDP [^{14}C] mannose but not from UDP [^{14}C]xylose or UDP [^{14}C]glucuronic acid. The implication of this is that the system under study was more likely to be involved in the synthesis of serum glycoproteins (61) than mucopolysaccharides (63). A detailed report of this work is now awaited.

2.4.4 Other lipid effects

Since glycosyl transferases are bound to lipid-rich membranes it is not surprising that their activity can be affected by the presence of detergents and by exogenous lipids or by the modification of endogenous lipids. Detergent effects have been observed in most of the previous sections and in fact the presence of a detergent, commonly Triton X-100, is often essential for maximal transferase activity and appears in most standard incubation media. The dependence of the glycosyl transferase activity of solvent-extracted particulate preparations upon exogenous phospholipid is well-described in bacterial systems[3] and has not gone unnoticed in eukaryotic organisms. Recently, a different aspect of this dependence upon phospholipids has been highlighted by the work of Mookerjea's group. They reported[141] that the transfer of galactose from UDP galactose to endogenous and exogenous proteins by rat liver microsomes was markedly stimulated by CDP choline in a synergistic manner with Triton X-100. It was suggested that the process of regeneration of membrane lecithin was activating the membrane galactosyl transferase, possibly by initiating conformational changes and that Triton X-100 aided this by stimulating protein agitation. Exogenous CDP choline has also been shown[142] to stimulate N-acetylglucosaminyl transfer to exogenous protein by rat liver microsomes, but it appears to be ineffective in mannosyl transfer to endogenous lipid by microsomes of mouse myeloma tumour (Heath, 1972, private communication). Conformational changes of membrane protein consequent on phospholipid changes, have also been proposed[143] to explain the stimulation of UDP glucuronic acid: p-nitrophenol glucuronyl transferase activity in bovine liver microsomes by phospholipase A and phospholipase C.

An awareness of these factors is clearly important in interpreting the results of studies with glycosyl transferases. In particular it has raised the necessity of workers in this field to distinguish between a role for lipids as acceptors for glycosyl transferases and one simply as stimulants of glycosyl transfer to lipid.

2.5 POLYPRENOLS AND GLYCAN BIOSYNTHESIS IN YEASTS AND FUNGI

2.5.1 Mannosyl transfer in yeast

Cell walls of *Saccharomyces cerevisiae* contain a polymannan as well as oligomannan linked to protein through an O-β glycosidic linkage with serine or threonine residues. The mannosidic bonds present are primarily a, 1–6 but side chains containing 1–2 and 1–3 linkages have also been reported. This organism also contains dolichols (Table 2.1) and investigations into a possible role of dolichol-linked mannose in mannan biosynthesis have been carried out.

The results of Tanner's group, using particulate preparation of *S.cerevisiae*[144-146], are consistent with the scheme in Figure 2.22. The [^{14}C]lipid (78) formed from GDP [^{14}C]mannose chromatographs as a single negatively-charged compound in a manner expected of dolichol phosphate mannose and is labile to mild acid but stable to mild alkali treatment (Section 2.2.1.7(c)). Exogenous dolichol monophosphate, isolated from pig liver or prepared by chemical phosphorylation of yeast dolichol, caused a four to fivefold increase in transfer of mannose from GDP mannose to the same compound and a 50% increase in transfer to polymer. Excess GDP, but not GMP, caused the reversal of step (*a*) which suggests a monophosphate bridge in (78). Addition of (78) labelled with [^{14}C]mannose led to the appearance of [^{14}C]mannose in polymer.

Step (b) was confirmed by taking advantage of a requirement for Mn^{2+} for mannan production whereas step (a) utilises either Mn^{2+} or Mg^{2+}. Radioactivity was allowed to build up in (78) by incubating with GDP [^{14}C] mannose in the presence of Mg^{2+} and upon addition of Mn^{2+} transfer of ^{14}C from (78) to (79) was observed. The scheme appeared to be concerned not only with biosynthesis of mannoprotein (Tanner, private communication) but also with side chain lengthening of polymannan. Other preliminary work[147] using preparations of S.cerevisiae supports the conclusions summarised in Figure 2.22 although evidence against the occurrence of lipid-linked intermediates in S.carlsbergensis has also been reported[148].

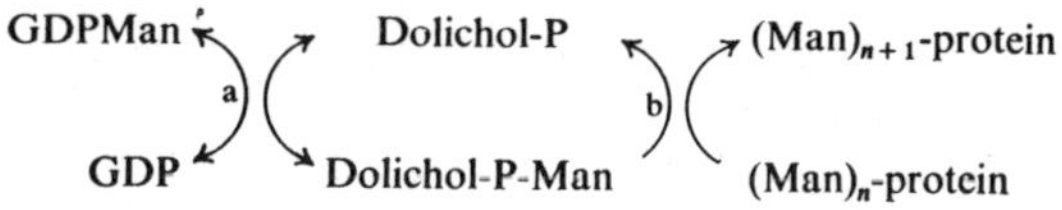

$$\text{GDPMan} \xrightleftharpoons[a]{} \text{Dolichol-P} \xrightleftharpoons[b]{} (Man)_{n+1}\text{-protein}$$
$$\text{GDP} \qquad \text{Dolichol-P-Man} \qquad (Man)_n\text{-protein}$$
$$(78)$$

Figure 2.22 The possible role of dolichol monophosphate in mannosyl transfer in S.cerevisiae (From Tanner et al.[144-146], by courtesy of Academic Press and Federation of American Societies for Experimental Biology)

A similar scheme appears to hold for the synthesis of mannan in particulate preparations of Hansenula holstii[149] although to date only step (a) has been demonstrated conclusively (R. K. Bretthauer, private communication). This step is again readily reversible by GDP and uses dolichol [^{32}P]monophosphate to form dolichol monophosphate mannose. The latter has been identified as a product with endogenous acceptor by virtue of its chromatographic properties and lability to acid (to give mannose) but stability to alkali. As with the other lipid-linked intermediates of yeast systems, results of direct chemical studies on the lipid portion have not yet been reported.

The secretion of a phosphorylated mannan containing repeating units (phosphate-man-man-man-man-man-) by H.holstii raises the possibility of transfer of mannose 1-phosphate from GDP mannose via a lipid-diphosphate-mannose intermediate but there is not yet any evidence for this (R. K. Bretthauser, private communication).

2.5.2 Mannosyl transfer in Aspergillus niger

The involvement of polyprenol phosphate in polymannan biosynthesis in fungal systems cannot be considered as established although it has been reported[35] that in Aspergillus niger a cell free system capable of catalysing the transfer of mannose from GDP mannose to endogenous mannan also catalyses its transfer to endogenous polyprenol monophosphate, presumably methylene-, hexahydro-, di*trans*-, poly*cis*-prenol monophosphate (see Table 2.1) by a reaction analogous to step (a) in Figure 2.22. The endogenous polyprenol monophosphate was identified chromatographically and on the basis of incorporation of [2-^{14}C]mevalonate and [^{32}P]inorganic phosphate from the growth medium. The [^{14}C]mannolipid formed during incubations with GDP [^{14}C]mannose was recognised chromatographically by its acid lability and by the increase in its formation in the presence of synthetic methylene-, hexahydro-, di*trans*-, poly*cis*-prenol monophosphate. The transferase also used the exogenous monophosphate of ficaprenol, but not that of cetyl alcohol as acceptor. The possibility that the mannolipid could be an intermediate in the formation of mannan has not been investigated further.

A cell free preparation of *Aspergillus oryzae* has been shown[150] to transfer mannose from GDP mannose to an endogenous glycoprotein but in this work lipid-soluble products were not studied. In a cell free preparation of another fungus, *Cryptococcus laurentii*, mannose was observed[151] to be transferred to both endogenous polymer containing a-1,2- and a-1,3-mannosyl bonds and to endogenous lipid. However the mannolipid was fairly stable to acid and labile to alkali treatment and in a pulse-chase experiment there was no evidence of the loss of [^{14}C]mannose from the lipid.

Evidence against the involvement of lipid linked intermediates in fungal systems has also been published[152].

2.6 POLYPRENOLS AND GLYCAN BIOSYNTHESIS IN GREEN PLANTS

2.6.1 General

Green plants produce a variety of polysaccharides ranging from starch in the plastids to cellulose, hemicelluloses, pectins and glycoproteins in the cell walls. The chemistry and biosynthesis of the wall glycans has been summarised by Northcote[153, 154]. There is good radioautographic and histological evidence for believing that much of the polymerisation of monosaccharides occurs in the Golgi apparatus and that this is transferred to plasma membranes as part of the development and movement of membrane from endoplasmic reticulum in a manner similar to that in mammalian systems discussed in Section 2.4.1. Whereas the synthetase systems for hemicelluloses and pectins appear to complete their activities before the membrane reaches the plasma membrane very little synthesis of cellulose occurs prior to this final location. The development and differentiation of many plant cells can be characterised by the nature of their cell wall and it has been suggested that control of the formation of nucleoside diphosphate glycoses and of their utilisation in glycan biosynthesis, is an important aspect of the process of differentiation. Because the glycosyl transferases concerned in the synthesis of cell wall glycans are membrane-bound and in view of the widespread occurrence in plant tissues of polyprenols (see Section 2.2.2.2) it is not surprising that several laboratories have been the site of investigations into the possible participation of prenol phosphates in the transfer process. In fact only in studies on mannosyl transfer and glucosyl transfer has evidence been put forward for the involvement of prenol phosphate glycose intermediates. A more detailed discussion of this evidence follows. Plant glycosyl transferases in general have been reviewed by Hassid[155, 156].

2.6.2 Mannosyl transfer

The transfer of ^{14}C mannose from GDP[^{14}C]mannose to an endogenous lipid by particulate fractions of young shoots of *Phaseolus aureus* has been observed in the laboratories of Kauss[157,158], Villemez[159,160], Hemming[161,162] and Hassid[156]. There is general agreement that the mannolipid yields mannose on mild acid treatment and the formation of some mannose 1-phosphate has been reported. It is stable to mild alkali and behaves in a manner very similar to undecaprenol monophosphate mannose in several chromatographic systems. These data suggest the formation of an allylic polyprenol monophosphate mannose by a reaction similar to step (*a*) in Figure 2.22 and the observed reversal of this by GDP[156,157,160] (but not by GMP) confirms this. It has been shown[163], using [^3H]betulaprenol monophosphate, that the transferase will use a prenol phosphate as acceptor but since dolichol monophosphate is also effective as acceptor, this does not prove the identity of the lipid moiety. The incorporation of [^3H]mevalonate into the partially purified lipid has been reported[157] and a very preliminary mention of a mass spectrum in favour of a polyisoprenoid structure has been made[160]. However, if a polyprenol phosphate is involved, the concentration must be extremely small for a careful search, using methods successful with other tissues, failed to reveal the presence of chemically detectable quantities of polyprenol or polyprenol phosphate derivatives in large quantities of *P.aureus* shoots[162]. In fact the failure to detect a lipid-linked intermediate *in vivo* is reminiscent of the situation in *P.aureus* with regard to GDP glucose. This compound is widely believed to be the precursor of cellulose in this plant but it has never been detected *in vivo*. It has been argued that this is probably due to its high rate of turnover[155]. The precise nature of the endogenous mannolipid must still be regarded as uncertain. A chromatographically identical mannolipid is formed by particulate preparations of *Pisum sativum*[156] and of tomato roots and of the alga *Codium fragile*[162].

While transferring mannose to lipid, the preparation from *P.aureus* also catalyses the transfer of mannose to glycoprotein and to polysaccharide. Neither of these products has been characterised. The glycoprotein has been reported[162] to be the major product but it has also been suggested[160] that this might well be an intermediate in the formation of the polysaccharide (presumably including mannans, glucomannans and galactoglucomannan). Whether or not the mannolipid is an obligatory intermediate in the formation of either glycoprotein or polysaccharide is still uncertain. There is evidence consistent with such a role and analogy with bacterial and mammalian systems provides a persuasive argument but the evidence is not yet firm. The kinetics of transfer of mannose to lipid and polymer have been reported on one hand to be in favour of such a scheme[158,159] and on the other hand to be inconclusive[162]. The results of pulse-chase experiments[157] would also allow this role as would the fact that *p*-hydroxymercuribenzoate causes an increase in recovery of mannolipid and a drop in transfer of mannose to polymer that can be reversed if mercaptoethanol is also present[162]. However, attempts to demonstrate the transfer of [^{14}C]mannose from exogenous [^{14}C]mannolipid to polymer have resulted in the recovery of pitiably small quantities of ^{14}C in the polymer fraction[156,157,159,162]. It has also been reported that endogenous [^{14}C]mannolipid, formed by incubating particulate preparations of *P.aureus* with GDP [^{14}C]mannose and then removing the latter by centrifugation and washing, failed to pass ^{14}C on to polymer during further incubation although the washed preparation was still able to transfer [^{14}C]mannose to polymer from GDP [^{14}C]mannose[156].

A complicating factor in these studies is that the polymer fraction is usually heterogeneous containing, for example both glycoprotein and polysaccharide. Very recently a solubilised mannosyl transferase, from *P.aureus* particulate preparations, has been described[164] which catalyses the formation of a (β1-4)-linked polymannan. The same activity was observed in a particulate preparation that had been extracted with polar organic solvents and thus presumably did not require the presence of any lipid. Whether one or more of the other polymers present *in vivo* is formed via a lipid-linked intermediate is still an open question.

2.6.3 Glucosyl transfer

Glucose is present in the polysaccharides of plant cells in the form of three glucans, namely cellulose (β-1,4 linkages), starch (a-1,4 and a-1,6 linkages) and callose (β-1,3 linkages). It is also present in the hemicelluloses as part of the glucomannans and galactoglucomannans linked to mannose residues by β-1,4-glucosidic bonds. The donor nucleotide derivatives are probably GDP glucose for forming the β-1,4-glucosidic bonds in cellulose and hemicelluloses, UDP glucose for forming the β-1,3-glucosidic bonds of callose and ADP glucose for forming the a-1,4-and a-1,6-glucosidic linkages[155, 156].

Over 10 years ago Colvin observed[165] the synthesis of cellulose microfibrils from precursors which could be extracted from *Avena sativa* and *Pisum sativum*, but not from *Phaseolus vulgaris*, by treatment with ethanol–water (4:1 v./v.). The system appeared to be exactly analogous to that described in *Acetobacter xylinum* (see Section 2.3.3). However, despite a number of investigations the case for lipid-linked intermediates in glucosyl transfer has not been pressed further than this. Incubation of particulate enzymes of *P. aureus* with UDP [^{14}C]glucose led to the formation of two [^{14}C]glucolipids but these were stable to mild acid treatment and no evidence was presented for the transfer of glucose from the lipid to polymer[166]. In fact Kauss observed[167] that 90–98% of the glucolipid formed in this system consisted of steryl glucosides which did not show any turnover in pulse-chase experiments. Using GDP [^{14}C]glucose in the same experimental system, Villemez has reported[160] that [^{14}C]glucolipid is not formed but that a [^{14}C]glucoprotein which is formed is probably an intermediate in glucan biosynthesis. Since polyprenols have been observed to accumulate in chloroplasts (see Section 2.2.2.3) it is surprising that no investigations into the possible role of polyprenol phosphates in the transfer of glucose from ADP glucose to starch or in the formation of galactolipids have been reported.

2.7 CONCLUDING DISCUSSION

It has been shown in Section 2.2 that the formation of several wall glycans in bacteria involves polyprenol phosphate derivatives as obligatory intermediates in the transfer of glycose(s) to glycan. The detection of these lipid-linked intermediates has resulted from careful incubations of cell-free particulate preparations with the appropriate sugar nucleotides. It may be that the slowing down of the transferase reaction by damage during preparation of the membrane system in which the enzymes are bound or by creating a shortage of acceptor is crucial to the detection of the intermediates. Indeed even then, the occurrence of radioactive lipid-linked intermediates in experimentally useful amounts has frequently required blocking of subsequent steps in the biosynthetic sequence or slowing the process down by lowering the temperature of incubation. This is not surprising for the free-energy change for the transfer of glycose from the lipid phosphate to polymer is probably very similar to that from nucleotide diphosphate, i.e. $ca - 3500$ cal mol^{-n}, that is, much in favour of transfer to polymer. That this is the situation in bacteria in which metabolism is geared very much to producing wall components, presents at least one reason why difficulty in detecting lipid intermediates in eukaryotic cells should not be unexpected. Nevertheless a role for polyprenol phosphates in mammalian, glucosyl and mannosyl transfer and in mannose transfer in yeast, appears well established. It may well be that a similar system is operative in plants. Perhaps other eukaryotic glycosyl transferases will be shown to be concerned as well.

The glycosyl transferases of bacteria are generally assumed to be members of a membrane-bound synthetase complex. These involve sugars linked to undecaprenol monophosphate or to the diphosphate. It appears that di- or oligo-saccharides are built up on the diphosphate intermediates and then transferred to polymer (which may also be linked to polyprenol diphosphate) whereas only monosaccharides are transferred via monophosphate intermediates. The biosynthesis of teichoic acids represents an exception to this generalisation (see Figures 2.13–2.15) but since these involve the transfer of glycose phosphates this does not invalidate the generalisation for normal glycans. There is evidence for a similar arrangement in glucosyl transfer in mammalian liver. It is possible also that the formation of a polyprenol diphosphate N-acetylglucosamine in mammalian liver (Section 2.4.2.3) represents the first step in the formation of lipid-linked oligosaccharides prior to their transfer to protein in the formation of serum glycoproteins. However, against this at present is the strong body of evidence[113] that these glycoproteins are glycosylated in a stepwise manner and also (Section 2.4.2.2) that dolichol monophosphate mannose appears to donate its mannose directly to protein.

Attention has been drawn[168] to the relatively low polarity of many membrane proteins and presumably this is related to their close proximity to and interaction with lipids within the membrane. It seems likely that one function of polyprenol phosphates is to anchor sugars in this hydrophobic environment and so facilitate the activity of those glycosyl transferases that are located there. Dolichol phosphate is a much more hydrophobic compound than is retinol phosphate and one possible rationalisation of the apparent involvement of these compounds in glycosyl transfer in mammalian liver is that dolichol phosphate is effective in the transfer of proximal N-acetylglucosamine and mannose residues and that the introduction of these into the protein (precursor of plasma glycoproteins) renders it less hydrophobic. Retinol phosphate may then be utilised in the introduction of further mannose and galactose residues. The glycoprotein is now so much more hydrophilic that further introduction of the distal N-acetylglucosamine, galactose and sialic acid, or fucose, residues can be carried out without the aid of lipid-linked intermediates. The fact that the transferases for introduction of these distal residues are more easily solubilised than the others is consistent with this idea but at present it must be considered highly speculative with little firm experimental support.

An important aspect of the anchorage of the sugar to lipid is that the transfer potential of the sugar intermediate should be retained. This is obviously achieved by the diphosphate intermediates and it has been suggested that the β-double bond ensures this in the case of the allylic polyprenol monophosphate glycose derivatives also. Although the β-double bond has a major influence on the prenol-phosphate bond its effect on the glycose-phosphate bond in these compounds appears to be much less as witnessed by the extreme acid lability of dolichol monophosphate glycose derivatives to yield dolichol monophosphate and glycose and by the ready reversibility of, for example, step (a) in Figure 2.20. It seems that much of the transfer potential of nucleoside diphosphate sugars is retained in the polyprenol monophosphate and diphosphate sugars, irrespective of the presence or absence of a β double bond in the polyprenol moiety.

The specificity of both bacterial and mammalian glycosyl transferase systems to polyprenol phosphates as acceptors is normally quite loose for it has been shown that although the stereochemistry and range of chain length of endogenous polyprenol phosphates used is constant exogenous polyprenol phosphates of varying chain length, stereochemistry and degree of saturation will substitute quite effectively. The specificity to the polyprenol moiety of the lipid-linked intermediate by enzymes catalysing the further transfer of glycose to polymer seems to be much tighter although in view of the range of chain length of endogenous polyprenols it presumably is still not absolute to one particular structure.

Observation of the anomeric configuration of the glycose moieties of nucleotide diphosphate sugars, lipid-linked intermediates and final polymers in some cases (e.g. the transfer of glucose to lipopolysaccharide (Section 2.3.5) and to teichoic acid (Section 2.3.7) lends support to the idea that an inversion occurs during transfer to the intermediate and again during transfer from the intermediate to polymer. This is consistent with each step being an S_N2 displacement. However, it is unlikely to be a general mechanism for the anomeric configuration of glycose residues in some bacterial polymers, formed via lipid linked intermediates, is opposite to that in the sugar nucleotide donor.

Some attention has been paid to the question of movement of polyprenol phosphate derivatives within the lipid milieu of the membrane. Baddiley[104] and co-workers suggest that the functioning of these compounds as vectorial intermediates transferring material from one side of the membrane to the other is unlikely and that polymers are more likely to be formed by a fairly static synthetase complex and to be extruded from the membrane as the polymer chain is gradually lengthened. The suggestion was also made that once the polyprenol phosphate is taken up by a particular synthetase complex it stays with the complex for several cycles of the synthetic process until the polymer is complete. It is not yet clear whether or not the polyprenol phosphate that is released at this stage becomes available to the complexes synthesising other polymers. It is possible that the trapping of polyprenol phosphate within a complex is dependent upon the pool size of the appropriate nucleoside diphosphate sugar (which in turn is subject to feed-back control [101, 169]), for the first step of the cycle[104]. Sutherland[98] makes the point that there is probably priority of some synthetase systems over others when competing for the limited amount of polyprenol phosphate available. It is observed that peptidoglycan synthesis has priority over capsular polysaccharide biosynthesis in growing cells but that the priority is absent, or reversed in cells that have ceased growing. The results of Wright (Section 2.3.5) suggest that in the 'O'-antigen synthetase 'complex' the endogenous lipid-linked intermediates are freely interchangeable with exogenous material and thus are not bound very tightly to the enzymes of the complex. Probably one of the most interesting current aspects of the role of polyprenol phosphates in glycosyl transfer is the potential they offer for controlling the rate of the process.

It was surmised (see Section 2.3.4.2) that the rate of glycosyl transfer in bacteria may be controlled by adjustment of the level of polyprenol monophosphate in the system and that in at least some bacteria this might be achieved not only through changing the rate of its biosynthesis but also by altering the balance between polyprenol monophosphate phosphatase and polyprenol phosphokinase. A polyprenol monophosphate phosphatase has been described (see Section 2.2.1.7(c)) in mammalian tissues but it has not yet been demonstrated if this will use dolichol monophosphate or retinol monophosphate, as a substrate. The presence or absence of dolichol phosphokinase in mammalian cells is still an open question but there is evidence for the presence of a retinol phosphokinase (see Section 2.4.3.2). It is, therefore, possible that a similar method of control could operate in mammalian cells.

Acknowledgements

The author is grateful to the following workers, and their colleagues, for providing information in advance of its publication: J. Baddiley, R. K. Bretthauer, T. Chojnacki, L. F. Leloir, L. deLuca, B. E. Ryman, I. W. Sutherland, K. Takayama, W. Tanner, K. J. I. Thorne, C. D. Warren, A. Wright.

Note added in proof

This review was completed during the summer of 1972. Several relevant papers have been published since then especially on studies with eukaryotic systems. Some of the more important reports are summarised below.

Leloir's group have published[170, 171] further evidence from experiments with rat liver preparations elaborating Figure 2.19 (p. 76). Results suggest that the first step in the formation of (70) is the transfer from UDPNAcGlc of

NAcGlc-P to dolichol monophosphate. The dolichol diphosphate-NAcGlc so formed then accepts a second NAcGlc residue from UDPNAcGlc to form dolichol diphosphate $(NAcGlc)_2$. Mannose can then be added to the disaccharide moiety. At least some of the mannose is derived from dolichol phosphate mannose. The final dolichol diphosphate oligosaccharide probably contains two glucose residues. The transfer of mannose to a lipid with the properties of (70) in pig liver microsomes has also been observed recently in the author's laboratory. The formation of dolichol monophosphate mannose in microsomal preparations has been extended to include calf pancreas[172]. The complete characterisation of dolichol monophosphate from pig liver preparations has now been published[173,174]. Of possible relevance as a rich source of dolichol is the report[175] that human pituitary contains extremely high levels (1.4g/kg) of dolichols, dolichol-19 predominating.

Work with yeast systems has continued in Tanner's laboratory where it has been shown[176] that intestinal alkaline phosphatase in large excess will remove the phosphate from the monophosphate of yeast dolichol. This step was used in confirming, by mass spectrometry, the nature of the endogenous lipid involved in mannosyl transfer in *S. cerevisiae*.

Retinol has been reported[177] to form a mannoside, galactoside and a glucoside, each of which is free from phosphate, when it is incubated with homogenates of rat thyroid and appropriate nucleotide diphosphate sugars.

Villemez[178] has suggested that phytanol phosphate should be a useful artificial acceptor for studying GDPmannose: lipid phosphate transmannosylase in *P.aureus*. This compound was found to be a much more efficient acceptor of mannose than phytol phosphate but less efficient than the endogenous acceptor. Interestingly the mannosylated phytanol phosphate was more polar than the mannosylated endogenous lipid when run on silica gel M with chloroform/methanol/water (12/6/1 by volume) as solvent.

References

1. Osborn, M. J. (1969). *Ann. Rev. Biochem.*, **38**, 501
2. Heath, E. C. (1971). *Ann. Rev. Biochem.*, **40**, 29
3. Rothfield, L. and Romeo, D. (1971). *Bact. Rev.*, **35**, 14
4. Ellar, D. J. (1970). *Organisation and Control in Prokaryotic and Eukaryotic Cells*, 167. (H. P. Charles and C. J. G. Knight, editors) (London: Cambridge University Press)
5. Leloir, L. F. (1971). *Science*, **172**, 1299
6. Lennarz, W. J. and Scher, M. G. (1972). *Biochim. Biophys. Acta*, **265**, 417
7. Nikaido, K. and Nikaido, H. (1971). *J. Biol. Chem.*, **246**, 3912
8. Getz, G. S. (1970). *Advan. in Lipid Research*, **8**, 175
9. Hemming, F. W. (1970). *Biochem. Soc. Sympos.*, **29**, 105. (T. W. Goodwin, editor) (London: Academic Press)
10. I.U.P.A.C. Commission for the Nomenclature of Biological Chemistry (1967). *Biochem. J.*, **102**, 15
11. Lindgren, B. O. (1965). *Acta Chem. Scand.*, **19**, 1317
12. Wellburn, A. R. and Hemming, F. W. (1966). *Nature (London)*, **212**, 1364
13. Thorne, K. J. I. and Kodicek, E. (1966). *Biochem. J.*, **99**, 123
14. Gough, D. P., Kirby, A. L., Richards, J. B. and Hemming, F. W. (1970). *Biochem. J.*, **118**, 167
15. Higashi, Y., Strominger, J. L. and Sweeley, C. C. (1967). *Proc. Nat Acad. Sc . USA*, **57**, 1878

16. Dunphy, P. J., Kerr, J. D., Pennock, J. F., Whittle, K. J. and Feeney, J. (1967). *Biochim. Biophys. Acta*, **136**, 136
17. Richards, J. B. and Hemming, F. W. (1972). *Biochem. J.*, **128**, 1345
18. Burgos, J., Hemming, F. W., Pennock, J. F. and Morton, R. A. (1963). *Biochem. J.*, **88**, 470
19. Walton, M. J. and Pennock, J. F. (1972). *Biochem. J.*, **127**, 471
20. Wellburn, A. R., Stevenson, J., Hemming, F. W. and Morton, R. A. (1967). *Biochem. J.*, **102**, 313
21. Stone, K. J., Wellburn, A. R., Hemming, F. W. and Pennock, J. F. (1967). *Biochem. J.*, **102**, 325
22. Fukawa, H., Toyada, M., Shimizu, T. and Murohashi, M. (1966). *Tetrahedion Lett.* **49**, 6209
23. Stone, K. J., Butterworth, P. H. W. and Hemming, F. W. (1967). *Biochem. J.*, **102**, 443
24. Barr, R. M. and Hemming, F. W. (1972). *Biochem. J.*, **126**, 1193
25. Kofler, M., Langemann, A., Ruegg, R. Gloor, U., Schweiter, U., Würsch, J., Wiss, O. and Isler, O. (1959). *Helv. Chim. Acta*, **42**, 2252
26. Hemming, F. W., Morton, R. A. and Pennock, J. F. (1963). *Proc. Roy. Soc. B*, **158**, 291
27. Wellburn, A. R. and Hemming, F. W. (1966). *J. Chromatog.*, **23**, 51
28. Feeney, J. and Hemming, F. W. (1967). *Analyt. Biochem.*, **20**, 1
29. Pennock, J. F. (1966). *Vitamins and Hormones*, **24**, 307
30. Morton, R. A. (1971). *Biol. Rev.*, **46**, 47
31. Pennock, J. F., Hemming, F. W. and Morton, R. A. (1972). *Nature (London)*, **186**, 470
32. Behrens, N. H. and Leloir, L. F. (1970). *Proc. Nat. Acad. Sci. USA*, **66**, 153
33. Popják, G., Cornforth, J. W., Cornforth, R. H., Ryhage, R. and Goodman, de W. S. (1962). *J. Biol. Chem.*, **237**, 56
34. Cramer, F. and Böhm, W. (1959). *Angew. Chem.*, **71**, 775
35. Barr, R. M. and Hemming, F. W. (1972). *Biochem. J.*, **126**, 1203
36. Warren, C. D. and Jeanloz, R. W. (1972). *Biochem Phy*, **11**, 2565
37. Khwaja, T. A., Reese, C. B. and Stewart, J. C. M. (1970). *J. Chem. Soc. C*, 2092
38. Lynen, F., Eggerer, H., Henning, U. and Kessel, J. (1959). *Angew Chem.*, **71**, 657
39. De Wolfe, R. H. and Young, W. G. (1956). *Chem. Rev.*, **56**, 753
40. Nikaido, K. and Nikaido, H. (1971). *J. Biol. Chem.*, **246**, 3912
41. Dawson, R. M. C. (1967). *Lipid Chromatogr. Anal.*, **1**, 163
42. Wright, A. (1971). *J. Bacteriol.*, **105**, 927
43. Capon, B. (1969). *Chem. Rev.*, **69**, 407
44. Hancock, J. C. and Baddiley, J. (1972). *Biochem. J.*, **127**, 27
45. Kent, J. L. and Osborn, M. J. (1968). *Biochemistry*, **7**, 4419
46. Allen, C. M., Alworth, W., Macrae, A. and Bloch, K. (1967). *J. Biol. Chem.*, **242**, 1895
47. Goldman, R. and Strominger, J. L. (1972). *J. Biol. Chem.*, **247**, 5116
48. Willoughby, E., Highasi, Y. and Strominger, J. L. (1972). *J. Biol. Chem.*, **247**, 5113
49. Tsai, S. C. and Gaylor, J. L. (1966). *J. Biol. Chem.*, **241**, 4043
50. Epstein, W. W. and Rilling, H. (1970). *J. Biol. Chem.*, **245**, 4597
51. Kurokawa, T., Ogura, K. and Seto, S. (1971). *Biochem. Biophys. Res. Commun.*, **45**, 251
52. Thorne, K. J. I. and Kodicek, E. (1962). *Biochim. Biophys. Acta*, **59**, 273
53. Thorne, K. J. I. and Kodicek, E. (1962). *Biochim. Biophys. Acta*, **59**, 280
54. Durr, J. F. and Habbol, M. Z. (1972). *Biochem. J.*, **127**, 345
55. Christenson, J. G., Gross, S. K. and Robbins, P. W. (1969). *J. Biol. Chem.*, **244**, 5436
56. Richards, J. B., Evans, P. J. and Hemming, F. W. (1971). *Biochem. J.*, **124**, 957
57. Alam, S. S. and Hemming, F. W. (1971). *FEBS Letters*, **19**, 60
58. Ruegg, R., Gloor, U., Langemann, A., Kofler, M., von Planta, C., Ryser, G. and Isler, O. (1960). *Helv. Chim. Acta*, **43**, 1745
59. Upper, C. D. and West, C. A. (1967). *J. Biol. Chem.*, **242**, 3285
60. Butterworth, P. H. W. (1964). *Ph.D. Thesis, University of Liverpool*.
61. Wellburn, A. R. and Hemming, F. W. (1966). *Phytochem.*, **5**, 969
62. Butterworth, P. H. W. and Hemming, F. W. (1968). *Arch. Biochem. Biophys.*, **128**, 503
63. Stone, K. J. and Hemming, F. W. (1968). *Biochem. J.*, **109**, 877
64. Dallner, G., Behrens, N. H., Parodi, A. J. and Leloir, L. F. (1972). *FEBS Letters*, **24**, 315

65. Wellburn, A. R. and Hemming, F. W. (1966). *Biochemistry of Chloroplasts* **1**, 173. (T. W. Goodwin, editor) (London: Academic Press)
66. Wellburn, A. R. and Hemming, F. W. (1967). *Biochem. J.*, **104**, 173
67. Gough, D. P. and Hemming, F. W. (1970). *Biochem. J.*, **117**, 309
68. Gough, D. P. and Hemming, F. W. (1970). *Biochem. J.*, **118**, 163
69. Stone, K. J. and Hemming, F. W. (1969). *Biochem. J.*, **104**, 43
70. Sandermann, H. and Strominger, J. L. (1971). *Proc. Nat. Acad. Sci. USA*, **68**, 2441
71. Costes, C. (1966). *Phytochem.*, **5**, 311
72. Thorne, K. J. I. and Barker, D. C. (1971). *Biochem. J.*, **122**, 45p
73. Momose, K. and Rudney, N. (1972). *J. Biol. Chem.*, **247**, 3930
74. Colvin, J. R. (1959). *Nature (London)*, **183**, 1135
75. Andersen, J. S., Matsuhashi, M., Haskin, M. A. and Strominger, J. L. (1965). *Proc. Nat. Acad. Sci. USA*, **53**, 881
76. Weiner, I. M., Hiyashi, T., Rothfield, L., Saltmarsh-Andrew, M., Osborn, M. J. and Horecker, B. L. (1965). *Proc. Nat. Acad. Sci. USA*, **54**, 228
77. Wright, A., Dankert, M., Fennessey, P. and Robbins, P. W. (1966). *Proc. Nat. Acad. Sci. USA*, **57**, 1798
78. Hiyashi, Y., Strominger, J. L. and Sweeley, C. C. (1967). *Proc. Nat. Acad. Sci. USA*, **57**, 1878
79. Scher, M. and Lennarz, W. J. (1969). *J. Biol. Chem.*, **244**, 2777
80. Lahav, M., Chiu, T. H. and Lennarz, W. J. (1969). *J. Biol. Chem.*, **244**, 5890
81. Scher, M., Lennarz, W. J. and Sweeley, C. C. (1968). *Proc. Nat. Acad. Sci. USA*, **59**, 1313
82. Dankert, M., Garcia, R. and Recondo, E. (1972). *The Biochemistry of the Glycosidic Linkage*, 199. (R. Piras and H. G. Pontis, editors) (New York: Academic Press)
83. Strominger, J. L., Higashi, Y., Sandermann, H., Stone, K. J. and Willoughby, E. (1972). *The Biochemistry of the Glycosidic Linkage*, 135. (R. Piras and H. G. Pontis, editors) (New York: Academic Press)
84. Struve, W. F., Sinha, R. K. and Neuhaus, F. C. (1966). *Biochemistry*, **5**, 82
85. Reynolds, P. E. (1971). *Biochim. Biophys. Acta*, **237**, 239
86. Stone, K. J. and Strominger, J. L. (1971). *Proc. Nat. Acad. Sci. USA*, **68**, 3223
87. Schechter, N., Momose, K. and Rudney, N. (1972). *Biochem. Biophys. Res. Commun.*, **48**, 833
88. Stone, K. J. and Strominger, J. L. (1972). *J. Biol. Chem.*, **247**, 5107
89. Wright, A., Dankert, M. and Robbins, P. W. (1965). *Proc. Nat. Acad. Sci. USA*, **54**, 235
90. Robbins, P. W., Bray, D., Dankert, M. and Wright, A. (1967). *Science*, **158**, 1536
91. Bray, D. and Robbins, P. W. (1967). *Biochem. Biophys. Res. Commun.*, **28**, 334
92. Kanegasaki, S. and Wright, A. (1970). *Proc. Nat. Acad. Sci. USA*, **67**, 951
93. Osborn, M. J. and Weiner, J. M. (1968). *J. Biol. Chem.*, **243**, 2631
94. Wright, A. (1969). *Fed. Proc.*, **28**, 658
95. Nikaido, H., Niakido, K., Nakoe, T. and Makela, P. H. (1971). *J. Biol. Chem.*, **246**, 3902
96. Jankowski, W. and Chojnacki, T. (1972). *Biochim. Biophys. Acta*, **260**, 93
97. Jankowski, W. and Chojnacki, T. (1972). *Acta Biochemica Polonica*, **19**, 51
98. Sutherland, I. W. (1972). *Advan. Microbiol. Physiol.*, **8**, 143
99. Troy, F. A., Frerman, F. E. and Heath, E. C. (1971). *J. Biol. Chem.*, **246**, 118
100. Sutherland, I. W. and Norval, M. (1970). *Biochem. J.*, **120**, 567
101. Baddiley, J. (1970). *Accounts Chem. Res.*, **3**, 98
102. Brooks, D. and Baddiley, J. (1969). *Biochem. J.*, **115**, 307
103. Watkinson, R. J., Hussey, H. and Baddiley, J. (1971). *Nature (London)*, **229**, 57
104. Anderson, R. G., Hussey, H. and Baddiley, J. (1972). *Biochem. J.* **127**, 11
105. Stow, M., Starkey, B. J., Hancock, S. C. and Baddiley, J. (1971). *Nature (London)*, **229**, 56
106. Brooks, D., Mays, L. L., Hatefi, Y. and Young, F. E. (1971). *J. Bact.*, **107**, 223
107. Kent, P. W. (1967). *Essays in Biochemistry*, **3**, 105
108. Taillard, H. (1970). *Molecular Biology*, 223 (A. Haidemanakis, editor) (New York: Gordon and Breach)
109. Spiro, R. G. (1970). *Annu. Rev. Biochem.*, **39**, 599
110. Watkins, W. M. (1966). *Science*, **152**, 172

111. Ginsburg, V. (1972). *The Biochemistry of the Glycosidic Linkage*, 387 (R. Piras and H. G. Pontis, editors) (New York: Academic Press)
112. Kaufman, B., Basu, S. and Roseman, S. (1968). *J. Biol. Chem.*, **243**, 5804
113. Roseman, S. (1970). *Chem. Phys. Lipids*, **5**, 270
114. Redman, C. M. and Cherian, M. G. (1972). *J. Cell. Biol.*, **52**, 231
115. Morré, D. J., Mollenhauer, H. H. and Bracker, C. E. (1971). *Origin and Continuity of Cell Organelles* 82. (J. Reinert and H. Ursprung, editors) (Heidelberg: Springer Verlag)
116. Dauwalder, M., Whaley, W. G. and Kephart, J. E. (1972). *Subcell. Biochem.*, **1**, 225
117. Whalley, W. G., Dauwalder, M. and Kephart, J. E. (1971). *Origin and Continuity of Cell Organelles*, **82**, 1
118. Wagner, R. R. and Cynkin, M. A. (1969). *Biochem. Biophys. Res. Commun.*, **35**, 139
119. Caccam, J. F., Jackson, J. J. and Eylar, E. H. (1969). *Biochem. Biophys. Res. Commun.*, **35**, 505
120. Behrens, N. H., Parodi, A. J., Leloir, L. F. and Krisman, C. (1971). *Arch. Biochem. Biophys.*, **143**, 375
121. Behrens, N. H., Parodi, A. J. and Leloir, L. F. (1971). *Proc. Nat. Acad. Sci. USA*, **68**, 2857
122. Parodi, A. J., Behrens, N. H., Leloir, L. F. and Dankert, M. (1972). *Biochim. Biophys. Acta*
123. Parodi, A. J., Behrens, N. H., Leloir, L. F. and Carminatti, (1972). *Proc. Nat. Acad. Sci. USA*
124. Krisman, C. (1972). *Biochem. Biophys. Res. Commun.*, **46**, 1206
125. Bosmann, H. B. and Martin, S. S. (1969). *Science*, **164**, 190
126. Bosmann, H. B. and Case, K. R. (1969). *Biochem. Biophys. Res. Commun.*, **36**, 830
127. Zatz, M. and Barondes, S. H. (1969). *Biochem. Biophys. Res. Commun.*, **36**, 511
128. Richards, J. B., Evans, P. J. and Hemming, F. W. (1972). *The Biochemistry of the Glycosidic Linkage*, 207. (R. Piras and H. G. Pontis, editors) (New York: Academic Press)
129. Richards, J. B. and Hemming, F. W. (1972). *Biochem. J.*, **130**, 77
130. Tetas, M., Chao, H. and Molnar, J. (1970). *Arch. Biochem. Biophys.*, **138**
131. Baynes, J. W. and Heath, E. C. (1972). *Fed. Proc.*, **31**, 1239
132. Molnar, J., Chao, H. and Ikehara, Y. (1971). *Biochim. Biophys. Acta*, **239**, 401
133. Bosmann, H. B. (1970). *Europ. J. Biochem.*, **14**, 33
134. Wasserman, R. H. and Corradino, R. A. (1971). *Annu. Rev. Biochem.*, **40**, 501
135. De Luca, L., Schumacker, M. and Wolf, G. (1970). *J. Biol. Chem.*, **245**, 4551
136. De Luca, L., Maestri, N., Rosso, G. and Wolf, G. (1972). *J. Biol. Chem.*
137. De Luca, L., Rosso, G. and Wolf, G. (1970). *Biochem. Biophys. Res. Commun.*, **41**, 615
138. Griffin, M. J., Dickenson, M., Chiba, N. and Johnson, B. C. (1971). *Fed. Proc.*, **30**, 1134
139. De Luca, L., Barber, A. J. and Jamieson, G. A. (1972). *Fed. Proc. Fed. Amer. Soc. Exp. Biol.*, **31**, 242
140. Helting, T. and Peterson, P. A. (1972). *Biochem. Biophys. Res. Commun.*, **46**, 429
141. Mookerjea, S., Cole, D. E. C. and Chow, A. (1972). *FEBS Letters*, **23**, 257
142. Mookerjea, S. and Chow, A. (1970). *Biochem. Biophys. Res. Commun.*, **39**, 486
143. Vessey, D. A. and Zakim, D. (1971). *J. Biol. Chem.*, **246**, 4649
144. Tanner, W. (1969). *Biochem. Biophys. Res. Commun.*, **35**, 144
145. Tanner, W., Jung, P. and Behrens, N. H. (1971). *FEBS Letters*, **16**, 245
146. Tanner, W., Jung, P. and Linden, J. C. (1972). *The Biochemistry of the Glycosidic Linkage*, 227. (R. Piras and H. G. Pontis, editors) (New York: Academic Press)
147. Sentandreu, R. and Lampen, J. O. (1971). *FEBS Letters*, **14**, 109
148. Behrens, N. H. and Cabib, E. (1968). *J. Biol. Chem.*, **243**, 502
149. Bretthauer, R. K. and Wu, S. (1972). *Amer. Chem. Soc. Meeting Abstr.*, 105
150. Richard, M., Letoublon, R., Louisot, P. and Got, R. (1971). *Biochim. Biophys. Acta* **230**, 603
151. Ankel, H., Ankel, E., Schutzbach, J. S. and Garancis, J. C. (1970). *J. Biol. Chem.*, **245**, 3985
152. McMurrough, I. and Bartnicki-Garcia, S. (1971). *J. Biol. Chem.*, **246**, 4008
153. Northcote, D. H. (1969). *Essays in Biochem.*, **5**, 89
154. Northcote, D. N. (1972). *Annu. Rev. Plant Physiol*
155. Hassid, W. Z. (1969). *Science*, **165**, 137

156. Hassid, W. Z. (1972). *The Biochemistry of the Glycosidic Linkage*, 315 (R. Piras and H. G. Pontis, editors) (New York: Academic Press)
157. Kauss, H. (1969). *FEBS Letters*, **5**, 81
158. Kauss, H. (1972). *The Biochemistry of the Glycosidic Linkage*, 221 (R. Piras and H. G. Pontis, editors) (New York: Academic Press)
159. Villemez, C. L. and Clark, A. F. (1969). *Biochem. Biophys. Res. Commun.*, **36**, 57
160. Villemez, C. L. (1970). *Biochem. Biophys. Res. Commun.*, **40**, 636.
161. Alam, S. S., Barr, R. M., Richards, J. B. and Hemming, F. W. (1971), *Biochem. J.*, **121**, 19P
162. Alam, S. S. and Hemming, F. W. (1973). *Phytochem*, in the press
163. Alam, S. S. and Hemming, F. W. (1972). *FEBS Letters*, **19**, 60
164. Heller, J. S. and Villemez, C. L. (1972). *Biochem. J.*, **128**, 243
165. Colvin, J. R. (1961). *Canad. J. Biochem. Physiol.*, **39**, 1921
166. Villemez, C. L., Vodak, B. and Albersheim, P. (1968). *Phytochem.*, **7**, 1561
167. Kauss, H. (1968). *Z. Naturforschung*, **236**, 1522
168. Capaldi, R. A. and Vanderkooi, G. (1972). *Proc. Nat. Acad. Sci. USA*, **69**, 930
169. Gottschalk, A. (1969). *Nature (London)*, **222**, 452
170. Leloir, L. F., Staneloni, F. J., Carminatti, H. and Behrens, N. H. (1973). *Biochem. Biophys. Res. Commun.*, **52**, 1285
171. Behrens, N. H., Carminatti, H., Staneloni, R. J., Leloir, L. F. and Cantarella, I. H. (1973). *Proc. Nat. Acad. Sci.*, in the press
172. Tkacz, J. S., Herscovics, A., Warren, C. D. and Jeanloz, R. W. (1973). *Biochem. Soc. Transactions*, in the press
173. Evans, P. J. and Hemming, F. W. (1973). *FEBS Lett.*, **31**, 335
174. Warren, C. D. and Jeanloz, R. W. (1973). *FEBS Lett.*, **31**, 332
175. Carroll, K. K., Vilim, A. and Woods, M. C. (1973). *Lipids*, **8**, 246
176. Tanner, W. (1973). *Eur. J. Biochem.*, in the press
177. Rodriguez, P., Bello, O. and Gaede, K. (1972). *FEBS Lett.*, **28**, 133
178. Clark, A. F. and Villemez, C. L. (1973). *FEBS Lett.*, **32**, 84

3
Biosynthesis of Saturated Fatty Acids

P. R. VAGELOS
Washington University, St. Louis, Missouri

3.1	INTRODUCTION	100
3.2	ACETYL-COA CARBOXYLASE	100
	3.2.1 *Molecular properties*	101
	3.2.2 *Subunit functions*	102
	3.2.3 *Control*	105
	3.2.3.1 *Allosteric regulation*	105
	3.2.3.2 *Roles of enzyme synthesis and degradation*	107
3.3	FATTY ACID SYNTHETASE	109
	3.3.1 *Acyl carrier protein*	111
	3.3.1.1 *Prosthetic group turnover*	117
	3.3.2 *Enzymes of* Escherichia coli *fatty acid synthetase*	118
	3.3.2.1 *Acetyl-CoA-ACP transacylase*	119
	3.3.2.2 *Malonyl-CoA-ACP transacylase*	120
	3.3.2.3 *β-Ketoacyl-ACP synthetase*	121
	3.3.2.4 *β-Ketoacyl-ACP reductase*	123
	3.3.2.5 *β-Hydroxyacyl-ACP dehydrase*	123
	3.3.2.6 *Enoyl-ACP reductase*	124
	3.3.3 *Multi-enzyme complexes*	124
	3.3.3.1 *Yeast*	124
	3.3.3.2 *Liver*	128
	3.3.3.3 *Mammary gland*	129
	3.3.3.4 *Mycobacterium phlei*	131
	3.3.3.5 *Euglena gracilis*	131
	3.3.4 *Control*	132
	3.3.4.1 *Allosteric regulation*	133
	3.3.4.2 *Adaptive changes in enzyme content*	134

3.1 INTRODUCTION

The *de novo* synthesis of saturated fatty acids, studied in a variety of biological systems, is catalysed by two enzyme systems which function sequentially, acetyl-CoA carboxylase and fatty acid synthetase. These enzyme complexes are found in the cell cytoplasm when cells are ruptured and the cellular components are fractionated by usual techniques. All the carbon atoms of the fatty acids are derived from acetyl-CoA, and equation (3.1) indicates the stoichiometry for the synthesis of palmitate from acetyl-CoA:

$$8\text{MeCO-S-CoA} + 7\text{ATP} + 14\text{NADPH} + 14\text{H}^+ \rightarrow$$
$$\text{Me(CH}_2)_{14}\text{CO}_2\text{H} + 8\text{CoA-SH} + 7\text{ADP} + 7\text{P}_i + 14\text{NADP}^+ + 6\text{H}_2\text{O} \quad (3.1)$$

Palmitate is the major fatty acid produced by most biosynthetic systems, and all of the long chain saturated, unsaturated (except for the essential fatty acids), and hydroxy fatty acids found in animal tissues can be derived from palmitate. In this review the acetyl-CoA carboxylase and fatty acid synthetase, which together account for the synthesis of palmitate from acetyl-CoA as summarised in equation (3.1), will be considered in detail.

3.2 ACETYL-CoA CARBOXYLASE

Acetyl-CoA carboxylase catalyses the first committed step in the synthesis of palmitate from acetyl-CoA. This reaction involves the biotin prosthetic group of the enzyme, ATP and bicarbonate, and it can be considered as two partial reactions which are indicated in equations (3.2) and (3.3).

$$\text{ATP} + \text{HCO}_3^- + \text{biotin-E} \underset{}{\overset{\text{Me}^{2+}}{\rightleftharpoons}} \text{CO}_2^- \text{-biotin-E} + \text{ADP} + \text{P}_i \quad (3.2)$$

$$\text{CO}_2^- \text{-biotin-E} + \text{MeCO-S-CoA} \rightleftharpoons {}^-\text{O}_2\text{CCH}_2\text{CO-S-CoA} + \text{biotin-E} \quad (3.3)$$

$$\text{Sum: ATP} + \text{HCO}_3^- + \text{MeCO-S-CoA} \underset{}{\overset{\text{Me}^{2+},\text{ biotin-E}}{\rightleftharpoons}} \text{ADP} + \text{P}_i + {}^-\text{O}_2\text{CCH}_2\text{CO-S-CoA} \quad (3.4)$$

The mechanism of action of acetyl-CoA carboxylase is similar to that elucidated for other biotin enzymes by Lynen and co-workers[1]. These investigators demonstrated that the enzymes contain covalently bound biotin as a prosthetic group and that a carboxybiotin intermediate is formed during the reaction[2-4]. The mode of binding of the carboxyl group to the biotin was demonstrated by studying a model reaction, the carboxylation of free biotin by β-methylcrotonyl-CoA carboxylase; the product of this reaction was methylated and identified as 1'-N-carboxymethylbiotin methyl ester[2,5]. When the reaction was carried out with substrate quantities of the enzyme, ATP and H^{14}CO$_3$, the carboxyenzyme formed was methylated and digested with Pronase, and 1'-N-carboxymethylbiocytin was identified[4]. Isolation of the biocytin derivative from the enzyme indicated that the biotin binding site on the enzyme is the ε-amino group of a lysine residue. Waite and Wakil[6] similarly identified biocytin (ε-N-biotinyl-L-lysine) in a peptic digest of acetyl-CoA carboxylase, confirming that the biotin of this enzyme is bound to lysine.

3.2.1 Molecular properties

Homogenous acetyl-CoA carboxylase has been obtained from avian liver[6-10], bovine adipose tissue[11] and rat liver[12, 13]. As originally demonstrated with the adipose tissue enzyme[14], acetyl–CoA carboxylase exists in at least two form, an active polymer and an inactive protomer. The inactive protomeric forms of the avian liver enzyme[6,7,10] has a molecular weight of 410 000; whereas the active polymeric form has a molecular weight of 4 000 000–8 000 000. The equilibrium toward the protomeric form is favoured by low protein concentration, Cl^-, $pH > 7.5$, carboxylation of the enzyme to produce carboxybiotin and palmityl–CoA. Aggregation and formation of the polymer are favoured by citrate (or isocitrate), high protein concentration, or pH 6.5–7.0. The electron microscopic appearance of the enzyme varies markedly with the state of aggregation[6]. The inactive protomeric form appears as particles with a minimum dimension of 70–150 Å. The active polymeric form appears as a network of filaments 70–100 Å in width and up to 4000 Å in length.

Dissociation of avian liver acetyl-CoA carboxylase has been carried out in 0.1 % sodium dodecyl sulphate[10, 15]. Subunit molecular weight, determined by sedimentation equilibrium, was 114 000 and by electrophoresis in sodium dodecyl sulphate polyacrylamide gels, 110 000. A single biotin group and single binding sites for citrate and acetyl-CoA were present per 410 000 molecular weight protomer. Thus, the subunits were not identical. The subunit functions could not be elucidated since they were inactivated by the dissociation procedure. In view of the information concerning subunit function that has been derived from the *Escherichia coli* enzyme, it is possible that the four subunits include a biotin carboxylase, a biotin carboxyl carrier protein, a transcarboxylase and a regulatory subunit with a binding site for citrate. Recent studies of rat liver acetyl-CoA carboxylase[13] suggest that it differs in structure from the avian enzyme. Analysis of the rat enzyme by sodium dodecyl sulphate gel electrophoresis has revealed protein bands with estimated molecular weights of 215 000, 125 000 and 118 000, and it appeared that the latter two peptides were derived from the 215 000 species. Thus this enzyme contains pairs of unlike subunits. One mole of biotin was found per 215 000 daltons; this suggests that there is twice as much biotin in the rat enzyme as in the avian enzyme. Another interesting feature of the rat-liver enzyme was the presence of 2.1 moles of phosphate per 215 000 daltons. The phosphate was not further identified, so it is not apparent whether it is present as a component of a prosthetic group such as AMP, 4'-phosphopantetheine, etc. However, its presence raises the possibility that enzyme activity might be regulated by phosphorylation and dephosphorylation, similar to the pyruvate dehydrogenase complex[16] and phosphorylase[17].

Yeast acetyl-CoA carboxylase has been dissociated in 0.1 M ammonia[18]. The native enzyme had a sedimentation coefficient of 15.5 S. After dissociation inactive subunits with sedimentation coefficients of 12 S, 9 S, and 6 S were observed. Sucrose density-gradient analysis of enzyme labelled with [^{14}C]biotin revealed constant specific radioactivity across the sedimentation pattern, suggesting identical 6 S subunits which were present as monomers, dimers, trimers, etc. Behaviour of the dissociated enzyme on gel electrophoresis was also consistent with an aggregating system. These data suggest that yeast acetyl-CoA carboxylase exists as a tetramer of 6 S subunits, each containing biotin. Since the enzyme molecular weight was estimated as 600 000, a subunit size of 150 000 would be consistent with the observations, and this would be similar to that reported for the rat and avian-liver enzymes.

3.2.2 Subunit function

Studies of animal and yeast acetyl-CoA carboxylase have defined the molecular and subunit structure of the enzymes. However, the functions of the various subunits could not be identified since the dissociating conditions that produced the subunits also denatured the proteins. Acetyl-CoA carboxylase of *E. coli*, on the other hand, was easily dissociated into functional subunits that have been investigated in detail. Protein-fractionation procedures initially gave rise to two protein fractions, E_a and E_b, that were required for the overall carboxylation reaction. Fraction E_a was found to contain biotin and to function in the reaction of equation (3.2), the formation of carboxybiotin–protein, while E_b contained no biotin and functioned in the reaction of equation (3.3), the formation of malonyl-CoA from acetyl-CoA and carboxybiotin–protein[19]. Further studies of E_a were facilitated by three additions to the experimental approach[20]. First it was discovered that E_a, usually very labile, was stable when maintained in 20% glycerol. Secondly, E_a was labelled in the biotin moiety by growing a biotin auxotroph of *E. coli* in the presence of [^{14}C]biotin. The covalently-bound biotin could then be followed even under conditions where catalytic activity was lost. Thirdly, E_a was found to catalyse a model reaction, the ATP-dependent carboxylation of free (+)-biotin, as indicated in equation (3.5), and this was designated biotin carboxylase activity.

$$ATP + HCO_3^- + (+)\text{-biotin} \xrightarrow{Me^{2+}} \text{carboxybiotin} + ADP + P_i \quad (3.5)$$

Purified E_a labelled with [^{14}C]biotin was subjected to disc gel electrophoresis at pH = 9. Two protein bands were observed and each was eluted from the gel. The slower migrating band catalysed free (+)-biotin carboxylation but contained no [^{14}C]biotin; the faster migrating band contained all the [^{14}C]biotin of the E_a preparation but it was inactive in the biotin carboxylase assay. Thus it was apparent that E_a was dissociated into two protein components and that *E. coli* acetyl-CoA carboxylase consists of three protein components. One, biotin carboxylase, catalyses biotin carboxylation as indicated in the model reaction of equation (3.5). This protein contains the sites for ATP, Mn^{2+} and bicarbonate. Lane and co-workers[21] later purified and crystallised biotin carboxylase and demonstrated that it has a molecular weight of *ca.* 100 000 and is composed of two subunits of 51 000 molecular weight. The second protein component of acetyl-CoA carboxylase contains covalently-bound biotin, and it was named biotin carboxyl carrier protein

(BCCP). The third protein is the transcarboxylase component (E_b), and that catalyses the transfer of the carboxyl group from carboxybiotin to acetyl-CoA, forming malonyl-CoA. Functions of the three protein components are demonstrated in equations (3.6) and (3.7):

$$\text{ATP} + \text{HCO}_3^- + \text{BCCP} \xrightleftharpoons{\text{biotin carboxylase}} \text{CO}_2^-\text{BCCP} + \text{ADP} + P_i \quad (3.6)$$

$$\text{CO}_2^-\text{BCCP} + \text{acetyl-CoA} \xrightleftharpoons{\text{transcarboxylase}} \text{malonyl-CoA} + \text{BCCP} \quad (3.7)$$

BCCP plays a central role in the carboxylation of acetyl-CoA. It is carboxylated to CO_2^--BCCP in the reaction catalysed by biotin carboxylase. The carboxyl group of CO_2^--BCCP is transferred to acetyl-CoA forming malonyl-CoA in the reaction catalysed by the transcarboxylase component of acetyl-CoA carboxylase. Several species of BCCP have been isolated from extracts of *E. coli*[22-25]. These differed in molecular weight and activity in the carboxylase and transcarboxylase reactions. The first species that was purified to homogeneity and crystallised[22] had a molecular weight of 9100, $\text{BCCP}_{(9100)}$, and contained one mole of biotin and 82 amino acid residues per mole. The apparent K_m of $\text{BCCP}_{(9100)}$ in the biotin carboxylase reaction was 2.5×10^{-5} M. $\text{BCCP}_{(10400)}$, which had a molecular weight of 10 400, was another form of BCCP that was isolated, and this had a similar K_m in the biotin carboxylase reaction as $\text{BCCP}_{(9100)}$. In addition to these two small peptides, limited quantities of a larger BCCP were isolated from *E. coli*, and it was therefore assumed that the smaller species of BCCP were derived from the largest one by proteolytic cleavage. The large BCCP was found to be the only biotin-containing protein present in crude extracts of *E. coli*[23]. Most significantly, this BCCP exhibited K_m values in the biotin carboxylase and transcarboxylase reactions of 2×10^{-7} M and 4×10^{-7} M respectively. These values were about 100 times lower than those obtained with $\text{BCCP}_{(9100)}$ and BCCP $_{(10400)}$.

More detailed studies of large BCCP were made possible by use of an isolation procedure that avoided proteolysis[24]. The polypeptide chain molecular weight of this presumably native BCCP was established as 22 500 by amino acid analysis, sodium dodecyl sulphate polyacrylamide gel electrophoresis, and gel filtration in the presence of 6 M guanidine–HCl. There was one biotin per 22 500 g of protein. Native BCCP probably exists as a dimer of two identical polypeptide chains, each containing one biotin, since gel filtration of BCCP found in crude extracts and homogeneous preparations indicated a molecular weight of 45 000. With mild treatment, such as heating or dialysis, the dimer readily dissociated to the monomer form. This dissociation could be prevented by the addition of *o*-phenanthroline, a known inhibitor of several metalloproteases. This suggests that the dissociation of BCCP might be due to a limited proteolytic event, induced by an endogenous or contaminating metalloprotease, resulting in loss of only a few amino acid residues and a monomer fraction of very similar molecular weight to the unaltered polypeptide chain. Although isolated native BCCP appears to be a dimer *in vitro*, its subunit structure *in vivo* and the nature of its interaction with the other components of acetyl-CoA carboxylase, the biotin carboxylase and the transcarboxylase, are unclear.

An explanation for the isolation of smaller forms of BCCP ($BCCP_{(9100)}$ and $BCCP_{(10400)}$) became apparent after studies of proteolytic modification of native BCCP[25]. BCCP was found to be susceptible to hydrolysis by the alkaline protease, subtilisin Carlsberg. Limited proteolysis of crude or purified preparations of native BCCP by subtilisin resulted in a mixture of smaller BCCP species analogous to those isolated earlier. More extensive proteolysis by subtilisin converted all of the intermediate species to a small form of BCCP ($BCCP_{(SC)}$). $BCCP_{(SC)}$ was indistinguishable from the smallest form of BCCP isolated previously, $BCCP_{(9100)}$. $BCCP_{(SC)}$ and $BCCP_{(9100)}$ were similar in size and charge (isoelectric point, 4.5). The amino acid composition of $BCCP_{(SC)}$ was identical to that of $BCCP_{(9100)}$ except that the latter contained two additional alanine residues. Each protein contained one biotin residue per molecule, and they had similar K_m values in the biotin carboxylase and transcarboxylase reactions. The relation of these small forms of BCCP to $BCCP_{(10400)}$ was demonstrated by showing that treatment of the latter with subtilisin Carlsberg resulted in its quantitative conversion into $BCCP_{(SC)}$. $BCCP_{(SC)}$ and $BCCP_{(9100)}$ are resistant to further proteolysis by high concentrations of subtilisin, as well as trypsin, chymotrypsin and thermolysin; thus this small BCCP represents a stable 'core' peptide of native BCCP. Presumably, during the initial stages of BCCP isolation, proteolysis of the subtilisin type occurred, accounting for the early isolation of BCCP $_{(9100)}$ and $BCCP_{(10400)}$.

Fraction E_b, the transcarboxylase component of acetyl-CoA carboxylase, was purified 1700-fold from *E. coli* extracts[26]. This component was shown to have a minimal molecular weight of 90 000 by gel filtration and by sodium dodecyl sulphate gel electrophoresis to be composed of two subunits of about 45 000 daltons. The purified transcarboxylase catalysed the carboxylation of BCCP by [^{14}C]malonyl-CoA in the absence of biotin carboxylase (equation (3.7)). The [^{14}C]carboxy-BCCP was separated from the radioactive substrate by gel filtration on Sephadex G-50 and identified by its characteristic acid lability. Transcarboxylase also catalysed a BCCP-dependent exchange between [^{14}C]-acetyl-CoA and malonyl-CoA. In addition, an acetyl-CoA and malonyl-CoA binding site was demonstrated on this protein by showing that incubation of transcarboxylase with [^{14}C]acetyl-CoA or [^{14}C]malonyl-CoA gave rise to [^{14}C]acetyl-CoA transcarboxylase or [^{14}C]malonyl-CoA transcarboxylase, respectively. Addition of unlabelled acetyl-CoA decreased the amount of [^{14}C]malonyl-CoA bound to the protein, suggesting that acetyl-CoA and malonyl-CoA were bound at the same site. These experiments established that the transcarboxylase component of acetyl-CoA carboxylase catalyses the reversible transfer of carboxyl groups from CO_2^-–BCCP to acetyl-CoA, as indicated in equation (3.7).

The acetyl-CoA carboxylase of plants has also been shown to be composed of similar active subunits. The wheat germ enzyme was separated into two active components, corresponding to E_a (9.4 S) and E_b (7.3 S)[27]. This enzyme is highly aggregated, and it appears to be intermediate in molecular properties between the animal and bacterial systems. Spinach leaf chloroplasts, on the other hand, which have procaryotic properties, were shown to have an acetyl-CoA carboxylase system similar to that of *E. coli*. Intact chloroplasts of spinach and lettuce incorporate [^{14}C]acetate into long-chain fatty acids in the presence of bicarbonate, ATP, CoA and light. Incubation of intact chloroplasts with [^{14}C]bicarbonate led to the binding of $^{14}CO_2$ to a protein component which could be separated from substrates by Sephadex G-50 filtration. Chloroplast stromal and lamellar fractions together could bind $^{14}CO_2$, but the individual fractions were inactive. When the stromal and lamellar fractions were separated after they had been incubated together with [^{14}C]bicarbonate, it was found that the $^{14}CO_2$ was associated with the lamellar fraction, indicating that the biotin carboxyl carrier protein of these organelles is membrane-bound. The carboxyl group of $^{14}CO_2^-$–BCCP, isolated from the chloroplast system, could be transferred to acetyl-CoA by the transcarboxylase component of the *E. coli* acetyl-CoA carboxylase or by a transcarboxylase that was purified 1000-fold from the stroma of the chloroplast. Biotin carboxylase activity was also demonstrated in the chloroplast stroma. Thus, although the chloroplast BCCP appears to be membranous, the acetyl-CoA carboxylase of this organelle resembles that of *E. coli* in that it has been resolved into three active components: biotin carboxylase, BCCP, and transcarboxylase.

3.2.3 Control

Acetyl-CoA carboxylase is the first committed reaction in the biosynthesis of fatty acids, and this enzyme is known to be subject to two types of control, allosteric regulation and adaptive changes in enzyme content.

3.2.3.1 Allosteric regulation

Acetyl-CoA carboxylase is considered to be the rate-limiting enzyme in fatty acid synthesis. The initial observations of Ganguly[29] and Numa *et al.*[30] indicated that the activity of the carboxylase is very much lower than that of the fatty acid synthetase in liver extracts. However, when the carboxylase is fully activated and assayed under optimal conditions, its activity is similar to that of the fatty acid synthetase[31]. Thus it is apparent that the degree to which the carboxylase is activated *in vivo* may play an important role in determining the extent that this enzyme directs the rate of fatty acid synthesis.

Two potential allosteric regulators that have been studied extensively are tricarboxylic acids and long chain fatty acyl-CoA derivatives. The early observation, that citrate stimulates fatty acid synthesis from acetate[32], was explained by the discovery that citrate and other Krebs-cycle intermediates cause polymerisation and activation of rat adipose tissue acetyl-CoA carboxylase[14, 33]. This basic observation was confirmed by studies of acetyl-CoA carboxylase of other animal sources, such as rat liver[34], avian liver[6-8, 15], bovine adipose tissue[35] and rat mammary gland[36]. In contrast, citrate has no effect on acetyl-CoA carboxylase of *E. coli*[19] or yeast[37]. Citrate activation of animal acetyl-CoA carboxylase is associated with polymerisation of inactive protomers to active polymeric filaments, a process which reflects clearly a conformational change of the enzyme[6-8, 33, 34]. Kinetic analyses of the enzyme of rat liver[38], avian liver[9, 39] and adipose tissue have established that the major effect of the activator is on the maximal velocity of the enzyme rather than on the K_m values for substrates. Lynen and co-workers[37] showed that both half reactions of the carboxylase (equations (3.2) and (3.3)) are activated by citrate. The first partial reaction was measured as an exchange reaction between ATP and either [^{32}P]orthophosphate or [^{14}C]ADP. The second partial reaction was measured as an exchange reaction between malonyl-CoA and [^{14}C]acetyl-CoA. Citrate activated both partial reactions catalysed by the rat liver enzyme. Lane and co-workers[7,9] confirmed these observations by showing that citrate stimulated both partial reactions catalysed by the avian liver enzyme. These investigators studied the carboxylation of free (+)-biotin as a measure of the first partial reaction, and they studied the carboxylation of acetyl-pantetheine by carboxybiotin-enzyme as a measure of the second partial reaction. Since both partial reactions were stimulated by citrate and the biotinyl group is involved in both of the half reactions, it has been concluded that the prosthetic group is involved in the citrate-induced conformational changes[15]. Moss and Lane[40] showed by kinetic analyses of the avian liver enzyme that citrate renders the biotinyl moiety of the carboxylase less accessible to avidin, a specific biotin inhibitor. In addition, citrate increased the rate of decarboxylation of the carboxylated biotinyl prosthetic group[41]; thus the reactivity of the biotinyl moiety is enhanced by the activator. All these experiments suggest that a conformational change in the vicinity of the biotin prosthetic group is associated with the allosteric activation of the enzyme by citrate.

The other allosteric effector of acetyl-CoA carboxylase is palmityl-CoA which, like other long-chain fatty acyl-CoA derivatives, severely inhibits this enzyme. Inhibition by palmityl-CoA is competitive with citrate[38,42]. The K_m value for citrate activation of the enzyme is 2–6 mM, while the K_i of palmityl-CoA is 0.8–1.1µM. Over 70% of the citrate in mammalian liver is associated with mitochondria[43]; estimates of concentrations of citrate in the cytoplasmic fraction are 0.1–0.2 mM[44]. This suggests that acetyl-CoA carboxylase may not be fully activated *in vivo*, although the extent of compartmentalisation of citrate within the cytoplasm is unknown. Concentrations of long-chain fatty acyl-CoA derivatives in mammalian and avian liver vary with the nutritional state of the animal, and values from 15–140 µM have been recorded[45-47]. Although these values are within the concentration range that causes inhibition of the carboxylase *in vitro*, the concentrations to which the carboxylase is exposed *in vivo* is entirely unknown. Perhaps of most importance, the degree of binding of these derivatives to other cellular proteins and the resulting concentration of acyl-CoA derivatives available for inhibition of the carboxylase is unknown. It should be emphasised that long-chain fatty acyl-CoA derivatives were found in increased concentrations in livers of starved and fat-fed rats[45,46]. Since increase in liver fatty acyl-CoA's was presumed to be secondary to adipose tissue lipolysis, it was reasonable to suspect that various physiological conditions associated with increased lipolysis and decreased fatty acid synthesis might be explained by inhibition of acetyl-CoA carboxylase by long chain acyl-CoA derivatives in the liver. However, this proposal was questioned because long-chain fatty acyl-CoA compounds are detergents which inhibit many enzymes non-specifically[48].

The physiological significance of the inhibition of chick liver acetyl-CoA carboxylase by long-chain fatty acyl-CoA compounds has recently been re-examined by Goodridge[49]. It was recognised that both the concentration of palmityl-CoA and the relative affinity of the carboxylase for this thioester are important determinants. Palmityl-CoA (100 µM), in the presence of bovine serum albumin (25 mg ml^{-1}), was shown to inhibit the incorporation of [^{14}C]citrate into fatty acids by a 100 000 × g supernatant fraction of chick liver homogenate and to inhibit the activity of purified chick liver acetyl-CoA carboxylase. Under similar conditions, palmityl-CoA (200 µM) did not inhibit other lipogenic enzymes, e.g. fatty acid synthetase, ATP-citrate lyase, malic enzyme and NADP-linked isocitrate dehydrogenase, or two other unrelated enzymes, pyruvate kinase and glutamate dehydrogenase. This 'specific' inhibition of acetyl-CoA carboxylase could be reversed by increasing the concentration of serum albumin, increasing the concentration of citrate, or addition of (+)-palmityl-carnitine. Inhibition by palmityl-CoA was competitive with citrate. The mechanism of the effect of palmityl-carnitine was unclear, but the possibility was considered that this derivative acted as a structural analogue which bound to the enzyme at the same site as palmityl-CoA but did not inhibit enzyme activity. Non-specific detergent inactivation of the carboxylase by palmityl-CoA was considered unlikely because the inhibition did not increase with time of incubation, it was relatively specific, and the inhibition was reversible and competitive with a potentially physiological activator, i.e. citrate. However, despite these *in vitro* results, the presence of inactive enzyme has not been demonstrated adequately in liver under conditions of fasting or high fat feeding when activity of carboxylase is low and the concentration of long-chain fatty acyl-CoA derivatives is high. In fact, all changes in carboxylase activity noted with various nutritional changes have been associated with changes in content of enzyme. However, it is possible that the adaptive changes in enzyme content serve in long-term regulation while allosteric control by citrate and long-chain acyl-CoA compounds serves for rapid regulatory changes.

3.2.3.2 *Roles of enzyme synthesis and degradation*

Acetyl-CoA carboxylase activity is depressed markedly in rats fasted, fed a high fat diet or made alloxan-diabetic, and it is markedly elevated in rats fed a fat-free diet[30, 50, 51]. Immunochemical techniques have been utilised to determine whether these changes reflect changes in catalytic efficiency or content of enzyme[12, 52]. Rat liver acetyl-CoA carboxylase preparations that were treated with citrate (50 S), palmityl-CoA (12 S) or untreated controls (12 S) were similarly inactivated and precipitated by antibody made against pure acetyl-CoA carboxylase[52]. Thus, all forms of the enzyme, protomeric or polymeric, of potential physiological importance could be detected by this antibody. Quantitative precipitin analyses were carried out with crude extracts from fasted, high-fat-fed, fat-free-fed or chow-fed rats; the acetyl-CoA carboxylase specific activities in these extracts varied *ca.* 25-fold. All the enzyme preparations had the same equivalence points in the quantitative precipitin analyses. Thus, there was a constant and equal amount of immunoprecipitable enzyme per unit of activity in each preparation, and, therefore, differences in specific activity reflected differences in content of enzyme. Nakanishi and Numa[12] confirmed these data and also demonstrated that the depressed carboxylase activity in diabetic rats was due to decreased content of enzyme.

The Schimke technique[53] of pulse labelling the enzyme with [^{3}H]leucine and isolating the radioactive carboxylase by immunoprecipitation was utilised to determine whether the differences noted in content of acetyl-CoA carboxylase reflected changes in enzyme synthesis[52]. The relative rate of enzyme synthesis was determined by relating incorporation of radioactivity into enzyme to incorporation into total soluble protein. A 4- to 10-fold increase in enzyme synthesis was noted with fat-free feeding[12,52]. Relative rates of enzyme synthesis decreased with fasting, high fat feeding, and in alloxan-diabetes. Thus, changes in rate of enzyme synthesis plays an important role in both nutritional and hormonal alterations in carboxylase activity.

To determine whether changes in the rate of enzyme degradation contributed to the alterations of carboxylase activity, the enzyme was labelled with [^{3}H]leucine and decay of radioactivity in the carboxylase with time was determined by the immunoprecipitation method. The half-life of the enzyme did not vary significantly during normal, fat-free or high-fat feeding or in alloxan-diabetic animals. Majerus and Kilburn[52] reported a $t_{\frac{1}{2}}$ of 48 h, Nakanishi and Numa[12] a $t_{\frac{1}{2}}$ of 55–59 h. However, during fasting enzyme degradation was definitely accelerated; Majerus[52] noted a $t_{\frac{1}{2}}$ of 18 h while Numa reported a $t_{\frac{1}{2}}$ of 3 h. During fasting, therefore, decreased acetyl-CoA carboxylase content was due to both decreased synthesis and accelerated degradation of the enzyme.

A marked increase in acetyl-CoA carboxylase activity occurs in the mouse at the time of weaning, when the nutritional intake changes from high-fat milk to lower-fat chow[54] and in the chick after hatching when the animals begin to feed[55]. It is not yet known if these increases reflect changes in content of enzyme related to changes in rates of enzyme synthesis or degradation. It might be predicted from the data cited above, relative to the effects of fat-free feeding, that enzyme content increase secondary to increase in enzyme synthesis will be an important regulatory mechanism in these systems. Moreover, increase in the synthesis of liver fatty acid synthetase in the rat at the time of weaning[56] and the synthesis of malic enzyme in the chick after hatching[57] have been reported; these enzymes, of course, are also important in lipogenesis.

Further insight into the regulation of acetyl-CoA carboxylase has been gained through studies of a genetic mutant, the obese hyperlgycaemic mouse, and a minimal deviation hepatoma. Genetically obese mice (C57BL/GJ-ob) carry a single recessive gene for obesity. These animals have elevated basal hepatic acetyl-CoA carboxylase[31], fatty acid synthetase and ATP-citrate lyase. By use of immunochemical techniques, Nakanishi and Numa[58] showed that the increased acetyl-CoA carboxylase level was due to increased content of enzyme and that the latter resulted from a 7.7-fold higher rate of enzyme synthesis in the obese animals and a 1.7-fold lower rate of degradation. Thus, in this mutant the primary derangement of acetyl-CoA carboxylase is due to acceleration in the rate of enzyme synthesis.

A notable feature of minimal deviation hepatomas is the loss of control of fatty acid synthesis following dietary alterations[59]. Majerus et al.[60] showed that neither acetyl-CoA carboxylase nor fatty acid synthetase of the tumour responds to dietary manipulations as does the host liver. The failure to respond was not related to differences of the enzymes, at least with respect to activation of the carboxylase by citrate or inhibition by palmityl-CoA. Preliminary experiments suggested that the increase in enzyme synthesis in host liver does not occur in neoplastic liver[52]. The mechanism whereby the malignant state results in loss of control of enzyme synthesis remains unclear. It was demonstrated, however, that the lack of dietary regulation of acetyl-CoA carboxylase and fatty acid synthesis in hepatoma was not due to the fact that the tumours do not receive nutrients from the portal blood flow, since these enzymes continued to respond to dietary manipulations in normal liver transplanted away from the portal circulation[61].

3.3 FATTY ACID SYNTHETASE

The enzyme system which catalyses the synthesis of saturated long-chain fatty acids from malonyl-CoA is called the fatty acid synthetase, and the requirements of this reaction include acetyl-CoA and NADPH, as indicated in equation (3.8):

$$\text{MeCO—S—CoA} + 7\text{HO}_2\text{CCH}_2\text{CO—S—CoA} + 14\text{NADPH} + 14\text{H}^+ \rightarrow \text{Et}(\text{CH}_2\text{CH}_2)_6\text{CH}_2\text{CO}_2\text{H} + 7\text{CO}_2 + 14\text{NADP}^+ + 8\text{CoA—SH} + 6\text{H}_2\text{O} \quad (3.8)$$

This enzyme system was discovered by Lynen and his associates in yeast, where it is present as a multi-enzyme complex[62, 63]. Comparable multi-enzyme complexes were later isolated from avian liver[64, 65], rat liver [66], mammary gland[67, 68] and *Mycobacterium phlei*[69]. In 1961 Lynen[62] proposed that the acyl intermediates in fatty acid biosynthesis are bound in thioester linkage to a component of the multi-enzyme complex. This proposal was based upon the susceptibility of the enzyme complex to inactivation by sulphydryl binding agents, the absence of free intermediates during the reaction and the identification of protein-bound acetoacetate. The intermediates proposed for the reactions were based on observations carried out with model compounds.

The nature of the protein-bound intermediates as well as the details of the intermediate reactions were elucidated in studies of bacterial systems, initially *Clostridium kluyveri* and then *Escherichia coli*[70-72]. In contrast to the tightly associated multi-enzyme complexes listed above, the fatty acid synthetase of these micro-organisms was found non-associated when the cells were disrupted and the individual proteins of the synthetase have been isolated and studied. Although these bacterial systems have facilitated the understanding of fatty acid biosynthesis, they have posed an interesting biological problem: Are the bacterial fatty acid synthetase proteins associated *in vivo* in a typical multi-enzyme complex? All attempts to isolate a complex from *E. coli* have failed thus far. Preliminary studies, however, have shown that acyl carrier protein (ACP), a central component of the fatty acid synthetase, is localised on or near, the inside surface of the plasma membrane of *E. coli*[73]. The fact that ACP is not randomly distributed in the cell suggests a degree of organisation which was not appreciated before, and this finding suggests that the *E. coli* fatty acid synthetase may be organised *in vivo*, perhaps in a typical multi-enzyme complex which includes ACP.

Studies of the individual enzymes of *E. coli* fatty acid synthetase have indicated the following reactions which explain the biosynthetic sequence from malonyl-CoA to the major saturated product, palmitate[70-72, 74, 75].

$$\text{MeCO—S—CoA} + \text{HS—ACP} \rightleftharpoons \text{MeCO—S—ACP} + \text{CoA—SH} \quad (3.9)$$

$$\text{MeCO—S—ACP} + \text{HS-E}_{cond} \rightleftharpoons \text{MeCO—S—E}_{cond} + \text{ACP—SH} \quad (3.10)$$

$$\text{HO}_2\text{CCH}_2\text{CO—S—CoA} + \text{HS—ACP} \rightleftharpoons \text{HO}_2\text{CCH}_2\text{CO—S—ACP} + \text{CoA—SH} \quad (3.11)$$

$$\text{HO}_2\text{CCH}_2\text{CO—S—ACP} + \text{MeCO—S—E}_{cond} \rightleftharpoons \text{MeCOCH}_2\text{CO—S—ACP} + \text{CO}_2 + \text{HS—E}_{cond} \quad (3.12)$$

$$\text{MeCOCH}_2\text{CO—S—ACP} + \text{NADPH} + \text{H}^+ \rightleftharpoons \text{MeCHOHCH}_2\text{CO—S—ACP} + \text{NADP}^+ \quad (3.13)$$

$$\text{MeCHO·HCH}_2\text{CO—S—ACP} \rightleftharpoons \text{H}_2\text{O} + \text{MeCH}=\text{CHCO—S—ACP} \quad (3.14)$$

$$\text{MeCH}=\text{CHCO—S—ACP} + \text{NADPH} + \text{H}^+ \rightleftharpoons \text{EtCH}_2\text{CO—S—ACP} + \text{NADP}^+ \quad (3.15)$$

In the initial reaction (equation (3.9)) an acetyl group is transferred by acetyl-CoA—ACP transacylase from the sulphydryl group of CoA to the single sulphydryl group of ACP to form acetyl-ACP. The acetyl group is then transferred to a sulphydryl group of the condensing enzyme (E_{cond}) to form an acetyl-enzyme intermediate and liberate ACP (equation (3.10)). Malonyl-CoA—ACP transacylase catalyses the transfer of a malonyl group from CoA to ACP in equation (3.11). This is followed by the condensation reaction (equation (3.12)), which takes place between malonyl-ACP and acetyl-enzyme to produce acetoacetyl-ACP, CO_2 and the free condensing enzyme. Acetoacetyl-ACP is then reduced by NADPH to form specifically D-(−)-β-hydroxybutyryl-ACP (equation (3.13)); the latter is dehydrated to form the *trans* unsaturated thioester, crotonyl-ACP (equation (3.14)); and crotonyl-ACP is reduced by NADPH to form the saturated thioester, butyryl-ACP (equation (3.15)). This can then react with the condensing enzyme to form the acyl-enzyme intermediate (equation (3.10) with butyryl-ACP substituting for acetyl-ACP), liberating ACP to react with another malonyl group and thus initiate another sequence of condensation, reduction, dehydration and reduction reactions. The synthetase of *E. coli*, like that of the animal and plant systems, results *in vitro* in free palmitate as the final product due to the action of a long chain thioesterase. However, it appears likely that *in vivo* in *E. coli* palmityl-ACP reacts directly with glycerol-3-phosphate and a specific membranous acyltransferase to form monoacylglycerol-3-phosphate, the first intermediate in the pathway of phospholipid biosynthesis[76-78]. The additional observation that ACP in *E. coli* is localised on, or very near, the cell membrane strongly suggests that the product of the synthetase reaction is well positioned for direct incorporation into the phospholipids of the cell membrane[73].

3.3.1 Acyl-carrier protein

A central component of all fatty acid synthetases is the acyl-carrier protein (ACP). The critical function of this protein in fatty acid synthesis is illustrated in the reactions of the enzyme system (equations (3.9)–(3.15)). The discovery, isolation and demonstration of some important structural and functional properties of *E. coli* ACP have been recently reviewed[79] and will not be reiterated here. However, it should be recalled that the most significant feature of all ACP compounds is the 4'-phosphopantetheine prosthetic group that was initially discovered in *E. coli* ACP[70]. The structure of the prosthetic group and its attachment to the protein as a phosphodiester through a hydroxyl group of a serine residue are illustrated in Figure 3.1 which demonstrates the structure of a peptide containing the 4'-phosphopantetheine that was isolated from a peptic digest of *E. coli* ACP. The acyl groups that are biosynthetic intermediates are bound in thioester linkage to the sulphydryl group of the 4'-phosphopantetheine.

While the majority of studies attempting to delineate the structure and function of ACP have been carried out with *E. coli* ACP, ACP or a protein which functions like ACP has been identified in every biological system which catalyses the *de novo* synthesis of fatty acids. Homogeneous preparations of ACP have been obtained from *Arthrobacter*, avocado, spinach[80], *Mycobacterium phlei*[81], *Clostridium butyricum*[82] and yeast[83]. The amino acid compositions of ACP compounds from these sources are shown in Table 3.1. The most striking features of these compositions is the presence of 1 mole each of taurine (the oxidation product of 2-mercaptoethylamine) and β-alanine, representing the 4'-phosphopantetheine prosthetic group. Five of the seven ACP compounds have only a single sulphydryl group, that contributed by the prosthetic group. Avocado and yeast ACP both have an additional sulphydryl group contributed by a cysteine residue; however, Simoni, Criddle and Stumpf[80] have shown conclusively that the cysteine of avocado ACP does not function as an acyl carrier. Molecular weights of the ACP compounds range from *ca.* 8600 for *C. butyricum* ACP[82] to 16 000 for yeast ACP[83]. The report of the isolation of yeast ACP represents the first isolation of ACP from one of the multi-enzyme complexes. Despite some similarities in amino acid composition and even primary structure around the site of attachment of the 4'-phosphopantetheine moiety (see below), important differences among various ACP compounds exist. Thus, although ACP compounds from plants and *E. coli* function interchangeably in the respective fatty acid synthetase systems, the products of fatty acid synthesis vary according to the ACP used[80].

It should be noted that all preparations of ACP studied so far, except yeast, have been obtained from organisms which contain fatty acid synthetase systems that are found non-associated when the cell membrane is ruptured. Preliminary studies of yeast ACP have been reported[83], but ACP has not been isolated from other organisms that have a fatty acid synthetase multi-enzyme complex. However, the presence of protein-bound 4'-phosphopantetheine has been shown in multi-enzyme complexes isolated from adipose tissue[84], pigeon liver[85], rat liver[66], lactating rat mammary gland[86], and *M. phlei*[87], and this

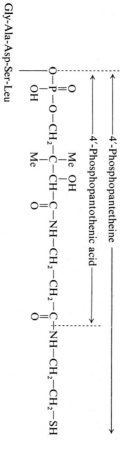

Figure 3.1 The structure of peptide isolated from a peptic digest of *E. coli* ACP

Table 3.1 Amino acid composition of ACP from several sources*

Amino acid	E. coli[†]	Arthrobacter[‡]	Avocado[‡]	Spinach[‡]	M. phlei[§]	C. butyricum[‖]	Yeast[¶]
Cysteic acid	0	0	1	0	0	0	1
Taurine	1	1	1	1	1	1	1
β-Alanine	1	1	1	1	1	1	1
Aspartic	9	14	12	12	11	13	14
Threonine	6	2	7	6	5	1	8
Serine	3	6	10	5	5	3	12
Glutamic	18	10	22	16	20	14	19
Proline	1	1	3	2	4	1	7
Glycine	4	5	7	4	6	1	13
Alanine	7	12	11	9	13	6	14
Valine	7	6	10	7	8	7	8
Methionine	1	1	1	1	1	4	2-3
Isoleucine	7	6	5	5	6	8	7
Leucine	5	6	9	7	8	7	13
Tyrosine	1	0	1	0	2	1	4-5
Phenylalanine	2	3	3	2	2	3	6
Lysine	4	5	10	9	5	4	10
Histidine	1	1	1	1	0	2	2-3
Arginine	1	1	1	0	3	0	5

Values for amino acid composition of yeast ACP were calculated assuming 1 mole of β-alanine/mole protein, and rounding to the nearest integer.
* Nearest integer values. † Vanaman, Wakil, and Hill[**] ‡ Simoni, Criddle, and Stumpf[⁹⁰] § Matsumura[⁹¹] ‖ Ailhaud, Vagelos, and Goldfine[⁴²] ¶ Willecke, Ritter, and Lynen[⁸³]

implies the presence of ACP or a protein that functions as ACP in each of these systems.

The complete amino acid sequence of *E. coli* ACP has been determined[88] and is illustrated in Figure 3.2. Several features of the amino acid composition and sequence are noteworthy. The molecule contains 14 residues of glutamic acid and eight residues of aspartic acid out of a total of 77 residues. There are only four lysine residues and one each of arginine and histidine. Therefore, ACP is quite acidic in nature with an isoelectric point of pH = *ca.* 4.2, and at this pH the protein is probably least soluble[88]. As is apparent in Figure 3.2, the acidic residues occur throughout the sequence, while the basic residues appear clustered at both amino and carboxyl ends. Residues 47–49 are in the sequence Glu-Glu-Glu and 56–58 are Asp-Glu-Glu. In fact, 9 out of 14 residues between residues 47 and 60 are acidic, while there are not extended sequences rich in hydrophobic side chains. The possible exception is the sequence from residue 62 to 69. The prosthetic group is attached to serine 36 and thus lies midway along the sequence. In view of the preponderance of charged residues and the fact that ACP behaves as a typical globular protein[89], the three-dimensional structure may show a preponderance of charged residues on the surface. This may have important implications for the interactions of acyl-ACP compounds with the various enzymes of the fatty acid synthetase system.

Detailed information on the structure of ACP compounds isolated from other sources is lacking. As mentioned above, all biological systems that catalyse *de novo* biosynthesis of fatty acids contain protein-bound 4'-phosphopantetheine, and in all cases that have been examined the 4'-phosphopantetheine is linked to a serine residue of the protein which has the function of ACP. The amino acid sequence around the 4'-phosphopantetheine has been studied in *Arthrobacter* and spinach ACP compounds (Table 3.2). Based on the results of compositional analyses and partial Edman degradations of peptides, Matsumura and Stumpf[90] suggest that nine residues around the prosthetic group appear to be identical in *Arthrobacter* ACP, spinach ACP and *E. coli* ACP. The similarity in the primary structure about the prosthetic group is perhaps not surprising since in all three systems the fatty acid synthetase is the non-associated variety.

There is less structural information concerning ACP compounds which are components of the tightly-associated fatty acid synthetase complexes of yeast and animals. Yeast ACP has been isolated, and the amino acid sequence[91] around the prosthetic group (Table 3.2) appears to have no resemblance to the corresponding segment of *E. coli* ACP. The animal fatty acid synthetase is more resistant to dissociation, and the protein-containing 4'-phosphopantetheine has not yet been isolated. Because of the difficulties encountered in the isolation of ACP from animal fatty acid synthetase, the amino acid sequence of rat-liver fatty acid synthetase peptides containing 4'-phosphopantetheine was determined. The isolation of these peptides was facilitated by the use of fatty acid synthetase labelled with either ^{3}H or ^{14}C in the prosthetic group[92]. The structure around the 4'-phosphopantetheine group of rat liver fatty acid synthetase is shown in Table 3.2. Five of the 13 residues in the region of the prosthetic group of liver fatty acid synthetase are identical with those found in the corresponding region of *E. coli* ACP. In particular,

```
                  20                                30
Gln-Glu-Val-Thr-Asp-Asn-Ala-Ser-Phe-Val-Glu-Asp-Leu-Gly-Ala-Asp-
         |
        (P)-Pantetheine-SH
                                 44                         50
Ser-Leu-Asp-Thr-Val-Glu-Leu-Val-Met-Ala-Leu-Glu-Glu-Phe-Asp-Thr-
36
              55                  60                            70
Glu-Ile-Pro-Asp-Glu-Glu-Ala-Glu-Lys-Ile-Thr-Thr-Val-Gln-Ala-Ala-Ile-Asp-
                                  ___
                  77
Tyr-Ile-Asn-Gly-His-Gln-Ala-$CO_2H$
```

Figure 3.2 The complete amino acid sequence of *E. coli* ACP. (From Vanaman, Wakil, and Hill[88], by courtesy of American Society of Biological Chemists, Inc.)

Table 3.2 Primary sequence of 4.-phosphopantetheine-peptides†

Source	Sequence of peptides
E. coli‡	-Asn-Ala-Ser-Phe-Val-Glu-Asp-Leu-Gly-Ala-Asp-*Ser-Leu-Asp-Thr-Val-Glu-
Arthrobacter§	Gly-Ala-Asp-*Ser, Leu, Asp, Thr, Val, Glu
Spinach§	Lys-Gly-Ala,Asp,*Ser,Leu,Asp,Thr,Val,Glu
Rat liver‖	Leu-Gly-Ser-Leu-Asn-Leu-Gly-Glx-Gly-Glu-Asp-*Ser-Leu
Yeast††	Lys-Gly-*Ser-Val-Pro-Ala

† The asterisk denotes the serine to which the prosthetic group, 4'-phosphopantetheine, is bound: underlined amino acids are identical with the amino acids in the corresponding sequence of *E. coli ACP*.
‡ From Vanaman, Wakil and Hill[88]. § From Matsumura and Stumpf[90]. ‖ From Roncari, Bradshaw and Vagelos[a2]. ¶ From Ayling, Pirson and Lynen[91].

four of the five residues around the 4′-phosphopantetheine are identical. Since the synthetases of yeast and animals behave as stable complexes, strong heteromeric contacts between the constituents of these complexes are implied and substantial differences in the primary structure of the corresponding subunits relative to the corresponding bacterial and plant subunits are expected. It is not known whether the regions of the mammalian and yeast fatty acid synthetase which differ in amino acid sequence from those of *E. coli* ACP form stronger heteromeric contacts with the component enzymes of the complex.

Both ACP and CoA contain 4′-phosphopantetheine as the component which carries acyl groups linked as thioesters. The fact that thioesters of ACP are obligatory intermediates in fatty acid biosynthesis, whereas thioesters of CoA are relatively inactive with the biosynthetic enzymes, suggests that the protein structure of ACP is important in its biological function. With the elucidation of the primary sequence of ACP, investigations were initiated to delineate those parts of the protein structure that are important for the activity of ACP with the ten enzymes with which it is known to react. These studies have indicated that few alterations of the polypeptide chain of ACP are tolerated by the enzymes of fatty acid biosynthesis or by the enzyme, holo-ACP synthetase, which catalyses the synthesis of holo-ACP from apo-ACP and CoA equation (3.16). The most critical feature of the amino acid sequence (Figure 3.2) is the serine residue to which the prosthetic group is attached at position 36, approximately the middle of the 77 amino acid peptide. The initial approaches to the understanding of structure–function relationships of ACP involved chemical-modification studies and proteolytic fragment analysis. Acetylation of the four lysine residues and the amino group of the terminal serine residue had no effect on the ability of ACP to function in fatty acid synthesis[93, 94]. Nitration of the single tyrosine at position 71 did not affect function of ACP in fatty acid synthesis, though the reaction of the palmityl derivative of this modified ACP with glycerol-3-phosphate acyltransferase was decreased[95]. Alkylation of the methionine at position 44 did not affect the activity of ACP in the malonyl-CoA—CO_2 exchange reaction, which depends upon the malonyl-CoA–ACP transacylase and β-ketoacyl-ACP synthetase reactions (equations (3.10–(3.12)) but it did decrease activity of both ACP in fatty acid synthesis and palmityl-ACP in the glycerol-3-phosphate acyltransferase reaction[95]. Amidisation of 11 of the 22 carboxyl functions resulted in total loss of biological function[95]. Not all modifications were accompanied by conformational changes (e.g. alkylation of methionine), and not all conformational changes were accompanied by activity changes (e.g. nitration of tyrosine). Several peptide fragments have been prepared after treatment of native ACP with trypsin, carboxy peptidase A or cyanogen bromide[93, 94]. ACP-(1–74) (carboxypeptidase A) was fully active in the malonyl-CoA-CO_2 exchange reaction as was the apopeptide in holo-ACP synthetase reaction. ACP-(1–44) (cyanogen bromide) and ACP-(19–61) (trypsin) were inactive in these reactions. Removal of the amino terminal hexapeptide (trypsin treatment of acetylated ACP; cleavage occurred only at the arginine at position 6) resulted in a peptide, ACP-(7–77), that had undergone a drastic conformational change and was inactive in the malonyl-CoA–CO_2 exchange and the holo-ACP synthetase reactions. Thus these experiments showed that the peptide structure of ACP is very important for activity both with enzymes of fatty acid synthesis and with holo-AcP synthetase.

An important advance in the understanding of structure-function relationships of ACP was made possible through the recent synthesis of apo-ACP-(1–74)[96,97] by the solid-phase method of Merrifield[98]. Holo-ACP-(1–74) was prepared enzymatically from synthetic apo-ACP-(1–74) by reaction with CoA in the presence of holo-ACP synthetase. The synthetic product was indistinguishable from native holo-ACP-(1–74) on the basis of ion-exchange chromatography, gel filtration on Sephadex G-50 and precipitation by specific antibody prepared against native ACP. A number of analogues of apo-ACP were synthesised by the solid phase method, and these were assayed for biological activity with the enzymes malonyl-CoA-ACP transacylase (equation (3.11)) and β-ketoacyl-ACP synthetase (equations (3.10) and (3.12)) and with specific antiserum prepared against native ACP[97]. When the single methionine at position 44 and the three lysines at positions 8, 9 and 18 were replaced with norleucine in the analogue, $[Arg(NO_2)^6, Nle^{8,9,18,44}]$-apo-ACP-(1–74), no loss in biological activity of the apo- or holo-proteins was noted. This demonstrated that these residues play a non-essential role in the biological function of ACP. This observation made further syntheses easier, since these vulnerable groups could be replaced by norleucine. The significance of the carboxy terminus was evaluated by synthesising $[Arg(NO_2)^6, Nle^{8,9,18,44}]$-apo-ACP-(1–74), subjecting that analogue to trypsin digestion (cleavage would occur only at the remaining lysine at position 61), and isolating and testing the resulting ACP fragment, $[Arg(NO_2)^6, Nle^{8,9,18,44}]$-apo-ACP-(1–61). The latter fragment was found to possess low but significant biological activity. Thus, a significant portion of the carboxy terminus (17 residues) does not participate directly in the activity of the protein. In contrast, the amino terminal hexapeptide is essential, as demonstrated by the inactivity of holo-ACP-(7–77) obtained by trypsin digestion of acetylated ACP. To examine the significance of this hexapeptide more closely, the following synthetic analogues were prepared: $[Arg(NO_2)^6, Nle^{8,9,18,44}]$-apo-ACP-(1–74); $[Arg(NO_2)^6, Nle^{8,9,18,44}]$-apo-ACP-(2–74); $[Arg(NO_2)^6, Nle^{8,9,18,44}]$-apo-ACP-(3–74); $[Arg(NO_2)^6, Nle^{8,9,18,44}]$-apo-ACP-(4–74); $[Arg(NO_2)^6, Nle^{8,9,18,44}]$-apo-ACP-(5–74); $[Arg(NO_2)^6, Nle^{8,9,18,44}]$-apo-ACP-(6–74); and $[Nle^{8,9,18,44}]$-apo-ACP-(7–74). A study of these analogues[97] showed a progressive decrease in activity as more of the amino terminus was omitted. Thus there was no single residue in the amino terminal hexapeptide that was absolutely critical for ACP function. The results suggest that each amino acid, by either ionic or hydrophobic interactions, contributes to the maintenance of some structural feature which is essential to the biological function of ACP. This result is consistent with earlier studies by Abita *et al.*[95] which demonstrated that the amino terminal hexapeptide plays an important role in maintaining the structure of ACP.

3.3.1.1 *Prosthetic group turnover*

4'-Phosphopantetheine is a component of both CoA and ACP, and CoA is the 4'-phosphopantetheine donor in the synthesis of holo-ACP[99]. The enzyme, holo-ACP synthetase, catalyses the synthesis of holo-ACP from apo-ACP and CoA according to equation (3.16):

$$\text{Apo-ACP} + \text{CoA} \xrightarrow{Mg^{2+}} \text{holo-ACP} + 3', 5'\text{-adenosine diphosphate} \quad (3.16)$$

This enzyme, which has been purified from extracts of *E. coli*, has rigid structural requirements for both CoA and the apoprotein. Neither dephospho-CoA nor oxidised CoA can substitute for reduced CoA as 4'-phosphopantetheine donor with this enzyme. A number of apopeptides of ACP have been tested[100], including apo-ACP-(1–74), apo-ACP-(19–61), apo-ACP-(1–44), and apo-ACP-(7–77). Only apo-ACP-(1–74) is active, and its activity approaches that of apo-ACP-(1–77).

Another *E. coli* enzyme, ACP hydrolase[101], catalyses the hydrolysis of ACP to yield the prosthetic group and the apoprotein according to equation (3.17):

$$\text{Holo-ACP} + H_2O \xrightarrow{Mn^{2+}} \text{apo-ACP} + 4'\text{phosphopantetheine} \quad (3.17)$$

Studies of this enzyme indicate that it is a highly specific phosphodiesterase; although it catalyses cleavage of 4'-phosphopantetheine from *E. coli* acetyl-ACP and ACP of *C. butyricum*, it is completely inactive with the trypsin peptide, holo-ACP-(19–61), of *E. coli* ACP.

The *in vivo* function of both holo-ACP synthetase and ACP hydrolase have been demonstrated by both pulse and pulse-chase experiments with labelled pantothenate in a pantothenate auxotroph of *E. coli*[102]. These studies established that CoA is the immediate precursor of the prosthetic group of ACP, that this 4'-phosphopantetheine moiety turns over rapidly, and that the rate of ACP turnover is four times the rate of growth of the ACP pool. This rapid prosthetic group turnover was suspected to have a regulatory role, but this has not yet been established.

Tweto *et al.*[103] studied prosthetic group turnover in rat liver by pulse labelling with [³H]pantothenate and determining specific radioactivity of 4'-phosphopantetheine of purified fatty acid synthetase and CoA. Maximum specific radioactivity of CoA was reached before that of the synthetase, supporting a precursor role of CoA for the 4'-phosphopantetheine group of ACP in rat liver. Specific radioactivity of the 4'-phosphopantetheine of fatty acid synthetase remained low until *ca.* 6 h after injection when it rose rapidly to reach a maximum 8 h later. The turnover rate of the prosthetic group was at least an order of magnitude faster than that of the whole complex which has a $t_\frac{1}{2}$ of *ca.* 70 h. As in *E. coli* this raises the distinct possibility that the turnover of the prosthetic group of ACP plays a regulatory role in fatty acid synthetase activity. Recently, Tweto and Larrabee[104] showed that this prosthetic group exchange in liver stops when the animal is fasted for 16–24 h. However, the physiological importance of this phenomenon has not been elucidated. It should be mentioned that the enzymes, holo-ACP synthetase and ACP hydrolase, have not yet been demonstrated in liver.

3.3.2 Enzymes of *Escherichia coli* fatty acid synthetase

As mentioned above, the fatty acid synthetase of most bacteria and plants is found completely non-associated, with ACP and all the enzymes of this biosynthetic system free in solution, when the cell membranes of the organisms are ruptured. Although this may not reflect the *in vivo* state of the fatty acid synthetase in these organisms, the non-associated nature of the various fatty acid synthetase components in these systems has permitted their isolation and study. The components of the *E. coli* fatty acid synthetase have been investigated most intensively and have led to the present understanding of the mechanism of fatty acid biosynthesis, therefore the individual enzymes of the *E. coli* fatty acid synthetase will be discussed in some detail.

3.3.2.1 Acetyl-CoA—ACP transacylase

Acetyl-CoA-ACP transacylase catalyses the transfer of acetyl groups from CoA to ACP forming acetyl-ACP as shown in equation (3.9). It has been purified *ca.* 230-fold from *E. coli* extracts[105-107], but the preparation was not pure at that stage. The enzyme is relatively unstable, and this has delayed further purification. The enzyme preparation is relatively specific for the acetyl moiety. Longer chain fatty acyl thioesters up to C_8 can replace acetyl CoA, but with slower rates. Thus the enzyme can transacylate CoA esters of propionic, butyric and hexanoic acids at rates of 23.4, 9.8 and 4.5% that of acetyl-CoA, respectively. Malonyl-CoA is inactive. Pantetheine can replace, ACP, and acetyl-pantetheine readily substitutes for acetyl-CoA. 2-Mercaptoethanol does not substitute for ACP.

Acetyl transacylase was inhibited 89% by 0.1 mM *N*-ethylmaleimide and 83% by 0.1 mM iodoacetamide, and this inhibition was prevented by prior incubation of the enzyme with acetyl-CoA[107]. This prompted the suggestion that the reaction occurs as two partial reactions with an acyl-enzyme intermediate as indicated in equations (3.18) and (3.19):

$$\text{Acetyl-CoA} + \text{ENZ} \rightleftharpoons \text{acetyl-ENZ} + \text{CoA} \qquad (3.18)$$

$$\text{Acetyl-ENZ} + \text{ACP} \rightleftharpoons \text{acetyl-ACP} + \text{ENZ} \qquad (3.19)$$

In equation (3.18) acetyl-CoA reacts with acetyl transacylase to form a stable acetyl-enzyme, and the latter reacts with ACP in equation (3.19) to form acetyl-ACP and the free enzyme. Williamson and Wakil[107] reported that incubation of the enzyme with [^{14}C]acetyl-CoA led to the formation of [^{14}C]acetyl-enzyme, which was separated from the reaction mixture by filtration through Sephadex G-25. The [^{14}C]acetyl-enzyme could transfer the [^{14}C]acetyl group to either CoA or ACP, supporting the role of an acetyl-enzyme intermediate in equations (3.18) and (3.19). Based upon the inhibitory effects of alkylating agents, Williamson and Wakil proposed that the enzyme intermediate is a thioester[107]. However, the nature of the acetyl-enzyme was not investigated. In light of the information concerning the *E. coli* malonyl-CoA—ACP transacylase (Section 3.3.2.2) and also the demonstration that the yeast acetyl-CoA—ACP transacylase forms an acetyl-*O*-serine enzyme intermediate (Section 3.3.3.1), it is very likely that the *E. coli* acetyl transacylase contains a serine in its active site and that the acetyl group is bound to the enzyme through the hydroxyl group of a serine residue.

3.3.2.2 *Malonyl-CoA—ACP transacylase*

Malonyl-CoA—ACP transacylase catalyses the transfer of malonyl groups from the sulphydryl group of CoA to the sulphydryl group of ACP, as shown in equation (3.11)[107-110]. This transacylase, purified to homogeneity from *E. coli* extracts, has a molecular weight of 36 700, obtained by sedimentation equilibrium measurements, sodium dodecyl sulphate gel electrophoresis, and by Sephadex G-100 column chromatography[110]. A value of $S_{20,w} = 2.31$ was determined for the enzyme by sedimentation-velocity measurements. Amino acid analysis and determination of the isoelectric point (pH 4.65) showed the enzyme to be acidic in nature. The presence of six half-cystine residues was indicated by the performic acid oxidation method of Moore[111].

Although the physiological thiols involved in the reaction are CoA and ACP, the enzyme also catalyses malonyl transfer to panteteheine, *N*-(*N*-acetyl-β-alanyl)cysteamine, and *N*-acetylcysteamine. Although acetyl-CoA is inactive as a substrate, it is a competitive inhibitor of malonyl-CoA with a K_i value of 115 µM[109]. Kinetic studies of the transacylase suggested the formation of a malonyl-enzyme intermediate during the reaction and were consistent with a Ping-Pong Bi Bi kinetic scheme for malonyl transacylation. The following partial reactions have been deduced:

$$\text{Malonyl-CoA} + \text{ENZ} \rightleftharpoons \text{malonyl-ENZ} + \text{CoA} \quad (3.20)$$

$$\text{Malonyl-ENZ} + \text{ACP} \rightleftharpoons \text{malonyl-ACP} + \text{ENZ} \quad (3.21)$$

The nature of the malonyl-enzyme intermediate was elucidated recently[110]. Although earlier work with sulphydryl inhibitors suggested a role for a thioester bond in the acyl-enzyme intermediate, inhibition of the transacylase by phenylmethanesulphonylfluoride suggested an active serine residue[109]. The nature of the enzyme intermediate was demonstrated in the following experiment. Incubation of enzyme with [^{14}C]malonyl-[^{3}H]CoA yielded a malonyl-enzyme intermediate which was isolated by gel filtration chromatography. The intermediate contained ^{14}C only, indicating that CoA was displaced during the formation of the intermediate. The isolated intermediate was not sensitive to performic acid oxidation, indicating that the malonate was not bound as a thioester. In order to identify the malonyl binding site of this transacylase [^{14}C]malonyl-enzyme was subjected to digestion by thermolysin. An isolated labelled tetrapeptide (alanine, glycine, histidine, and serine) was degraded further by pronase to yield malonyl-*O*-serine, which was identified by co-electrophoresis and co-chromatography with synthetic malonyl-*O*-serine. The nature of the malonyl attachment in the malonyl transacylase was thereby established.

Although it has been established that the malonyl group is bound to a serine of the transacylase, sulphydryl groups seem intimately involved in the enzymic reaction. Stimulation of the enzyme by dithiothreitol and inhibition by a variety of sulphydryl binding agents have been demonstrated[108, 110]. The inhibition was pH-dependent, indicating that the reactive sulphydryl residue is ionisable. Binding of malonate to the enzyme prevented inhibition. These data indicate that the reactive sulphydryl residue is either located within the active site of the transacylase, or it is located at a site removed from the catalytic region but structurally proximal to it.

3.3.2.3 β-Ketoacyl-ACP synthetase

β-Ketoacyl-ACP synthetase catalyses the condensation reaction of fatty acid biosynthesis (equations (3.10) and (3.12)). The enzyme has been purified to homogeneity and crystallised[112-114]. The molecular weight of the enzyme, determined by equilibrium sedimentation studies, is 66 000. Treatment of the enzyme with 6 M guanidine—HCl or 0.1 % sodium dodecyl sulphate causes it to dissociate into two, apparently identical subunits of 35 000 dalton[113]. The amino acid composition of the enzyme has been reported[112]. Amino acid analysis after performic acid oxidation or titration of the protein with 5,5'-dithiobis-(2-nitrobenzoic acid) indicated the presence of 8 moles of cysteine per mole of protein.

Experiments with the pure condensing enzyme[112] indicate that the reaction proceeds in two partial reactions with an acyl-enzyme intermediate, as indicated in equations (3.22) and (3.23).

$$\text{Acetyl—S—ACP} + \text{HS—E}_{cond} \rightleftharpoons \text{acetyl—S—E}_{cond} + \text{HS—ACP} \quad (3.22)$$

$$\text{Acetyl—S—E}_{cond} + \text{malonyl—S—ACP} \rightleftharpoons \text{acetoacetyl—S—ACP} + CO_2 + \text{HS—E}_{cond} \quad (3.23)$$

Kinetic studies of the enzyme[112] indicated that it was inactivated after reaction with one mole of iodoacetamide. Acetyl-ACP completely prevented this inhibition. When [^{14}C]iodoacetamide was reacted with the enzyme and the alkylated enzyme was subjected to acid hydrolysis, all the radioactivity of the protein was found associated with carboxymethylcysteine. These experiments indicated that the enzyme has a cysteine that is essential for catalytic activity and probably at the site of acetyl binding. This suspicion received support from the results of an experiment in which enzyme incubated with [^{3}H]acetyl-[^{14}C]ACP was shown to give rise to [^{3}H]acetyl-enzyme which was isolated by gel filtration and sucrose density centrifugation. The [^{3}H]acetyl-enzyme was active in transferring the [^{3}H]acetyl group either to ACP to form [^{3}H]acetyl-ACP (equation (3.22)) or to malonyl-ACP to form [^{3}H]acetoacetyl-ACP (equation (3.23)). The [^{3}H]acetyl group was released from the enzyme when the protein was treated with alkali, neutral hydroxylamine or performic acid. Thus, a thioester linkage was suggested. These findings, coupled with the isolation of carboxymethylcysteine at the active site of the enzyme, strongly suggest that the acetyl group is bound covalently as a thioester to a cysteine residue of the condensing enzyme.

The β-ketoacyl-ACP synthetase also catalyses two additional reactions which are probably models of partial reactions catalysed by this enzyme. The first is fatty acyl-CoA—ACP transacylation[115], as indicated in equation (3.24):

$$\text{RCO—S—CoA} + \text{ACP—SH} \rightleftharpoons \text{RCO—S—ACP} + \text{CoA—SH} \quad (3.24)$$

Although this reaction probably has no physiological significance, it has interesting substrate specificities that have been studied in detail. The second model reaction catalysed by this enzyme is malonyl-ACP decarboxylation, as shown in equation (3.25):

$$\text{Malonyl-ACP} \rightarrow \text{acetyl-ACP} + CO_2 \qquad (3.25)$$

The relationship of this decarboxylase activity to the physiological reaction is not clear.

The detailed study of substrate specificity of the β-ketoacyl-ACP synthetase has demonstrated the importance of the condensing enzyme in the termination of chain elongation and in the accumulation of specific saturated and unsaturated fatty acids in the cell[114]. The predominating saturated fatty acid in *E. coli* is palmitate whereas the predominating unsaturated fatty acids are palmitoleate and *cis*-vaccenate. The fatty acid synthetase of this organism synthesises both saturated and unsaturated fatty acids *in vitro*, and, as shown by Bloch and his co-workers[116], the critical reaction in the biosynthetic pathway that leads to unsaturated fatty acids is catalysed by β-hydroxydecanoyl thioester dehydrase. β-Hydroxydecanoyl-ACP is the intermediate at the branch point between the pathways to saturated and unsaturated fatty acids. Dehydration of this compound by β-hydroxydecanoyl thioester dehydrase gives rise to *cis*-3-decenoyl-ACP. This thioester presumably condenses with malonyl-ACP to initiate chain elongation of *cis*-unsaturated acyl-ACP intermediates. Other hypothetical intermediates expected to undergo condensation with malonyl-ACP in this pathway include *cis*-5-dodecenoyl-ACP, *cis*-7-tetradecenoyl-ACP, and *cis*-9-hexadecenoyl-ACP (palmitoleate). On the other hand, dehydration of β-hydroxydecanoyl-ACP by β-hydroxyacyl-ACP dehydrase give rise to *trans*-2-decenoyl-ACP, the normal intermediate in saturated fatty acid synthesis. The thioester intermediates expected to undergo chain elongation in the condensation reactions of the saturated pathway include decanoyl-ACP, dodecanoyl-ACP, and tetradecanoyl-ACP.

The availability of pure condensing enzyme permitted investigations to determine whether this enzyme catalyses all the condensations in the biosynthesis of both saturated and unsaturated fatty acids and whether the specificity of this enzyme can explain the accumulation of fatty acids of particular chain lengths in the cell[114]. As shown in Table 3.3, reactions with acetyl-ACP,

Table 3.3 Activity of β-ketoacyl-ACP synthetase with various intermediates in fatty acid synthesis in *E. coli**

Intermediate	K_m (μM)	V_{max} (μmol product min^{-1} mg^{-1})
Acetyl-ACP	0.52	2.8
Decanoyl-ACP	0.33	2.8
Dodecanoyl-ACP	0.27	0.97
Tetradecanoyl-ACP	0.28	0.31
Hexadecanoyl-ACP	—	N.A.
cis-3-Decenoyl-ACP	0.71	1.9
cis-5-Dodecenoyl-ACP	0.20	1.7
cis-9-Hexadecenoyl-ACP	0.37	0.37
cis-11-Octadecenoyl-ACP	—	N.A.

* Assays were carried out according to Greenspan *et al.*[144]. Kinetic constants were determined from Lineweaver–Burk plots of the data obtained.
N.A. = no activity.

decanoyl-ACP and dodecanoyl-ACP gave approximately similar results for both K_m and V_{max}. The enzyme is slightly less active with tetradecanoyl-ACP. However, it is completely inactive with hexadecanoyl-ACP, the C_{16} saturated fatty acid that accumulates in the cell. Assay of the enzyme with cis-3-decenoyl-ACP and cis-5-dodecenoyl-ACP, two early intermediates in the synthesis of unsaturated fatty acids, indicated that they are both as active as acetyl-ACP or decanoyl-ACP and that they have similar K_m values. Activity of cis-9-hexadecenoyl-ACP is decreased to approximately one-fifth of the rate of the cis-5-dodecenoyl-ACP. The cis-11-octadecenoyl-ACP is completely inactive. These results indicate that this condensing enzyme can function in both saturated and unsaturated fatty acid biosynthesis in E. coli. The specificity of this enzyme explains how chain elongation is terminated specifically at C_{16} in the saturated pathway and at C_{16} and C_{18} in the unsaturated pathway.

3.3.2.4 β-Ketoacyl-ACP reductase

The initial reaction in *de novo* fatty acid biosynthesis is catalysed by β-ketoacyl-ACP reductase according to equation (3.13)[117-119]. The reduction product of acetoacetyl-ACP has the D(—)-configuration. The enzyme can utilise CoA and pantetheine thioesters, although both are much less active than ACP thioesters. The enzyme is absolutely specific for NADPH. The enzyme has a broad substrate specificity for chain length of the acyl group of the substrate, and intermediates of both the saturated and unsaturated pathways are included[119]. It is equally active with β-ketoacyl-ACP derivatives of C_4–C_{16}[118].

3.3.2.5 β-Hydroxyacyl-ACP dehydrase

This enzyme catalyses the dehydration of D(—)-β-hydroxyacyl-ACP thioesters to form *trans*-2-enoyl-ACP thioesters according to equation (3.14)[117,120]. Like the β-ketoacyl-ACP synthetase, the dehydrase has an absolute specificity for thioesters of ACP; it is inactive with thioesters of CoA or pantetheine. The reaction is stereospecific; the D(—) isomer of β-hydroxybutyryl-ACP was dehydrated whereas the L(+) isomer was not.

The enzyme has been purified 2900-fold but it is not homogeneous at this stage of purification[120]. The substrate specificity of the purified preparation was found to include chain lengths from C_4 to C_{16}; however, there was a bimodal distribution with the lowest activity with the 10-carbon substrate, β-hydroxydecanoyl-ACP. This is of interest since β-hydroxydecanoyl-ACP is the last intermediate which is common to both the saturated and unsaturated biosynthetic pathways. Bloch and his co-workers[116] demonstrated that another enzyme, β-hydroxydecanoyl thioester dehydrase, catalyses the formation of cis-3-decenoyl-ACP, the initial step in the unsaturated pathway. The β-hydroxydecanoyl thioester dehydrase has a very narrow substrate specificity; it is only active with 10-carbon thioesters and essentially inactive with all other chain length thioesters. This specificity explains the distribution of double bonds in the unsaturated fatty acids of E. coli. Since β-hydroxyacyl-ACP dehydrase catalyses the conversion of β-hydroxydecanoyl-ACP into trans-2-decenoyl-ACP, an intermediate in saturated fatty acid synthesis, it is apparent that competition for β-hydroxydecanoyl-ACP by these two dehydrases could determine the relative abundance of saturated and unsaturated fatty acids synthesised by the cell. It is, therefore, of interest that *E. coli* extracts contain similar quantities of the two dehydrases and that the K_m and V_{max} values of the two enzymes with β-hydroxydecanoyl-ACP are very similar.

Earlier investigations suggested the presence in *E. coli* extracts of three different dehydrases, specific for short, medium and long chain β-hydroxyacyl-ACP derivatives[121,122]. Although the discovery that the specificity of the highly-purified β-hydroxyacyl-ACP dehydrase discussed above includes all the intermediates of both saturated and unsaturated fatty acid biosynthetic pathways, the specific involvement of additional dehydrases cannot be ruled out, and the interpretation of the possible functions of the various dehydrases must await their purification or perhaps the isolation of mutants lacking each of the individual dehydrases.

3.3.2.6 Enoyl-ACP reductase

An enoyl-ACP reductase preparation has been purified 250-fold from *E. coli* extracts[123]. Available evidence indicates that there are two different enzymes in the purified preparation that catalyse reduction of enoyl-ACP derivatives according to equation (3.15). An NADPH specific enzyme is inactive at pH > 8.0 and has absolute specificity for ACP derivatives. This enzyme also exhibits greatest activity with crotonyl-ACP as substrate. The presence of an essential sulphydryl group is indicated, since the enzyme is readily inhibited by *p*-hydroxymercuribenzoate, iodoacetate, and *N*-ethylmaleimide.

In contrast to the NADPH specific enzyme, an NADH specific enzyme exhibits activity over a wide range of pH values. This enzyme exhibits higher activity with decenoyl-ACP as substrate compared to crotonyl-ACP. Moreover, enoyl-CoA thioesters are active substrates with the NADH specific reductase, and a different pattern of inhibition by sulphydryl reagents was observed[123]. The two enoyl reductase activities have not yet been separated.

3.3.3 Multi-enzyme complexes

3.3.3.1 Yeast

The fatty acid synthetase of yeast was purified as a homogeneous protein with a molecular weight of 2 300 000[62,124]. Lynen's studies of this enzyme system provided the first major insight into the chemical details of fatty acid synthesis from acetyl-CoA and malonyl-CoA. The work also provided an understanding of the mode of action of the fatty acid synthetases that occur as tightly associated multi-enzyme complexes, i.e. those of animals and yeast. The overall reaction catalysed by the yeast enzyme is:

$$\text{MeCO—S—CoA} + n\text{CO}_2\text{HCH}_2\text{CO—S—CoA} + 2n\text{NADPH} + 2n\text{H}^+ \rightarrow$$
$$\text{Me(CH}_2\cdot\text{CH}_2)_n\text{CO—S—CoA} + n\text{CO}_2 + n\text{CoA} + 2n\text{NADP}^+ + n\text{H}_2\text{O}$$
(3.26)

The similarity to the reaction for *E. coli* (equation (3.8)) is obvious; the most important difference is the product obtained *in vitro*, i.e. acyl-CoA thioesters rather than free long-chain fatty acids as in bacteria and animals.

The yeast complex was visualised by electron microscopy by negative staining techniques with phosphotungstic acid[125]. Oval shaped particles were seen with a longitudinal diameter of 250 Å and a cross diameter of 210 Å. Lynen has proposed a model of the complex in which ACP is in the centre, enclosed by the various enzymes[63]. Small angle x-ray analysis confirms the existence of a hollow space in the rotationally symmetric complex[126]. Since the complex has been crystallised recently, more extensive x-ray analysis should now be possible[127].

Active subunits have not been isolated from the yeast fatty acid synthetase. Dissociation by 0.2 M deoxycholate or 6 M urea give subunits with an apparent weight-average molecular weight of 110 000 (Ref. 63). Seven different amino-terminal amino acids were found. Less drastic dissociation was carried out by freezing and thawing the complex repeatedly in 1 M NaCl and LiCl. The products of this partial dissociation had a molecular weight of 200 000–250 000 and were inactive in the fatty acid synthetase assay. However, re-association to an active complex occurred when the ionic strength was decreased[128].

Yeast fatty acid synthetase is the only one of the multi-enzyme complexes from which ACP has been isolated, as noted above. Analysis of the 4'-phosphopantetheine of the yeast enzyme complex indicated that it contains from 3.5 to 6.0 moles per mole of enzyme[129]. The 4'-phosphopantetheine was cleaved by treatment with mild alkali, indicating that it is probably bound to the ACP via a serine residue, as had been shown in *E. coli* ACP. [^{14}C]Pantetheine-labelled synthetase was subjected to treatment with guanidine–HCl and then alkylation with iodoacetamide. The radioactive material was isolated by preparative polyacrylamide gel electrophoresis. Molecular weight, calculated on the basis of β-alanine content, was 16 000.

The details of the reaction mechanism of yeast fatty acid synthetase were elucidated primarily by kinetic and inhibition studies, isolation of acyl intermediates and study of model reactions. Acetyl-CoA was shown to function as primer[62] although a number of saturated acyl-CoA derivatives could substitute. Stimulation by thiols and inhibition by sulphydryl binding agents indicated that sulphydryl groups were important for the overall reaction. These facts, coupled with the failure to demonstrate any free intermediates, led to the hypothesis that the acyl groups were transferred to sulphydryl groups on the enzymes[62]. The identification of acetoacetyl-enzyme proved this hypothesis. By use of model substrates, all the intermediate steps of fatty acid synthesis were demonstrated in yeast. The model compounds were the various intermediates which were tested as thioesters of either CoA, pantetheine or *N*-acetylcysteamine. The addition of FMN stimulated the second reduction, and removal of flavin by charcoal adsorption markedly decreased the activity of this step. This participation of FMN in the yeast synthetase is a distinguishing feature of this complex.

Several types of acyl binding sites, both thiol and non-thiol, have been characterised in the yeast synthetase. The yeast complex was shown to have two types of sulphydryl groups, the 'peripheral' thiol and the 'central' thiol, distinguished on the basis of reactivity with alkylating agents and specific substrates[63, 130]. The 'peripheral' thiol is much more sensitive to iodoacetamide than is the 'central' thiol. The alkylation of the 'peripheral' thiol can be prevented by prior reaction with acetyl-CoA. This 'peripheral' thiol has been identified as a cysteine residue and presumably belongs to the subunit involved in the catalysis of the condensation reaction[124, 130]. Lynen proposed that the 'central' thiol was arranged in the centre of the complex among the enzymically active subunits and served to hold the covalently bound acyl intermediates close to the various catalytic centres[63, 130]. This thiol was identified as the 4'-phosphopantetheine moiety of ACP[83]. Despite this intriguing concept of synthetase structure and function, it should be noted that the isolated condensing enzyme and ACP of *E. coli* also demonstrate differences in sensitivity to alkylating agents similar to those described for the yeast thiols. Thus, the differences in sensitivity to alkylating agents cannot be explained by the location of enzyme sites within a multi-enzyme complex.

Acetyl binding sites have been studied by investigations of [^{14}C]acetyl-enzyme formed upon incubation of [^{14}C]acetyl-CoA with yeast fatty acid synthetase. Treatment of the radioactive acetyl-enzyme with performic acid resulted in release of 50% of the label. Thus, 50% of the acetyl groups were bound in thioester linkage and 50% in non-thiol linkage. Peptic hydrolysis of [^{14}C]acetyl-enzyme gave rise to three classes of peptides. One class was stable to treatment with performic acid and presumably contained the active site of acetyl-CoA-ACP transacylase[63, 130]. Further experiments established the presence of serine in the active site of yeast acetyl-CoA-ACP transacylase and showed that the acetyl group is bound to the enzyme as acetyl-*O*-serine[131].

A second class of peptides contained the 4'-phosphopantetheine moiety of ACP. A third class of peptides contained acetyl groups in thioester linkage, and this class was not formed when enzyme was pretreated with iodoacetamide. A cysteine residue was shown to contribute the sulphydryl group involved, and this cysteine was presumably a component of the condensing enzyme.

Malonyl binding sites were studied in a similar manner[129]. Treatment with performic acid released only 25% of the protein-bound radioactivity. Thus, malonyl groups, like acetyl groups, were bound by non-thiol as well as thiol linkage. Peptic hydrolysates of [^{14}C]malonyl-enzyme resulted in two major classes of peptides. One class was labile to performic acid and contained the 4'-phosphopantetheine moiety of ACP. A second class of peptides, stable to performic acid oxidation, presumably contained the active site of the malonyl-CoA—ACP transacylase. The performic acid-stable acetyl peptides behaved differently from the analogous malonyl peptides on DEAE-Sephadex. This class of malonyl peptides was digested further with subtilisin S, and identified as overlapping penta- and hexa-peptides. Each contained a serine residue and no cysteine. Thus, it was concluded that malonate is bound to serine in an *O*-ester linkage at the active site of the malonyl-CoA—ACP transacylase. It should be mentioned that the amino acid sequence around the malonyl-*O*-serine in the presumed malonyl-CoA—ACP transacylase is different from the sequence around the acetyl-*O*-serine in the presumed acetyl-CoA—ACP transacylase. Thus these two transacylations are undoubtedly catalysed by separate enzymes in the yeast complex, as is the case in the *E. coli* fatty acid synthetase.

Incubation of yeast fatty acid synthetase with [^{14}C]palmityl-CoA gave rise to [^{14}C]palmityl-enzyme. Digestion of the latter with pepsin allowed the isolation of [^{14}C]palmityl-peptides that were recently identified[91]. At least three different sites for presumed covalent binding of palmitate were deduced. Two of the peptides were labile to performic acid oxidation and thus contained thioester linkages. One was derived from the 4'-phosphopantetheine moiety of ACP. The second peptide was absent when the complex was pretreated with iodoacetamide; thus, presumably it represented the cysteine of the condensing enzyme. A third class of peptides, stable to performic acid, contained a non-thiol site, presumably of the palmityl transacylase. The major peptide of this group had an amino acid composition identical to the active-site peptide of the malonyl transacylase. The latter results could be interpreted in either of two ways. Either the palmityl and malonyl transacylase activities of the fatty acid synthetase are catalysed by the same enzyme, or there are two different transacylases with common amino acids at their active sites.

A mechanism of action of the yeast fatty acid synthetase complex has been proposed by Lynen on the basis of the above data[63, 129, 130]. The primer acetyl group binds to a non-thiol acceptor, a serine of the acetyl CoA-ACP transacylase, and the malonyl group binds to a separate non-thiol acceptor, a serine residue of the malonyl-CoA—ACP transacylase. The acetyl group is transferred to the sulphydryl group of ACP and then to the sulphydryl group of a cysteine residue of the condensing enzyme. The malonyl group is transferred to the sulphydryl of ACP, and the condensation occurs between the acetyl-enzyme and malonyl-ACP. The product, acetoacetyl-ACP, undergoes reduction, dehydration and reduction. The saturated acyl group is transferred from ACP to the condensing enzyme, liberating ACP which can accept another malonyl group, and the cycle repeats itself until acyl chain lengths of 16–18 carbons are reached. The final reaction is the transfer of the palmityl or stearyl group from ACP to CoA by the fatty acyl transferase[62, 129]. The marked similarity of this mechanism of fatty acid synthesis to that discussed for *E. coli* fatty acid synthetase is immediately obvious.

The mechanism whereby chain termination occurs after chain lengths of 16 and 18 carbons are reached is still an unsolved problem of the yeast fatty acid synthetase. A clue that the condensing enzyme might be involved in chain termination was obtained by the observation that protection of the 'peripheral' thiol from inhibition by *N*-ethylmaleimide decreased with increasing chain length of the acyl-CoA preincubated with the enzyme[63]. Palmityl-CoA was only about one-third as effective as acetyl-CoA. More recently, a model has been proposed based upon two assumptions: (a) the probability of an acyl group bound to the enzyme being transferred to CoA depends on the relative velocities of the condensing and transferring reactions, and (b) the growing alkane chain interacts with the protein after a chain length of 13 carbon atoms has been reached and this interaction changes the relative velocities in favour of termination. Since the transferase reaction did not demonstrate specificity with thioesters between 6 and 20 carbons, it is apparent that the condensation reaction is the step affected. Thus, as in *E. coli*, chain termination in yeast is apparently related to specific properties of the condensing enzyme.

3.3.3.2 Liver

Fatty acid synthetase has been purified to homogeneity as a tightly-associated heteropolymeric complex from liver of the pigeon[64], rat[66] and chicken[65]. The enzyme from pigeon liver has a molecular weight of 450 000[132], from rat liver 540 000[66], and from chicken liver 508 000[65, 133]. Each of the complexes was inhibited readily by sulphydryl binding agents. One mole of 4′-phosphopantetheine per mole of complex has been demonstrated in the pigeon[134-136] and rat[66] enzymes. These two complexes contain no flavin[65, 66]. Products of both the avian[137] and mammalian[138] synthetases are free fatty acids, primarily palmitate, rather than thioesters of CoA as in yeast.

None of the multi-enzyme complexes of animals or yeast has been dissociated into active component enzymes. However, it has been possible to dissociate the complexes of pigeon[134, 139, 140], rat[66] and chicken[133] liver into approximately one-half molecular weight subcomplexes. This is accomplished most easily with the enzyme from pigeon liver, and the most detailed studies of this phenomenon have been carried out with this system by Porter and co-workers. The dissociation can be carried out by storage in 2-mercaptoethanol ('ageing'), reaction with carboxymethylsulphide, or exposure to low ionic strength, high pH or low temperature. Stability is enhanced by phosphate ions[141], fructose-1,6-diphosphate and NADPH[142]. The mechanism of the dissociative process is not understood. Dissociation can occur without oxidation of sulphydryl groups and inactivation by oxidation of sulphydryl groups can occur without dissociation[133, 141]. The relative contributions of changes in hydrophobic and electrostatic interactions remain unclear.

The approximately one-half molecular weight subcomplexes, formed by dissociation of the pigeon liver enzyme by low ionic strength, were shown to catalyse all the partial reactions of the intact complex, except for the condensation reaction[139]. Porter and co-workers[144] have shown that the sulphydryl binding site for acetyl groups, presumably a cysteine residue on the condensing enzyme, cannot be demonstrated upon incubation of subcomplex with [^{14}C]acetyl-CoA. The subcomplexes are considered to be non-identical, since only 1 mole of 4′-phosphopantetheine is present per mole of undissociated complex, and disc gel electrophoresis data, obtained with the dissociated chicken liver enzyme[133], indicate that the subcomplexes, although nearly identical in size, differ significantly in net charge.

Dissociation of the type just described is reversible. Irreversible dissociation occurred after treatment with maleate[134], or detergents like palmityl-CoA or sodium dodecyl sulphate[134, 143]. The pigeon liver enzyme was also dissociated by treatment with phenol-acetic acid-urea[132] and subjected to electrophoresis on polyacrylamide gels. At least eight bands were visible. Amino terminal analysis of the pigeon liver enzyme indicated at least five different DNP-amino acids[132].

The mechanism of action proposed for the animal fatty acid synthetase is similar to that proposed earlier by Lynen for the yeast multi-enzyme complex. The partial reactions of the liver synthetase were demonstrated with model substrates[139], as was done earlier with the yeast enzyme complex.

The binding sites for acetyl and malonyl groups on the liver fatty acid synthetase have been studied extensively in the laboratories of Porter and Wakil. Three groups of [^{14}C]acetyl-peptides and two groups of [^{14}C]malonyl-peptides were isolated[144, 145]. One group of peptides for each acyl group contained approximately one mole each of β-alanine, taurine, pantothenic acid and radioactive substrate[141]. Thus, one binding site for both the acetyl and malonyl groups is ACP. A second binding site for these two groups proved to be a non-thiol site since the [^{14}C]acyl groups were stable to performic acid oxidation[145]. Serine was contained in each peptide. Thus, it was concluded that the linkage was not a thioester bond but probably an oxygen ester of serine. Joshi et al.[146] demonstrated that a product of hydrolysis of [^{14}C]acetyl-peptide co-chromatographed with authentic acetyl-O-serine. The possibility of a common non-thiol site for acetyl and malonyl groups was suggested by competition studies[146, 147]. However, whether there is a common site with a single serine or two sites with the same specificities was not yet clear. In light of the discovery that the yeast fatty acid synthetase contains different non-thiol acyl binding peptides for malonyl and acetyl groups, as discussed above, it is likely that different peptides will also be found in the liver system. The third binding site for acetyl groups is a sulphydryl group. [^{14}C]Acetyl groups were released from this class of peptides by performic acid oxidation or hydroxylamine[145]. The purified peptides contained cysteine, and it is presumed that this cysteine residue is in the condensing enzyme. The analogy with the yeast complex is obvious.

Chain termination is perhaps the single reaction of the animal complex that differs from the yeast synthetase, since the products in animals are free fatty acids and in yeast, CoA thioesters. It is clear that saturated acyl-CoA derivatives of 12 and 14 carbons are deacylated poorly, and those of 16 and 18 carbons are deacylated readily by the liver fatty acid synthetase[148, 149]. It is unclear whether the activity of this enzyme determines termination of chain elongation by animal fatty acid synthetase.

3.3.3.3 Mammary gland

Fatty acid synthetase has been purified to homogeneity from the mammary gland of the lactating rat[67] and rabbit[68]. The multi-enzyme complex of rat mammary has a molecular weight of 478 000, similar to the liver synthetases, whereas synthetase of rabbit mammary has a molecular weight of 910 000. The rat enzyme contains one 4'-phosphopantetheine moiety per mole[67]. At optimal substrate concentrations both synthetases form predominately palmitic acid. Fatty acid synthetase of rat mammary is cold-labile, and the native form (13 S) dissociates into half-molecular weight (9 S) subunits upon ageing in the cold[150]. Fatty acid synthetase activity is lost with dissociation and is restored when the subunits re-associate at 20–30 °C.

Recent studies have shown that the primer for the fatty acid synthetase of mammalian liver and mammary gland may be butyryl-CoA rather than acetyl-CoA. Since the initial work on fatty acid synthetase of pigeon liver, yeast and bacteria, it has been assumed that the universal primer for the reaction is acetyl-CoA. Nandedkar and Kumar[151] demonstrated in the soluble fraction of lactating goat mammary enzymes that catalysed the reversal of the β-oxidation steps from butyryl-CoA to acetyl-CoA, i.e. acetoacetyl-CoA thiolase, NADH-linked acetoacetyl-CoA reductase, and crotonyl-CoA hydratase. The final enoyl-CoA reductase step was catalysed by the fatty acid synthetase. Lin and Kumar[152] showed that a partially-purified preparation of synthetase from rabbit mammary gland had a K_m for butyryl-CoA that was approximately one-sixth and a V_{max} for butyryl-CoA that was approximately sixfold higher than the corresponding values for acetyl-CoA. The latter observations are similar to those made by Smith and Abraham[153] for purified fatty acid synthetase of lactating rat mammary gland. Lin and Kumar[154] showed that the synthetase of rat and rabbit liver, as well as mammary gland, were more active with butyryl-CoA than with acetyl-CoA as primer. In addition, by (a) incubation of particle-free supernatant solution in a reaction mixture suitable for fatty acid synthesis and the β-reductive steps, containing [1-^{14}C]acetyl-CoA and unlabelled malonyl-CoA, and (b) subjecting the fatty acids formed to Kuhn–Roth oxidation, it was demonstrated that 90% of the acetyl-CoA was converted to butyryl-CoA prior to condensation with malonyl-CoA. Thus, under these conditions the important primer for mammalian liver and mammary gland was butyryl-CoA.

The mechanism of fatty acid chain termination in lactating mammary gland has received considerable attention because crude extracts of this organ produce fatty acids of various chain length from C_4 to C_{18}[155-158]. The major question is: is this difference in fatty acids produced related to a unique fatty acid synthetase system or to exogenous factors in the cells of mammary tissue? That the pattern of fatty acids synthesised is not due to a separate and unique enzyme system was demonstrated when antibody to rat mammary fatty acid synthetase was shown to inhibit synthesis of all chain length fatty acids by whole homogenates of mammary gland. That the fatty acid synthetase *per se* is not unusual was shown by the fact that the pure complexes of lactating rabbit and rat glands produce predominantly palmitate[67,68]. The choice of primer made no difference in the products[153]. The palmityl thioesterase activity of both the pure synthetase and crude extracts of rat mammary showed substrate specificity for long chain saturated acyl-CoA derivatives, especially C_{16}[153]. Thus, a thioesterase intrinsic or extrinsic to the synthetase is not responsible for the production of shorter chain fatty acids. However, the relative percentage of fatty acids of various chain lengths produced *in vitro* by the pure enzyme of rabbit and rat mammary gland and rat liver can be altered by varying the concentration of acetyl-CoA and malonyl-CoA[68,159,160]. Thus, a relative increase in malonyl-CoA resulted in the synthesis of more longer chain fatty acids. The physiological relevance of this phenomenon is unclear, since fatty acid composition of lipids of rat liver and rat and rabbit milk differ dramatically, whereas the change in chain length of fatty acids produced by changes in substrate concentration are similar for the fatty acid synthetases from the two tissues. At present, it appears that factors within the mammary cell, but exogenous to the fatty acid synthetase, cause the production of distinctive fatty acids. Dils and coworkers[161,162] have suggested that the rate of acetyl-CoA carboxylation might be a regulatory factor under these circumstances, and Abraham and co-workers[153] have suggested that chain-length specificity of the fatty acid activating thiokinases or the acyl transferases of phospholipid and triglyceride synthesis may be the critical steps involved.

3.3.3.4 Mycobacterium phlei

Mycobacterium phlei, an advanced prokaryote, contains a fatty acid synthetase of the multi-enzyme complex type[69]. This is the least advanced organism shown to contain a tightly-associated synthetase complex. The heteropolymer has a molecular weight of 1 700 000 and contains 4′-phosphopantetheine. It is very unstable in solutions of low ionic strength, dissociating into fragments of lower molecular weights. The products of the synthetase reaction are fatty acids ranging from C_{14} to C_{26}, with peaks at C_{18} and C_{24}. The complex elongates octanoyl- and stearyl-CoA as well as acetyl-CoA.

The activity of this synthetase was greatly stimulated by a heat-stable fraction from M. phlei extracts[69]. This fraction was separated into two subfractions[163]. One contained FAD and FMN, and addition of FMN to the reaction mixtures reproduced the stimulation by this subfraction. The second subfraction contained three groups of polysaccharides, one composed of mannose and 3-O-methylmannose, and the other two, glucose and 6-O-methylglucose. This subfraction stimulated the synthetase at low concentrations of acetyl-CoA and was shown to reduce the K_m for this substrate by ca. 50-fold. The polysaccharides did not influence the acetyl-CoA-pantetheine transacylase reaction or the acyl-CoA-dependent malonyl-CoA-CO_2 exchange reaction, and thus their effect could not be localised to either the acetyl transacylase step or the condensation reaction. The relation of the polysaccharide stimulation observed in vitro to the in vivo activity of this enzyme system remains unclear.

Maximal activity of the fatty acid synthetase of M. phlei was obtained when both NADH and NADPH were present[164]. NADH was preferentially utilised for the enoyl reductase step, the second reduction catalysed by the synthetase. FMN was shown to be involved in the second reduction catalysed by the yeast complex, and this may also be the case in M. phlei.

An unexpected finding with the Mycobacterial system was the isolation of free ACP from crude extracts[81, 165]. The purified ACP has a molecular weight of 10 600 and an amino acid composition similar to E. coli ACP (Table 3.1). It is of interest, in view of the free ACP, that a second fatty acid synthetase was demonstrated in M. phlei[69, 165]. This system has a molecular weight of less than 250 000 by Sephadex G-150 chromatography, requires added ACP for activity, elongates palmityl-CoA or stearyl-CoA but not acetyl-CoA, and thus produces only longer chain fatty acids. It is possible that this non-associated elongating system was a part of the large complex, accounting for the elongation of the C_{16} and C_{18} acids to the longer chain compounds, and that it was formed upon dissociation of the complex.

3.3.3.5 Euglena gracilis

The unicellular phytoflagellate, Euglena gracilis, is of particular interest because it displays animal as well as plant characteristics. Chloroplast formation occurs when the cells are grown in the light. Bloch and co-workers[166] have shown that, when grown in the dark, *E. gracilis* contains a typical fatty acid synthetase multi-enzyme complex with a molecular weight greater than 650 000. This synthetase produces predominantly palmitate. Like the heteropolymers of *M. phlei* and yeast, the complex was stimulated by FMN and was most active in the presence of both NADPH and NADH[164]. When the organism was grown in the light, two fatty acid synthetases could be isolated. One was clearly the complex present in etiolated cells. The other was smaller, had an absolute requirement for ACP, and produced primarily stearic and arachidic acids. This latter synthetase is comparable to that of *E. coli* and plants.

Synthesis of the ACP-dependent synthetase was dependent on the protein synthesising machinery of the chloroplast[167]. When etiolated cells were grown in the light in the presence of chloramphenicol, which inhibits protein biosynthesis in the chloroplast, appearance of the ACP-dependent synthetase was completely suppressed. Cycloheximide, which suppresses cytoplasmic protein synthesis, had no such effect, and, thus, cytoplasmic protein synthesis is not necessary for formation of this chloroplast fatty acid synthetase. It is noteworthy that in another phytoflagellate, the unicellular green alga, *Chlamydomonas reinhardi*, there is only one fatty acid synthetase which is dependent on ACP[168]. This organism differs from *E. gracilis* in that chloroplast formation occurs in the dark as well as in the light. Antibiotics that selectively inhibit transcription and translation in either the nuclear–cytoplasmic system or in the chloroplast were utilised to show that the chloroplast ribosomes are responsible for the *de novo* protein synthesis required for the synthesis of fatty acid synthetase in *C. reinhardi*.

3.3.4 Control

Control of fatty acid synthetase *in vivo* is a critical process, as indicated by the marked changes in fatty acid synthesis during various nutritional states and developmental stages of the animal. Changes in fatty acid synthesis are not mediated entirely through effects on acetyl-CoA carboxylase. The importance of nutritional factors in the regulation of fatty acid synthetase activity in liver of mature animals has been explored in a variety of studies with the rat[169-175] and mouse[176]. The effects observed were impressive increases with fat-free feeding and decreases with fasting or feeding a high-fat diet. The effectors of these responses are unclear. Gibson and his coworkers[170, 176, 177] demonstrated that the high hepatic fatty acid synthetase activity, accompanying fat-free feeding, returns to normal after gastric intubation of linoleate and linolenate, but not after administration of palmitate or oleate. Whether these polyunsaturated fatty acids are the primary effectors of the synthetase under these conditions is not yet known.

Fatty acid synthetase in the developing liver has been studied in greatest detail in the rat[178] and mouse[54]. In both animals, hepatic activity rises dramatically at the time of weaning when the animals change their diet from high-fat milk to lower-fat chow. A similar rapid rise in activity can be induced earlier than the natural time of weaning by prematurely weaning the suckling pups to a low-fat diet[54, 178]. These observations indicate the importance of nutritional factors in the control of the developmental changes of hepatic synthetase activity. It should also be noted in this context that in the studies of developing mouse liver[54] the capacity of liver slices to generate fatty acids from acetate correlated best with changes in fatty acid synthetase activity. Of all the lipogenic enzymes studied, including acetyl-CoA carboxylase, fatty acid synthetase responded most rapidly, most consistently, and to the greatest degree. Thus, this is additional evidence that the synthetase is one of the critical regulatory enzymes in fatty acid synthesis, particularly in developing tissues.

Recent studies of fatty acid synthetase of developing brain demonstrated that the highest level of synthetase activity occurs in late foetal life and that specific activity gradually falls as the animal matures[178]. The regulation of synthetase in brain is different from that in liver. Under conditions such as fasting, high-fat or fat-free feeding, profound changes in hepatic synthetase activity occurred without any effects observed in brain. Thus, nutritional factors do not seem important in regulation of brain fatty acid synthetase.

The possible importance of hormonal influences in the physiological control of synthetase activity in developing animals is suggested by recent observations[178]. Hydrocortisone administered daily in the first 6–10 days of life resulted in a decrease in hepatic activity to as low as one-third that of control animals. Tri-iodothyronine administered daily until clinical signs of hyperthyroidism appeared led to a 75–100% increase in hepatic synthetase activity. No change was noted in fatty acid synthetase activity in brain at any age. However, synthetase specific activity of the brain was consistently decreased in rats made hypothyroid by daily injections of methimazole from birth; the greatest diminution was *ca.* 50% when 27 days old. Hepatic synthetase activity was also consistently reduced in hypothyroid animals, it was *ca.* 50% the activity of control animals. Synthetase activity was also reduced in livers of alloxan-diabetic rats[179, 180], and administration of insulin to these rats either partially[180] or fully[179] restored activity. These observations raise important questions about the role of adrenal and thyroid hormones and insulin in regulation of fatty acid synthetase in these tissues.

3.3.4.1 *Allosteric regulation*

Fatty acid synthetases of pigeon liver and *E. coli* have been shown to be stimulated by phosphorylated sugars, especially fructose-1,6-diphosphate[181, 182]. At comparable concentrations (20 mM) lesser degrees of stimulation were observed with glucose 1-phosphate, glucose 6-phosphate, α-glycerol phosphate, pyrophosphate and orthophosphate. These observations were of considerable interest because they suggested a means whereby fatty acid synthesis could be attuned to glucose metabolism. Kinetic studies with the pigeon liver synthetase[182] suggested that the effect of phosphorylated sugars was related to inhibition of the synthetase by malonyl-CoA. This inhibition was competitive with respect to NADPH. The K_m for NADPH was decreased by fructose 1,6-diphosphate. It was suggested that hexose diphosphate acted either by competing for a regulatory site or by binding at a specific site to promote a conformational change of the enzyme rendering it insensitive to malonyl-CoA inhibition. The significance of these findings are difficult to interpret for several reasons. First, Porter was unable to show any effect of hexose diphosphates on the enzyme activities of purified pigeon or rat liver fatty acid synthetases[149]. In fact, the avian enzyme did not require any phosphate for activity, it is fully active in histidine buffers[183]. Second, the concentration of phosphorylated sugars tested *in vitro* were well above physiological concentrations[149]. Third, concentrations of malonyl-CoA as high as 100 μM had no effect on binding of NADPH to the pigeon liver fatty acid synthetase, as determined by fluorescence emission spectroscopy[183].

Feed-back inhibition of fatty acid synthetase was first considered a means of regulation on the basis of inhibition by palmityl-CoA of the enzyme complex of yeast[124], pigeon liver[143], rat liver, and rat brain[184]. The idea received support from findings that liver concentrations of long chain acyl-CoA derivatives are elevated in starvation and high fat feeding[45, 46] when hepatic synthetase activity is low. Lust and Lynen[185] demonstrated that the inhibition of the yeast complex by long chain acyl-CoA compounds was competitive with respect to malonyl-CoA. Addition of serum albumin diminished the inhibition. The effect of palmityl-CoA on the pigeon liver enzyme was evaluated critically by Dorsey and Porter[143]. Inhibition by the thioesters was found to depend on the molar ratio of palmityl-CoA to protein. The fact that the molar ratio was important for the inhibition of the synthetase suggested that palmityl-CoA acts as a detergent and that it is not a site-specific inhibitor. The inhibition was irreversible and similar to that produced by sodium dodecyl sulphate, another detergent. The physiological significance of feed-back inhibition by palmityl-CoA has also been questioned since it inhibits non-specifically a wide variety of enzymes[48] via its detergent properties.

3.3.4.2 Adaptive changes in enzyme content

In contrast to the questions surrounding the physiological importance of the allosteric effects demonstrated *in vitro*, dramatic adaptive changes in content of the fatty acid synthetase have been demonstrated. Porter and co-workers demonstrated that the content of the synthetase, determined by direct purification, was considerably lower in liver of fasted versus fat-free-fed pigeons[171] and rats[172]. Accelerated synthesis of enzyme upon re-feeding starved animals a fat-free diet was demonstrated by pulse labelling *in vivo* with [^{14}C]leucine and purifying the synthetase; a striking increase of incorporation of radioactivity into the enzyme was noted 6 h after re-feeding[172]. However, quantitative conclusions from these experiments were difficult to assess since not all of the preparations of fatty acid synthetase were homogeneous. By pulse labelling with [U-^{14}C]amino acids and isolating the synthetase by purification, Tweto and Larrabee[104] showed that the rate of synthesis of the complex in liver of starved rats fell to its lowest values after approximately 16 h of fasting. More recently, immunochemical techniques have been utilised to study the content, synthesis and degradation of fatty acid synthetase in both liver and brain in various nutritional states and during development[56].

The striking changes in hepatic fatty acid synthetase activity with fasting and fat-free feeding and during development are entirely related to changes in enzyme content[56]. This was determined by quantitative precipitin analyses, as described above for regulation of acetyl-CoA carboxylase. To delineate the mechanisms of the changes in content, rate constants of synthesis and degradation were derived from data obtained by isotopic-immunochemical experiments. A *ca.* 20-fold difference in synthetase activity per g of liver was noted between starved and fat-free-fed rats. The rate of synthesis of hepatic fatty acid synthetase per g of liver in the fat-free-fed rats was sixfold greater than that in liver of starved rats. The rate of degradation of the enzyme was nearly fourfold greater in the livers of starved rats. Thus the observed 20-fold difference in activity per g of liver is very similar to the difference predicted on the basis of changes in rates of enzyme synthesis and degradation[56]. This remarkable acceleration of degradation, suggested by previous determinations of the rate of decrease in enzyme activity after fasting[104], plays a major role in determining the lower hepatic synthetase activity in starved animals. This mechanism of enzyme regulation has been shown to be operative for two other lipogenic enzymes, acetyl-CoA carboxylase[12,52] and malic enzyme[57]. Rate of degradation of hepatic synthetase in animals fed a fat-free diet ($t_{\frac{1}{2}} = $ 2.7 days) was essentially identical to that for normally fed animals ($t_{\frac{1}{2}} = 2.8$ days). The increase in hepatic fatty acid synthetase activity and content in fat-free-fed animals was related entirely to an acceleration of enzyme synthesis. None of the nutritional states was accompanied by change in content, synthesis or degradation of brain synthetase. Finally, the rise in hepatic activity at the time of weaning was also shown to be due entirely to an increase in synthesis of enzyme. The 15-fold difference in activity per g of liver between suckling and weaned animals was accompanied by a 12-fold difference in rate of synthesis. Thus, the same mechanism for the increase in fatty acid synthetase content, i.e. increased synthesis of enzyme, is operative in developing animals at weaning as that accounting for the difference in enzyme content in adult animals that are on a chow versus a fat-free diet.

The mechanisms that cause the developmental changes of fatty acid synthetase in brain are different, at least in part, from those operative in liver. Quantitative precipitin analyses demonstrated that the changes in activity of brain synthetase with development are entirely related to changes in content of enzyme[56]. As in liver, changes in rate of synthesis play an important role. Thus, synthetase activity per g of brain was *ca.* twofold higher in 4-day-old versus adult animals. The rate of enzyme synthesis in the young animals was *ca.* sevenfold greater than in the adult animals. The rate of degradation of brain fatty acid synthetase in these young animals ($t_{\frac{1}{2}} = 1.9$ days) was greater than threefold faster than the rate in the adults ($t_{\frac{1}{2}} = 6.4$ days). It is apparent that the twofold difference in brain synthetase activity is explained by both an acceleration of synthesis and degradation of the enzyme. The enzyme complex undergoes rapid turnover in the young brain. The striking increase in rate of degradation of the brain synthetase is distinctly different from the situation with developing liver, and this suggests that this process may be important in regulation of the fatty acid synthetase in the developing nervous system.

Acknowledgements

Unpublished data from the author's laboratory have been obtained through support by Grant GB-5142X from the National Science Foundation and Grant 1-RO1-HE-10406 from the National Institutes of Health.

References

1. Lynen, F. (1967). *Biochem. J.*, **102**, 381
2. Lynen, F., Knappe, J., Lorch, E., Jutting, G., Ringelmann, E. and LaChance, J. P. (1961). *Biochem. Z.* **335**, 123
3. Himes, R. H., Young, D. L.,Ringelmann, E. and Lynen, F. (1963). *Biochem. Z.*, **337**, 48
4. Knappe, J., Wenger, B. and Wiegand, U. (1963). *Biochem. Z.*, **337**, 232
5. Knappe, J., Ringelmann, E. and Lynen, F. (1961). *Biochem. Z.*, **335**, 168
6. Gregolin, C., Ryder, E., Kleinschmidt, A. K., Warner, R. C. and Lane, M. D. (1966). *Proc. Nat. Acad. Sci. USA*, **56**, 148
7. Gregolin, C., Ryder, E., Warner, R. C., Kleinschmidt, A. K. and Lane, M. D. (1966). *Proc. Nat. Acad. Sci. USA*, **56**, 1751
8. Goto, T., Ringelmann, E., Reidel, B. and Numa, S. (1967). *Life Sci.*, **6**, 785
9. Gregolin, C., Ryder, E. and Lane, M. D. (1968). *J. Biol. Chem.*, **243**, 4227
10. Gregolin, C., Ryder, E., Warner, R. C., Kleinschmidt, A. K., Chang, H. C. and Lane, M. D. (1968). *J. Biol. Chem.*, **243**, 4236
11. Moss, J., Yamagishi, M., Kleinschmidt, A. K. and Lane, M. D. (1972). *Biochemistry*, **11**, 3779
12. Nakanishi, S. and Numa, S. (1970). *Europ. J. Biochem.*, **16**, 161
13. Inoue, H. and Lowenstein, J. M. (1972). *J. Biol. Chem.*, **247**, 4825
14. Vagelos, P. R., Alberts, A. and Martin, D. B. (1963). *J. Biol. Chem.*, **238**, 533
15. Lane, M. D., Edwards, J., Stoll, E. and Moss, J. (1970). *Vitamins and Hormones*, **28**, 345
16. Linn, T. L., Pettit, F. H. and Reed, L. J. (1969). *Proc. Nat. Acad. Sci. USA*, **62**, 234
17. Fischer, E. H., Pockes, A. and Sarri, J. (1970). *Essays Biochem.*, **6**, 23
18. Sumper, M. and Riepertinger, C. (1972). *Europ. J. Biochem.*, **29**, 237
19. Alberts, A. W. and Vagelos, P. R. (1968). *Proc. Nat. Acad. Sci. USA*, **59**, 561
20. Alberts, A. W., Nervi, A. M. and Vagelos, P. R. (1969). *Proc. Nat. Acad. Sci. USA* **63**, 1319
21. Dimroth, P., Guchait, R. B., Stoll, E. and Lane, M. D. (1970). *Proc. Nat. Acad. Sci. USA*, **67**, 1353
22. Nervi, A. M., Alberts, A. W. and Vagelos, P. R. (1971). *Arch. Biochem. Biophys.*, **143**, 401
23. Fall, R. R., Nervi, A. M., Alberts, A. W. and Vagelos, P. R. (1971). *Proc. Nat. Acad. Sci. USA*, **68**, 1512
24. Fall, R. R. and Vagelos, P. R. (1972). *J. Biol. Chem.*, **247**, 8005
25. Fall, R. R. and Vagelos, P. R. (1972). *J. Biol. Chem.*, in the press
26. Alberts, A. W., Gordon S. G. and Vagelos, P. R. (1971). *Proc. Nat. Acad. Sci. USA*, **68**, 1259
27. Heinstein, P. F. and Stumpf, P. K. (1969). *J. Biol. Chem.*, **244**, 5374
28. Kanangara, C. G. and Stumpf, P. K. (1972). *Arch. Biochem. Biophys.*, **152**, 83
29. Ganguly, J. (1960). *Biochim. Biophys. Acta*, **40**, 110
30. Numa, S., Matsuhashi, M. and Lynen, F. (1961). *Biochem. Z.*, **334**, 203
31. Chang, H. C., Seidman, J., Teebor, G. and Lane, M. D. (1967). *Biochem. Biophys. Res. Commun.*; **28**, 682
32. Brady, R. O. and Gurin, S. (1952). *J. Biol. Chem.*, **199**, 421
33. Martin, D. B. and Vagelos, P. R. (1962). *J. Biol. Chem.*, **237**, 1787
34. Matsuhashi, M., Matsuhashi, S. and Lynen, F. (1964). *Biochem. Z.*, **340**, 263
35. Kleinschmidt, A. K., Moss, J. and Lane, M. D. (1969). *Science*, **166**, 1276
36. Miller, A. L. and Levy, H. R. (1969). *J. Biol. Chem.*, **244**, 2334
37. Matsuhashi, M., Matsuhashi, S. and Lynen, F. (1964). *Biochem. Z.*, **340**, 243

38. Numa, S., Ringelmann, E. and Lynen, F. (1965). *Biochem. Z.*, **343**, 243
39. Ryder, E., Gregolin, C., Chang, H. C. and Lane, M. D. (1967). *Proc. Nat. Acad. Sci. USA*, **57**, 1455
40. Moss, J. and Lane, M. D. (1972). *J. Biol. Chem.*, **247**, 4944
41. Moss, J. and Lane, M. D. (1972). *J. Biol. Chem.*, **247**, 4952
42. Numa, S., Bortz, W. M. and Lynen, F. (1965). *Advan. Enzyme Regul.*, **3**, 407
43. Schneider, W. C., Striebich, M. J. and Hogeboom, G. H. (1956). *J. Biol. Chem.*, **222**, 969
44. Greenbaum, A. L., Gumaa, K. A. and McLean, P. (1971). *Arch. Biochem. Biophys.*, **143**, 617
45. Bortz, W. M. and Lynen, F. (1963). *Biochem. Z.*, **339**, 77
46. Tubbs, P. K. and Garland, P. B. (1964). *Biochem. J.*, **93**, 550
47. Yeh, Y. Y. and Leveille, G. A. (1971). *J. Nutrition.*, **101**, 803
48. Taketa, K. and Pogell, B. M. (1966). *J. Biol. Chem.*, **241**, 720
49. Goodridge, A. G. (1972). *J. Biol. Chem.*, **247**, 6946
50. Wieland, O., Eger-Neufeldt, J., Numa, S. and Lynen, F. (1963). *Biochem. Z.*, **336**, 455
51. Allmann, D. W., Hubbard, D. D. and Gibson, D. M. (1965). *J. Lipid Res.*, **6**, 63
52. Majerus, P. W. and Kilburn, E. (1969). *J. Biol. Chem.*, **244**, 6254
53. Schimke, R. T. and Doyle, D. (1970). *Annu. Rev. Biochem.*, **39**, 929
54. Smith, S. and Abraham, S. (1970). *Arch. Biochem., Biophys.*, **136**, 112
55. Ryder, E. (1970). *Biochem. J.*, **119**, 929
56. Volpe, J. J., Lyles, T. O., Roncari, D. A. K. and Vagelos, P. R. (1973). *J. Biol. Chem.*, in the press
57. Silpananta, P. and Goodridge, A. G. (1971). *J. Biol. Chem.*, **246**, 5754
58. Nakanishi, S. and Numa, S. (1971). *Proc. Nat. Acad. Sci. USA*, **68**, 2288
59. Elwood, J. C. and Morris, H. P. (1968). *J. Lipid Res.*, **9**, 337
60. Majerus, P. W., Jacobs, R., Smith, M. B. and Morris, H. P. (1968). *J. Biol. Chem.*, **243**, 3588
61. Bartley, J. C. and Abraham, S. (1972). *Biochim. Biophys. Acta*, **260**, 169
62. Lynen, F. (1961). *Fed. Proc.*, **20**, 941
63. Lynen, F. (1967). *Biochem., J.*, **102**, 381
64. Hsu, R. Y., Wasson, G. and Porter, J. W. (1965). *J. Biol. Chem.*, **240**, 3736
65. Hsu, R. Y. and Yun, S. L. (1970). *Biochemistry*, **9**, 239
66. Burton, O. N., Haavik, A. G. and Porter, J. W. (1968). *Arch. Biochem. Biophys.*, **126**, 141
67. Smith, S. and Abraham, S. (1970). *J. Biol. Chem.*, **245**, 3209
68. Carey, E. M. and Dils, R. (1970). *Biochim. Biophys. Acta*, **210**, 371
69. Brindley, D. N., Matsumura, S. and Bloch, K. (1969). *Nature (London)*, **224**, 666
70. Majerus, P. W., Alberts, A. W. and Vagelos, P. R. (1965). *Proc. Nat. Acad. Sci. USA*, **53**, 410
71. Vagelos, P. R., Majerus, P. W., Alberts, A. W., Larrabee, A. R. and Ailhaud, G. P. (1966). *Fed. Proc.*, **25**, 1485
72. Majerus, P. W. and Vagelos, P. R. (1967). *Advan. Lipid Res.*, **5**, 1
73. van den Bosch, H., Williamson, J. R. and Vagelos, P. R., (1970). *Nature (London)*, **228**, 338
74. Majerus, P. W., Alberts, A. W. and Vagelos, P. R. (1964). *Proc. Nat. Acad. Sci. USA*, **51**, 1231
75. Wakil, J. J., Pugh, E. L. and Sauer, F. (1964). *Proc. Nat. Acad. Sci. USA*, **52**, 106
76. Ailhaud, G. P. and Vagelos, P. R. (1966). *J. Biol. Chem.*, **241**, 3866
77. Goldfine, H., Ailhaud, G. P. and Vagelos, P. R. (1967). *J. Biol. Chem.*, **242**, 4466
78. van den Bosch, H. and Vagelos, P. R. (1970). *Biochim. Biophys. Acta*, **218**, 233
79. Prescott, D. J. and Vagelos, P. R. (1972). *Advan. Enzymol.*, **36**, 269
80. Simoni, R. O., Criddle, R. S. and Stumpf, P. K. (1967). *J. Biol. Chem.*, **242**, 573
81. Matsumura, S. (1970). *Biochem. Biophys. Res. Commun.*, **28**, 238
82. Ailhaud, G. P., Vagelos, P. R. and Goldfine, H. (1967). *J. Biol. Chem.*, **242**, 4459
83. Willecke, K., Ritter, E. and Lynen, F. (1969). *Europ. J. Biochem.*, **8**, 503
84. Larrabee, A. R., McDaniel, E. G., Bakerman, H. A. and Vagelos, P. R. (1965). *Proc. Nat. Acad. Sci. USA*, **54**, 267
85. Butterworth, P. H., Yang, P. C., Bock, R. M. and Porter, J. W. (1967). *J. Biol. Chem.*, **242**, 3508

86. Smith, S. and Dils, R. (1966). *Biochim. Biophys. Acta*, **116,** 23
87. Delo, J., Ernst-Fonberg, M. L. and Bloch, K. (1971). *Arch. Biochem. Biophys.*, **143,** 385
88. Vanaman, T. C., Wakil, S. J. and Hill, R. L. (1968). *J. Biol. Chem.*, **243,** 6420
89. Takagi, T. and Tanford, C. (1968). *J. Biol. Chem.*, **243,** 6432
90. Matsumura, S. and Stumpf, P. K. (1968). *Arch. Biochem. Biophys.*, **125,** 932
91. Ayling, J., Pirson, R. and Lynen, F. (1972). *Biochemistry*, **11,** 526
92. Roncari, D. A. K., Bradshaw, R. A. and Vagelos, P. R. (1972). *J. Biol. Chem.*, **247,** 6234
93. Majerus, P. W. (1968). *Science*, **159,** 428
94. Majerus, P. W. (1967). *J. Biol. Chem.*, **242,** 2325
95. Abita, J. P., Lazdunski, M. and Ailhaud, G. P. (1971). *Europ. J. Biochem.*, **23,** 412
96. Hancock, W. S., Prescott, D. J., Marshall, G. J. and Vagelos, P. R. (1972). *J. Biol. Chem.*, **247,** 6224
97. Hancock, W. S., Marshall, G. J. and Vagelos, P. R. (1972). *J. Biol. Chem.*, in the press
98. Merrifield, R. B. (1963). *J. Amer. Chem. Soc.*, **85,** 2149
99. Elovson, J. and Vagelos, P. R. (1968). *J. Biol. Chem.*, **243,** 3603
100. Prescott, D. J., Elovson, J. and Vagelos, P. R. (1969). *J. Biol. Chem.*, **244,** 4517
101. Vagelos, P. R. and Larrabee, A. R. (1967). *J. Biol. Chem.*, **242,** 1776
102. Powell, G. L., Elovson, J. and Vagelos, P. R. (1969). *J. Biol. Chem.*, **244,** 5616
103. Tweto, J., Liberati, M. and Larrabee, A. R. (1971). *J. Biol. Chem.*, **246,** 2468
104. Tweto, J. and Larrabee, A. R. (1972). *J. Biol. Chem.*, **247,** 4900
105. Alberts, A. W., Majerus, P. W., Talamo, B. and Vagelos, P. R. (1964). *Biochemistry*, **3,** 1563
106. Alberts, A. W., Goldman, P. and Vagelos, P. R. (1963). *J. Biol. Chem.*, **238,** 557
107. Williamson, J. P. and Wakil, S. J. (1966). *J. Biol. Chem.*, **241,** 2326
108. Alberts, A. W., Goldman, P. and Vagelos, P. R. (1963). *J. Biol. Chem.*, **238,** 557
109. Joshi, V. C. and Wakil, S. J. (1971). *Arch. Biochem. Biophys.*, **143,** 493
110. Ruch, F. (1972). *Ph.D. Thesis*, Washington University, St. Louis, Missouri
111. Moore, S. (1963). *J. Biol. Chem.*, **238,** 235
112. Greenspan, M. D., Alberts, A. W. and Vagelos, P. R. (1969). *J. Biol. Chem.*, **244,** 6477
113. Prescott, D. J. and Vagelos, P. R. (1970). *J. Biol. Chem.*, **245,** 5484
114. Greenspan, M. D., Birge, C. H., Powell, G. L., Hancock, W. S. and Vagelos, P. R. (1970). *Science*, **170,** 1203
115. Alberts, A. W., Bell, R. M. and Vagelos, P. R. (1972). *J. Biol. Chem.*, **247,** 3190
116. Bloch, K. (1969). *Accounts Chem. Res.*, **2,** 193
117. Majerus, P. W., Alberts, A. W. and Vagelos, P. R. (1965). *J. Biol. Chem.*, **240,** 618
118. Toomey, R. E. and Wakil, S. J. (1966). *J. Biol. Chem.*, **241,** 1159
119. Birge, C. H. and Vagelos, P. R. (1972). *J. Biol. Chem.*, **247,** 4921
120. Birge, C. H. and Vagelos, P. R. (1972). *J. Biol. Chem.*, **247,** 4930
121. Mizugaki, M., Weeks, G., Toomey, R. E. and Wakil, S. J. (1968). *J. Biol. Chem.*, **243,** 3661
122. Mizugaki, M., Swindell, A. C., and Wakil, S. J. (1968). *Biochem. Biophys. Res. Commun.*, **33,** 520
123. Weeks, G. and Wakil, S. J. (1968). *J. Biol. Chem.*, **243,** 1180
124. Lynen, F., Hopper-Kessel, I. and Eggerer, H. (1964). *Biochem. Z.*, **340,** 95
125. Hagen, A. and Hofschneider, P. H. (1964). *Proc 3rd European Regional conf. Electron Microscopy*, Vol. B, 69 (Prague: Czechoslovak Academy of Sciences)
126. Pilz, I., Herbst, M., Kratky, O., Oesterhelt, D. and Lynen, F. (1970). *Europ. J. Biochem.*, **13,** 55
127. Oesterhelt, D., Bauer, H. and Lynen, F. (1969). *Proc. Nat. Acad. Sci. USA*, **63,** 1377
128. Sumper, M., Oesterhelt, D., Riepertinger, C. and Lynen, F. (1969). *Europ. J. Biochem.*, **10,** 377
129. Schweizer, E., Piccinini, F., Duba, C., Günther, S., Ritter, E. and Lynen, F. (1970). *Europ. J. Biochem.*, **15,** 483
130. Lynen, F., Oesterhelt, D., Schweizer, E. and Willecke, K. (1968). *Cellular Compartmentalisation and Control of Fattty Acid Metabolism*, 1 (F. C. Gran, editor) (New York: Academic Press)
131. Ziegenhorn, J., Niedermeier, R., Nüssler, C. and Lynen, F. (1972). *Europ. J. Biochem.*, **30,** 285
132. Yang, P. C., Butterworth, P. H. W., Bock, R. M. and Porter, J. W. (1967). *J. Biol. Chem.*, **242,** 3501

133. Yun, S. L. and Hsu, R. Y. (1972). *J. Biol. Chem.*, **247**, 2689
134. Butterworth, P. H. W., Yang, P. C., Bock, R. M. and Porter, J. W. (1967). *J. Biol. Chem.*, **242**, 3508
135. Jacob, E. J., Butterworth, P. H. and Porter, J. W. (1968). *Arch. Biochem. Biophys.*, **124**, 392
136. Williamson, I. P., Goldman, J. K. and Wakil, S. J. (1966). *Fed. Proc.*, **25**, 340
137. Bressler, R. and Wakil, S. J. (1962). *J. Biol. Chem.*, **237**, 1441
138. Brady, R. O., Bradley, R. M. and Trams, E. G. (1960). *J. Biol. Chem.*, **235**, 3093
139. Kumar, S., Dorsey, J. A., Muesing, R. A. and Porter, J. W. (1970). *J. Biol. Chem.*, **245**, 4732
140. Yang, P. C., Bock, R. M., Hsu, R. Y. and Porter, J. W. (1965). *Biochim. Biophys. Acta*, **110**, 608
141. Kumar, S., Muesing, R. A. and Porter, J. W. (1972). *J. Biol. Chem.*, **247**, 4749
142. Kumar, S. and Porter, J. W. (1971). *J. Biol. Chem.*, **246**, 7780
143. Dorsey, T. A. and Porter, J. W. (1968). *J. Biol. Chem.*, **243**, 3512
144. Chesterton, C. J., Butterworth, P. H. W. and Porter, J. W. (1968). *Arch. Biochem. Biophys.*, **126**, 864
145. Phillips, G. T., Nixon, J. E., Abramovitz, A. S. and Porter, J. W. (1970). *Arch. Biochem. Biophys.*, **138**, 357
146. Joshi, V. C., Plate, C. A. and Wakil, S. J. (190). *J. Biol. Chem.*, **245**, 2857
147. Nixon, J. E., Phillips, G. T., Abramovitz, A. S. and Porter, J. W. (1970). *Arch. Biochem. Biophys.*, **138**, 372
148. Barnes, E. M. Jr., and Wakil, S. J. (1968). *J. Biol. Chem.*, **243**, 2955
149. Porter, J. W., Kumar, S. and Dugan, R. E. (1971). *Progr. Biochem. Pharm.*, **6**, 1
150. Smith, S. and Abraham, S. (1971). *J. Biol. Chem.*, **246**, 6428
151. Nandedkar, A. K. N. and Kumar, S. (1969). *Arch. Biochem. Biophys.*, **134**, 563
152. Lin, C. Y. and Kumar, S. (1971). *J. Biol. Chem.*, **246**, 3284
153. Smith, S. and Abraham, S. (1971). *J. Biol. Chem.*, **246**, 2537
154. Lin, C. Y. and Kumar, S. (1972). *J. Biol. Chem.*, **247**, 604
155. Abraham, S., Matthes, K. J. and Chaikoff, I. L. (1961). *Biochim. Biophys. Acta*, **49**, 268
156. Dils, R. and Popják, G. (1962). *Biochem. J.*, **83**, 41
157. Smith, S. and Dils, R. (1966). *Biochim. Biophys. Acta*, **116**, 23
158. Kumar, S., Singh, V. N., Leren-Paz, R. (1965). *Biochim. Biophys. Acta*, **98**, 221
159. Carey, E. M. and Dils, R. (1970). *Biochim. Biophys. Acta*, **210**, 388
160. Hansen, H. J. M., Carey, E. M. and Dils, R. (1970). *Biochim. Biophys. Acta*, **210**, 400
161. Hansen, H. J. M., Carey, E. M. and Dils, R. (1971). *Biochim. Biophys. Acta*, **248**, 391
162. Carey, D. M., Hansen, H. J. M. and Dils, R. (1972). *Biochim. Biophys. Acta*, **260**, 527
163. Ilton, M., Jevans, A. W., McCarthy, E. O., Vance, D., White, H. B. and Bloch, K. (1971). *Proc. Nat. Acad. Sci. USA*, **68**, 87
164. White, H. B. III, Mitsuhashi, A. and Bloch, K. (1971). *J. Biol. Chem.*, **246**, 4751
165. Matsumura, S., Brindley, D. N. and Bloch, K. (1970). *Biochem. Biophys. Res. Commun.*, **38**, 369
166. Delo, J., Ernst-Fonberg, M. L. and Bloch, K. (1971). *Arch. Biochem. Biophys.*, **143**, 384
167. Ernst-Fonberg, M. L. and Bloch, K. (1971). *Arch. Biochem. Biophys.*, **143**, 392
168. Sirevåg, R. and Levine, R. P. (1972). *J. Biol. Chem.*, **247**, 2586
169. Bortz, W., Abraham, S. and Chaikoff, I. L. (1963). *J. Biol. Chem.*, **238**, 1266
170. Allmann, D. W., Hubbard, D. O. and Gibson, D. M. (1965). *J. Lipid Res.*, **6**, 63
171. Butterworth, P. H. W., Guchwait, R. B., Baum, H., Olson, E. B., Margolis, S. A. and Porter, J. W. (1966). *Arch. Biochem. Biophys.*, **116**, 453
172. Burton, D. N., Collins, J. M., Kennan, A. L. and Porter, J. W. (1969). *J. Biol. Chem.*, **244**, 4510
173. Zakim, D. and Ho, W. (1970). *Biochim. Biophys. Acta*, **222**, 558
174. Hicks, S. E., Allman, D. S., and Gibson, D. M. (1965). *Biochim. Biophys. Acta*, **106**, 441
175. Craig, M. D., Dugan, R. E., Muesing, R. A., Slakey, L. L. and Porter, J. W. (1972). *Arch. Biochem. Biophys.*, **151**, 128
176. Allman, D. W. and Gibson, D. M. (1965). *J. Lipid. Res.*, **6**, 22
177. Muto, Y. and Gibson, D. M. (1970). *Biochem. Biophys. Res. Commun.*, **38**, 9
178. Volpe, J. J. and Kishimoto, Y. K. (1972). *J. Neurochem.*, **19**, 737

179. Gibson, D. M. and Hubbard, D. D. (1960). *Biochem. Biophys. Res. Commun.*, **3,** 531
180. Dahlen, J. V., Kennan, A. L. and Porter, J. W. (1968). *Arch. Biochem. Biophys.*, **124,** 51
181. Wakil, S. J., Goldman, J. K., Williamson, I. P. and Toomey, R. E. (1966), *Proc. Nat. Acad. Sci. USA*, **55,** 880
182. Plate, C. A., Joshi, V. C., Sedgwick, B. and Wakil, S. J. (1968). *J. Biol. Chem.*, **243,** 5439
183. Dugan, R. E. and Porter, J. W. (1970). *J. Biol. Chem.*, **245,** 2051
184. Robinson, J. D., Brady, R. O. and Bradley, R. M. (1963). *J. Lipid Res.*, **4,** 144
185. Lust, G. and Lynen, F. (1968). *Europ. J. Biochem.*, **7,** 68

4
The Dynamic Role of Lipids in the Nervous System

D. M. BOWEN, A. N. DAVISON and R. B. RAMSEY
The National Hospital, London

4.1	INTRODUCTION	142
4.2	LIPID DISTRIBUTION WITHIN THE NERVOUS SYSTEM	143
	4.2.1 *Distribution of lipid in the whole adult brain*	143
	4.2.2 *Regional lipid distribution*	143
	4.3.2 *Cellular*	145
	4.2.4 *Subcellular*	147
4.3	LIPID METABOLISM	147
	4.3.1 *Phospholipids—biosynthesis*	147
	4.3.2 *Metabolism of fatty acids*	150
	4.3.2.1 *Synthesis*	150
	4.3.2.2 *Incorporation into complex lipids*	151
	4.3.2.3 *Catabolism*	151
	4.3.2.4 *Free fatty acids*	152
	4.3.3 *Cholesterol*	152
	4.3.3.1 *Acetate to squalene*	153
	4.3.3.2 *Squalene to cholesterol*	153
	4.3.3.3 *Physiological role of steryl esters*	155
	(a) *Brain degeneration*	156
	(b) *Brain developent*	156
	4.3.4 *Ganglioside*	157
	4.3.4.1 *Biosynthesis*	157
	4.3.4.2 *Degradation*	159
	4.3.4.3 *Turnover*	161
	4.3.5 *Glycosphingolipids not containing sialic acid*	161
	4.3.6 *Cellular metabolism*	162
	4.3.6.1 *Whole tissue*	162
	4.3.6.2 *Neural cell-enriched fractions*	163

4.4	EXCHANGE OF BRAIN LIPIDS	164
	4.4.1 *Cholesterol exchange*	164
	4.4.1.1 *Experiments* in vitro	165
	4.4.2 *Phospholipid exchange*	166
4.5	LIPIDS AND TRANSMISSION IN THE NERVOUS SYSTEM	167
4.6	DEVELOPMENT AND AGEING	169
	4.6.1 *Development*	169
	4.6.2 *Ageing*	170
	4.6.2.1 *Lipofuscin*	171
	4.6.2.2 *Lysosomes and lipid peroxidation*	171
4.7	BRAIN DISEASE	172
	4.7.1 *Lipidoses*	172
	4.7.2 *Aminoacidurias*	172
	4.7.3 *Demyelinating diseases*	174
4.8	CONCLUSION	175
	ACKNOWLEDGEMENTS	175

4.1 INTRODUCTION

In the past, lipids have been thought to play a static role in the brain as uninteresting components of cerebral membranes, not significantly participating in any of the diverse metabolic reactions of the nervous system. This was because detailed research on the function of cerebral lipids had necessarily to await the isolation and identification of the very complex lipid species found in the tissue. Thus, despite the pioneer work of Thudicum in 1884 showing that there were phosphorus- and galactose-containing lipids in the brain, progress on their chemistry was largely held up until 1957 when an efficient and simple method of isolation of brain lipids was first reported[1]. Soon further work led to the discovery of the different kinds of kephalins and to the identification of new lipids such as the gangliosides and the inositol-containing phosphatides. This facilitated research on the physiological role of lipids in the brain. It has long been recognised from dietary experiments that lipids of the adult nervous system were relatively stable and early studies with deuterium and radioactive tracers confirmed this view. For example, it was suggested that much of the myelin sheath was relatively inert and even, possibly, part of the cerebral mitochondrial membrane was equally stable[2, 3]. Work from a number of laboratories has, however, made it necessary to re-evaluate completely current concepts of membrane lipid metabolism, for there is good evidence that lipid molecules can exchange between different brain membranes. It is now realised that lipids also have an important dynamic function in enzyme reactions, in membrane translocation processes and in intermediary metabolism within the nervous system. Although the brain lipids appear to be resistant to starvation, electrical stimulation and treatment with hormones can elicit fascinating alterations in their concentration,

Changes in lipid composition of the developing nervous system demonstrate the important participation of these components in newly-formed organelles and cells. In neurological diseases alterations in lipid profile provide an indication of the mechanism of brain pathology. It is, therefore, our object in this chapter to review the biochemistry of brain lipids with special reference to these more dynamic aspects of neural metabolism. Details of the structure, chemistry and methodology of both lipids and myelin have been recently reviewed[4-7].

4.2 LIPID DISTRIBUTION WITHIN THE NERVOUS SYSTEM

Adult nervous tissue has high lipid content in comparison to other tissues. Prior to myelination the lipid composition of foetal brain resembles that of other organs but, as the myelin–lipoprotein membranes are laid down round nerve fibres, there is a dramatic increase in lipid concentration, particularly in the white matter and peripheral nerve. However, even after the final stage of development of the brain has finished, changes in composition continue. There is an increase in total lipid of whole human brain up to about 30 years of age, and thereafter there is a decrease in total lipid[8]. This loss is seen particularly in the phospholipid content; there are only slight changes in cholesterol and cerebroside concentration. In addition to changes in the proportion of brain lipid, alterations are also found in fatty-acid profiles of the different lipids. These observations indicate the continually changing lipid levels in the human nervous system throughout life and emphasise the dynamic state of brain lipids.

4.2.1 Distribution of lipid in the whole adult brain

Analysis of different areas of the nervous system shows the expected relationship between anatomical structure and lipid distribution. Thus the white matter has a composition close to that of myelin, while cortical grey matter contains less of the typical myelin lipids (e.g. cerebroside, sulphatide and phosphatidal ethanolamine) and relatively more ganglioside (Table 4.1). Attention has also focused on examination of specific regions of the nervous system. This has added interest where it may have functional significance. For example, since the cerebellum is important in control of balance, and since this is one of the last major brain structures to develop, analysis of cerebellar lipids has been used as an index of structural changes in this part of the brain in relation to ataxia. Further work has been directed at analysis of regional cellular and subcellular lipids at which level functional correlates should be more meaningful.

4.2.2 Regional lipid distribution

Not a great deal of information is available on the lipid composition of various regions of the central nervous system; this lack of data is particularly evident

Table 4.1 Brain grey and white matter composition. The lipid content of adult human grey and white matter has been expressed both on a weight and molecular basis in terms of total dry weight of tissue. Data have been recalculated from that of Norton, Poduslo and Suzuki [9]. Molecular weights used in calculations: cholesterol 386; cerebroside 858; sulphatide 937; and phospholipids 775.

Lipid	White Matter		Grey Matter	
	(mg/100 mg dry wt.)	(μmol/100 mg dry wt.)	(mg/100 mg dry wt.)	(μmol/100 mg dry wt.)
Total lipid	54.9	95.4	32.7	53.9
Cholesterol	15.1	39.1	7.18	18.6
Total galactolipid	14.5	15.9	2.38	2.25
Cerebroside	10.9	12.7	1.77	2.07
Sulphatide	2.96	3.16	0.556	0.176
Total phospholipid	25.2	32.5	22.7	29.3
Phosphatidal-ethanolamine	8.18	10.6	7.40	9.55
Phosphatidal-choline	7.03	9.07	8.73	11.3
Sphingomyelin	4.23	5.46	2.26	2.92
Phosphatidal-serine	4.34	5.60	2.84	3.66
Phosphatidal-inositide	0.494	0.637	0.883	1.14
Plasmalogens	6.15	7.94	2.88	3.72
Water content (%)	71.6		81.9	

Table 4.2 Changes in lipids of developing sheep central nervous system [10,11]

	Lipid molar ratio (cholesterol : phospholipid : cerebroside)			
	Cerebrum	Cerebellum	Brain stem	Spinal cord
Early myelination (140 days post-conception)	1:1.36:0.18	1:1.25:0.29	1:0.94:0.53	1:0.85:0.48
Late myelination (180 days, 5 weeks old)	1:0.35:0.29	1:0.57:0.37	1:0.71:0.37	1:0.59:0.29
Mature	1:0.56:0.43	1:0.70:0.50	1:0.61:0.52	1:0.64:0.61
Adult myelin	1:0.89:0.40			

with regard to human tissue. However, an extensive study has been reported on the regional distribution of lipid in the adult and developing sheep brain (Table 4.2). Although the concentration of phospholipid in the cerebellum was found to be slightly higher than other parts of the central nervous system,

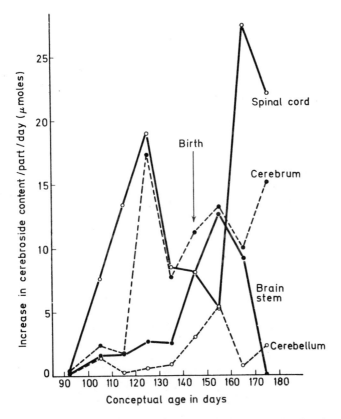

Figure 4.1 Changes in cerebroside concentration during foetal and postnatal development of the lamb central nervous system (From Patterson, Swensey and Hebert[10] by courtesy of *J. Neurochem.*)

little variation of molar lipid ratios was seen in the mature animal. Marked differences in cerebroside concentration in each region during development were apparent (Figure 4.1.1). These changes could be correlated with the varying rates of myelination in the ovine central nervous system.

4.2.3 Cellular

Although isolation of brain cell-enriched preparations has only been attempted in the last few years, considerable information is available regarding the lipid composition of these cell fractions (Table 4.3). Most notable are the lipid

Table 4.3 Lipids in neural cell types. Values for astroglia and neurones from rat have been calculated from data of Norton and Poduslo[12] except for the quantities of the individual galactolipids which were derived from the molar fractions of these compounds found by Hamberger and Svennerholm[13] with astroglia and neurones from rabbit cortex. Values for oligodendroglia have been calculated from the results of Podulso and Norton[14] with corpus callosum of calf brain as were all the CNPH activities. Myelin values were from Norton and Autilio[15].

Cell type	DNA (pg/cell)	CPNH*	Cholesterol (μmol/100 mg dry wt.)	Cerebroside (μmol/100 mg dry wt.)	Sulphatide (μmol/100 mg dry wt.)	Total Phospholipid (μmol/100 mg dry wt.)	Ganglioside (μmol/100 mg dry wt.)	Molar ratios Cholesterol : Cerebroside : Phospholipid			
Astroglia	8.18	41	14.1	0.689	0.142	35.6	0.582	1	: 0.05	:	2.5
Neurones	11.2	42	6.61	0.513	0.0897	22.4	0.223	1	: 0.075	:	3.5
Oligodendroglia	5.14	323	10.8	2.61	0.472	23.6	0.239	1	: 0.25	:	2.2
Myelin		372	54.9	22.0	2.89	41.8	0.0453	1	: 0.40	:	0.76

* 2′,3′-Cyclic nucleotide 3′phosphohydrolase (CPNH) activity has been expressed as μmol of Pi min^{-1} per 100 mg dry wt.

molar ratios of the cell enriched fractions, which indicate that oligodendroglia are much more enriched in cerebroside than are neurones or astrocytes. The much higher phospholipid content of the three cell types when compared to myelin would also suggest cell plasma membranes of composition quite different from myelin.

4.2.4 Subcellular

Many studies have been reported on the lipid composition of brain subcellular fractions, especially myelin (Table 4.4). While it would appear that the list of examined subcellular fractions is quite exhaustive, such subcellular components as lysosomes, the Golgi apparatus and lipofuscin have yet to be adequately characterised. The understanding of the role of lipids in these subcellular fractions may yield valuable insight into how these organelles may contribute to ageing and disease processes in the brain.

The lipid composition of synaptosomal and myelin membranes have been directly compared by various workers. For example, in the squirrel monkey, Sun and Sun[17] found very low levels of galactolipid in synaptic membranes whereas the molar ratio of galactolipid to phospholipid in myelin was 0.66. Synaptic membranes were characterised by a high choline phosphatide content (molar ratio cholesterol:phospholipid = 0.74:1) whereas myelin contained a much higher proportion of ethanolamine phosphoglycerides and relatively more cholesterol (molar ratio cholesterol:phospholipid = 1.24:1). Synaptosomal membranes contained a high proportion of $C_{22:6}$ $(n-3)$ fatty-acid esters whereas myelin was richer in the monoenoyl acyl groups; the latter may be necessary for a more stable configuration of myelin. The high level of polyunsaturated acyl groups in synaptic membranes renders them highly susceptible to attack by peroxidase or lipoxygenase (see p. 171).

4.3 LIPID METABOLISM

Research published in the last few years on the metabolism of lipids throws new light on their dynamic role in the brain. There is evidence of ready exchangeability of lipids between neural membranes and hence knowledge of the metabolism of these molecules assumes a new importance. The general features of lipid synthesis and catabolism in the nervous system are well described and in this section we will, therefore, only review more recent advances relevant to the main theme of this chapter.

4.3.1 Phospholipids—biosynthesis

As in other tissues the cytidine pathway described by Kennedy and his colleagues is of overriding importance for the synthesis of brain phospholipids. Table 4.5 gives an indication of the activity of some enzymes concerned in phospholipid metabolism. Clearly, there is considerable heterogeneity in the

Table 4.4 Lipids in subcellular fractions of rat brain cortex. Values are calculated from data of Lapetina, Soto and De Robertis[16]. Number of analyses of rat brain cortex are shown in parenthesis. The molecular weights were taken as 846 and 775 for galacto- and phospholipids respectively. Ganglioside is expressed as NANA µg g^{-1} fresh tissue. Values for mitochondrial galactolipid is taken from Cuzner, Davison and Gregson[2]. Relative specific concentration of ganglioside is shown as % of NANA/protein.

	Total protein(8) (µg mg^{-1} fresh wt.)	Proteolipid(8) (µg mg^{-1} fresh wt.)	Cholesterol(7) (µmol g^{-1} fresh wt.)	Galactolipid(6) (µmol g^{-1} wet wt.)	Total Phospholipid(7) (µmol g^{-1} fresh wt.)	Ganglioside(5) µg g^{-1} wet wt.$^{-1}$	Relative specific concentration %
Total cortex	123.7±0.49	4.15±0.66	36.6±3.0	6.35±1.6	56.5±4.5	680±5	—
Nuclei	8.6±1.87	0.35±0.06	3.1±0.4	1.3 ±0.28	3.7±1.4	20±3	0.49
Myelin	5.7±0.82	1.46±0.10	4.9±0.3	2.0 ±0.18	4.9±0.97	27±3	1.10
Mitochondria	14.1±1.83	0.90±0.07	1.55±0.1	0.01	4.9±0.78	15±3	0.24
Nerve ending membranes (M$_1$ 1,2)	3.5±0.85	0.24±0.03	1.3 ±0.13	0.35±0.08	1.8±0.42	26±4	1.72
Synaptic vesicles	1.7±0.24	0.13±0.01	0.77±0.05	0.12±0.03	1.3±0.19	9±2	0.56
Microsomes	18.2±2.15	0.41±0.02	5.4±0.65	0.35±0.67	11.3±1.07	168±32	1.00
Axoplasm	8.4±1.67	0.13±0.02	0.5±0.05	0.12±0.04	0.39±0.19	7±0.2	0.17
Supernatant	23.3±1.63	0.33±0.03	0.82±0.23	0.35±0.07	1.42±0.45	13±2	0.12

Table 4.5 Brain phospholipid synthetic enzyme activity[18-20]

	Activity (μmol h^{-1} g^{-1} fresh brain)
Choline kinase	15–20
Phosphorylcholine cytidyl transferase	8–12
Phosphorylcholine diglyceride transferase	0.86
Phosphorylethanolamine diglyceride transferase	0.45
Glycerol phosphate acyl transferase	0.4

potential rates of such enzyme-directed reactions, particularly if account is taken of their subcellular localisation.

Microsomes are the main site for the synthesis of phospholipid *de novo* within the cell. However the cellular heterogeneity of nervous tissue has always made it especially difficult to obtain pure subfractions. Miller and Dawson[21] have overcome some problems by correcting fractions for microsomal contaminants (using NADPH-cytochrome c reductase, NAD nucleosidase and RNA as markers). They demonstrated that mitochondria from adult guinea-pig do not synthesise phospholipid from CDP-[Me-^{14}C]choline or phosphoryl-[Me-^{14}C]choline. Unlike liver mitochondria[22] the organelles could synthesise phosphatidic acid and diphosphatidylcholine from a [^{32}P] phosphate precursor, but could not synthesise nitrogen-containing phosphoglycerides or phosphatidylinositol. Although the synaptosomal outer membrane and intraterminal mitochondria cannot incorporate CDP-[Me-^{14}C] choline into phosphatidylcholine this can be achieved by synaptic vesicles and by the intraterminal endoplasmic reticulum. The exchange of choline is catalysed by a calcium-dependent non-energy-requiring reaction by microsomal and brain synaptosomal fractions and there is fast entry of choline into the lecithin of the synaptic vesicles[23]. *In vivo*, ^{32}P shows maximal incorporation into mitochondrial and microsomal fraction, with synaptosomes appreciably less[21, 24]. Myelin phospholipids are hardly labelled up to 10 h after intraperitoneal injection. Some transport of phospholipid from the nerve cell body down the axon may be possible[25].

There has been considerable research on the mechanism of synthesis of phosphatidyl ethanolamine and the corresponding plasmalogen (the major brain phospholipid, e.g. see Ansell[26]).

Synthesis of ethanolamine plasmalogen (1-alkenyl-2-acyl-*sn*-glycero-3-phosphorylethanolamine) was thought to be formed mainly by reduction of the 1-acyl residue of phosphatidyl ethanolamine. There is now good evidence for the direct metabolic conversion of alkyl acyl ethanolamine phosphatides or alkyl ethanolamine lysophosphatides to the corresponding plasmalogens or lysoplasmalogens in mammalian brain and other tissues. [1-^{14}C]Hexadecanol is readily incorporated into the ethanolamine and choline phosphatides of myelinating rat brain[27]. Radioactivity can be shown to be present in alkyl and acyl moieties of the phosphatides but labelled alk-1-enyl moieties are present only in ethanolamine phospholipid[28]. In addition, DL-sphingosine (dehydrosphingosine) can be converted into the alkenyl ether chain of plasmalogen[29].

4.3.2 Metabolism of fatty acids

Since fatty acids are major components of membrane lipids, study of changes in the fatty-acid ester composition provides some insight into the physiological role of lipids in the nervous system. Very long-chain fatty acids linked to sphingosine are important in stabilising the myelin membrane. Polyunsaturated fatty acids are possibly essential to the nervous system and the long-chain 2-hydroxy fatty acids, such as cerebronic acid, $C_{24h:0}$, are characteristically present in cerebroside and sulphatides of the myelin sheath. Fatty acids may be taken up into the brain by the developing animal but this route is of less importance in the adult. Endogenous synthesis of fatty acid is rapid in developing and rather slower in mature brain.

4.3.2.1 Synthesis

The biosynthesis of fatty acids has been comprehensively reviewed elsewhere[30-33]. At least two fairly well documented cellular mechanisms exist for synthesis of fatty acids in brain (Figure 4.2). The first is the *de novo* synthesis of saturated acids, primarily palmitic acid, by cytoplasmic enzymes. The second system is present in microsomes and mitochondria and catalyses the elongation of preformed fatty acids by the addition of two-carbon units. A third synthetic mechanism for supplying nervous tissue with fatty acids may be by

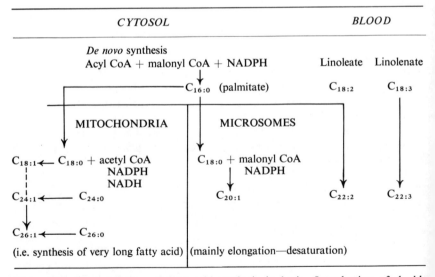

Figure 4.2 Representation of fatty acid synthesis in brain. Introduction of double bonds beyond $n-9$ (n from methyl end) is not possible

oxidation of long-chain alcohols. Schmid and his associates[34, 35] have shown that mono-, di- and tri-unsaturated long chain alcohols administered to myelinating rat brain are not only incorporated into both the O-alkyl and O-alk-1-enyl moieties of ethanolamine phosphatides but are also oxidised to the corresponding acid. It is salient that small amounts of saturated and mono-unsaturated alcohols have been detected in brain[36, 37]. The most likely route for the biosynthesis of longer-chain unsaturated fatty acids is by chain elongation of the medium-chain unsaturated fatty acids[30].

The chain elongation pathway has only recently been examined in detail. Aeberhard and Menkes[38] found that subcellular particles from rat brain homogenates only synthesised small amounts of the long-chain fatty acids that are characteristic of the myelin membrane. However, elongation of behenyl-CoA ($C_{22:0}$) to lignoceryl-CoA ($C_{24:0}$) by the microsomal fraction from rat brain demonstrates that enzymes are present in brain for the synthesis of very long-chain fatty acids[39]. A failure of chain elongation may cause the sphingolipid deficiency seen in the 'Quaking' mouse mutant[40]. This and other data[41] have led to the view that decreased chain elongation might be important in the aetiology of multiple sclerosis. Though information is lacking on the factors that control chain elongation the sites at which the *de novo* route is controlled are well established. Control of *de novo* synthesis is exerted at the level of acetyl CoA carboxylase and fatty acid synthetase, as well as on the supporting pathways which provide intermediate carbon precursors and reducing equivalents[42].

From *in vivo* experiments it seems likely that saturated and unsaturated non-hydroxy fatty acids give rise to the analogous hydroxy fatty acids. *In vitro* studies on the hydroxylation of non-hydroxy fatty acids by brain preparations have not yielded clear results, though non-enzymatic conversion can certainly be obtained.[43]

4.3.2.2 Incorporation into complex lipids

The products of chain elongation in brain are the very long chain $n - 3$ (ω 3) fatty acids. It is these fatty acids that are incorporated into the complex lipids characteristic of nervous tissue. The transacylation reactions involved in the synthesis of some complex brain lipids appear to have a specificity resembling the fatty acid composition of the respective lipid class[44].

4.3.2.3 Catabolism

When added to cerebral cortical slices β-hydroxybutyrate, a product of the β-oxidation of fatty acids, can increase the formation of lactate from glucose[45]. Since lactate can be converted into glucose the increase in lactate formation is equivalent to a sparing of glucose. This and other data demonstrate that ketone bodies derived from fatty acids by β-oxidation, can be utilised as an energy source by brain[46]. In addition to β-oxidation brain contains a 1-carbon degradation system that catalyses the oxidation of the a-carbon atom of fatty acids to carbon dioxide[32]. A recent communication suggests that ascorbic acid is the natural co-factor for this α-oxidation process. The results of assaying subcellular fractions indicate that the 1-carbon degradation system has a microsomal localisation[47, 48]. Data obtained in this laboratory (D.M. Bowen, unpublished observations) suggest that in crude homogenate of brain, microsomes become enriched with peroxisomes; a similar effect may occur in some tumour tissues[49]. Since hydrogen peroxide is a by-product of α-oxidation and as peroxisomes characteristically contain enzymes such as L-α-hydroxy acid oxidases and catalase, it is possible that peroxisomes contain part or all of the 1-carbon degradation system. A fascinating application of the concept of α-oxidation are the investigations carried out to elucidate the nature of Refsum's disease.

4.3.2.4 Free fatty acids

Free fatty acids (FFA) occur as trace components of brain lipids. FFA may be estimated satisfactorily only after minimising the time interval between death and homogenising and by separation from the relatively large complex lipid samples in a single chromatographic step. Using these techniques the FFA content of rat brain appears to be 37 μg g^{-1} fresh weight[50]. It seems likely that the published data[51] on the FFA contents of human brain represent a gross overestimate. The FFA of brain constitute a labile pool that is increased following very short periods of ischaemic and electroconvulsive shock[52] and following convulsions induced by pentylenetetrazol[53]. Since cholesterol esters are virtually absent from normal adult brain and as the kinetics of FFA appearance and triglyceride disappearance are not comparable the FFA are believed to originate mainly from phospholipids[52].

4.3.3 Cholesterol

Although it has been clearly established that cholesterol is synthesised readily in developing brain, it has always been difficult to conclusively demonstrate cholesterol synthesis in the adult nervous system. However, adult nervous tissue contains the full complement of enzymes necessary for sterol synthesis and their localisation within the cell is similar to that found for liver and other organs. No failure has been detected in the ability of the brain to utilise a whole range of water-soluble precursors. The main biosynthetic pathway would appear to contain no unusual intermediates, yet there must be something about the nature of cholesterol biosynthesis in the brain which distinguishes it from all other parts of the body. The glia have the ability to generate great quantities of sterols (and other compounds) during development but after maturation synthesis becomes minimal. One possibility is that biosynthesis is tightly controlled by a feed-back inhibition mechanism. Alternatively, lack of a co-factor or a carrier protein or possibly a modification of the biosynthetic pathway to allow the channeling of potential intermediates into other pathways may explain the restricted synthesis. Although many of these questions are as yet unanswered our present knowledge of brain sterol biosynthesis provides clues to the solution to these questions.

4.3.3.1 Acetate to squalene

Examination of the biosynthetic steps between acetate and squalene has suggested the presence of the same basic pathway as has been described in liver cholesterol synthesis. Much discussion has been centred on what water-soluble precursors serve as the best substrates. Various reports have indicated glucose as most efficient in brain, while acetate and glucose were of comparable activity in brain stem and cord. After manipulation of co-factors, however, mevalonate could be shown to be most efficiently utilised[54]. The fact that mevalonate is a much better precursor of cholesterol than glucose or acetate in liver under comparable conditions does suggest differences in sterol biosynthesis between the two tissues. Since cholesterol biosynthesis in the intact tissues passes through acetate to mevalonate in both instances, differences observed in substrate administered studies must reflect more a problem of uptake and membrane-bound intermediates than one of differences of metabolic pathways.

Little is known about the possible control mechanism affecting the acetate to squalene sequence of sterol biosynthesis. It has been suggested that possibly there is a feed-back inhibition of brain β-hydroxy-β-methyglutaryl coenzyme A reductase (HMG-CoA) as there is in liver[55]. The results of Kandutsch and Saucier[56] would indicate that HMG-CoA reductase regulation controls brain sterol formation in the rat from eleven days after birth but not before. It has, on the contrary, been indicated that there is no active feed-back mechanism in CNS tissue sterol synthesis[57].

In contrast, Professor Popjak[58] has suggested an alternative pathway to explain the decreased cholesterol synthesis. He suggests a second pathway from mevalonate, not leading to sterol biosynthesis (Figure 4.3). This would involve the attack by a phosphatase on dimethylallyl pyrophosphate yielding dimethylallyl alcohol. This would then be metabolised by way of CO_2 fixation to dimethylacrylyl coenzyme A, which could then be broken down to HMG-CoA and ultimately to acetyl CoA. The operation of such a pathway in brain, as well as in skin, has indeed been shown (J. Edmund and G. Popják, personal communication). Certainly the knowledge of the functioning of such a pathway may give us more clues about the control of brain sterol synthesis.

4.3.3.2 Squalene to cholesterol

It would appear that whereas in liver the critical points in cholesterol biosynthesis lie between acetate and mevalonic acid the critical steps in central nervous system sterol formation are between squalene and cholesterol itself. The role of squalene as an intermediate is firmly established but there is uncertainty as to the reason for its relatively slow metabolism in brain compared to liver[59, 60]. It has been suggested that the relatively slow metabolism of squalene by brain may be due to the lack of a sterol carrier protein[61]. Indeed this has been shown to be the case[62]. As yet it is not known if the lack of sterol carrier protein is unique to nervous tissue. This absence of carrier

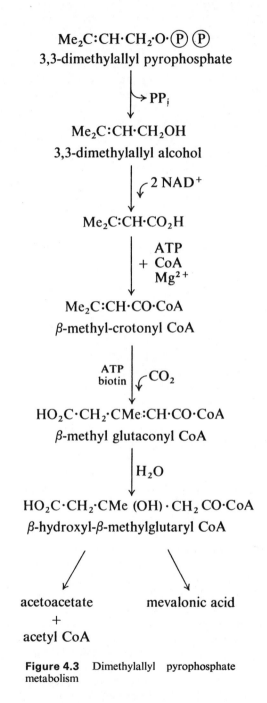

Figure 4.3 Dimethylallyl pyrophosphate metabolism

protein could certainly account for at least part of the slow biosynthesis of cholesterol.

The intermediate in cholesterol biosynthesis, now generally presumed to follow squalene, squalene epoxide, has not itself been isolated from brain. Squalene epoxide can, however, be converted to sterols in both developing and mature brain tissue[54,63]. Although earlier studies[64] involving rat, chick and human brain failed to show lanosterol as an active intermediate, later work on both developing and adult rat has shown it to be involved in brain cholesterol biosynthesis[63,65]. All the above studies have demonstrated the role of 4,4-dimethyl-5α-cholesta-8,24-dien-3β-ol in sterol formation and the relative amounts of this sterol present during metabolic studies would suggest that its demethylation at C-4 is possibly one of the slower steps between squalene and cholesterol. After the initial demethylation at C-14, there could be two possible substrates for the removal of the remaining methyl at C-4. One is the 4α-methyl sterol derived from the previously described demethylation 4α-methyl-5α-cholesta-8,24-dien-3β-ol, and the second is a sterol that has undergone isomerisation of the Δ^8 double bond to Δ^7, 4α-methyl-cholesta-7,24-dien-3β-ol. Weiss, Galli and Paoletti[64] found no evidence for the second of these but Ramsey et al.[54,63] have demonstrated the intermediary role of the former in developing rat brain and the latter in adult rat brain. If this is true, one of the factors affecting the rate of cholesterol formation in adult versus developing brain may be the isomerisation of the nuclear double bond at this point.

The demethylation of the 4α-methyl sterols should yield $\Delta^{8,24}$ and $\Delta^{7,24}$ 4-desmethylsterols, respectively. While the existence[65] of the C_{27} $\Delta^{8,24}$ as an intermediate in brain has not been shown specifically[66], it does occur in the central nervous system[64]. However, the brain contains a variety of sterols which are probably derived from minor pathways or dietary sources[67]. The C_{27} $\Delta^{7,24}$ sterol has, on the other hand, been found highly labelled in metabolic studies involving [2-^{14}C]mevalonic acid and at the same time being at very low endogenous levels in the tissue[63]. The presence of C_{27} $\Delta^{5,7,24}$ as an important precursor of cholesterol has been best shown by Scallen, Candie and Schroepfer[68] by use of the drug AY-9944 which blocks reduction of the Δ^7 double bond. With its use they demonstrated the accumulation in brain tissue of this sterol (C_{27} $\Delta^{5,7,24}$) as well as its C-24 dihydro analogue.

The penultimate and final steps of sterol biosynthesis presumably involve the reduction of the Δ^7 nuclear double bond and the subsequent reduction of the C_{27} $\Delta^{5,24}$ compound, desmosterol, to give cholesterol. The final reduction step is slow and has for a long time been felt to be a rate-limiting step in brain sterol biosynthesis, indeed its activity appears to be maximal during development and minimal after maturation[69]. If this were the limiting step, however, one would expect to find a significant quantity of desmosterol in the adult brain; this is not the case.

4.3.3.3 Physiological role of sterol esters

Although, during development, significant quantities of steryl ester are present in the brain, their role in brain metabolism has as yet remained a mystery. Histological examination shows Marchi-positive sudanophilic material to be present in human foetal brain with particularly elevated sudanophilia staining (mainly for cholesterol esters) at 6 and 7 months gestation[70]. This reaction does not appear in normal adult brain but Marchi-positive sudanophilic lipid is seen in demyelinated tissue. In neural degeneration involving multiple sclerosis, sudanophilic leucodystrophy, gangliosidosis or neural virus infections steryl esters have often been found in abundance. The same is true for tumours that are derived from neuronal or glial cell types. It is not known if these cholesterol esters are directly formed as a result of the degenerative processes or if they are even of neural origin. In the developing brain it has been suggested that cholesterol ester may be involved in lipid transport.

(a) *Brain degeneration*—As a result of *in vitro* experiments it was thought that a considerable portion of the sterol moiety of steryl esters were cholesterol precursors[63]. However, examination of ester formed by intact tissue, either during tissue maturation or during demyelination, has shown cholesterol to be the major sterol (R. B. Ramsey and A. N. Davison, unpublished work). It may be concluded that steryl ester formation *in vitro* is dissimilar to *in vivo* ester formation.

The fatty acid component of the ester is possibly of more interest. Normal developing human tissue contains as much as 20% of the total sterol as ester. The main fatty acid ester is oleate which represents approximately 50% of the total ester (R. B. Ramsey and A. N. Davison, unpublished work). The cholesterol ester fatty acid pattern of human brain has been shown to be remarkably uniform throughout life[71]. The fatty acid composition of cholesterol esters from the brain in three different instances of demyelinating diseases have been found to be similar with a relative enrichment of oleic acid ester, accounting for nearly 70% of the total fatty acids. It is proposed that the raised proportion of oleic acid ester is the result of increased acylation of cholesterol with free fatty acids released during myelin breakdown by the action of phospholipase A on phosphatidylcholine. It seems unlikely that macrophages are responsible for steryl ester formation for Werb and Cohn[72] have shown in extensive experiments that macrophages could not form steryl esters but rather absorbed and hydrolysed pre-formed esters. This evidence would therefore suggest that the steryl esters might be formed by some type of reactive endogenous cell such as the astroglia, which we know to multiply in areas of brain degeneration.

(b) *Brain development*—The enzymes of steryl ester formation and degradation in developing and mature tissue have been examined in the rat[73, 74] but this has shed little light on the reasons for elevated esters in development. In the rat, as in other species, the fatty acid pattern of the cholesterol esters of the developing and adult brain are similar but quite distinct from the fatty acid profile of serum and liver cholesterol esters. This does not suggest that cholesterol esters originate from extra-cerebral sources. Within the brain immediately prior to myelination cholesterol ester may serve as a convenient reservoir of cholesterol and fatty acid molecules. Alternatively, it is possible to speculate that the ester may provide a vehicle for fatty acid elongation. It seems unlikely that steryl esters are necessary intermediates for cholesterol synthesis, in fact they might well block sterol synthesis. Certainly there is no evidence for active sterol synthesis as a result of demyelination when steryl esters are elevated.

Steryl esters are not membrane bound in developing, normal mature or demyelinating brain tissue (R. B. Ramsey and A. N. Davison unpublished work). Indeed, in instances of elevated steryl ester well over 90% of the ester may be present in the soluble fraction of nervous tissue. Since it is in the cytosol, it could be proposed that cholesterol acts as a carrier of long chain fatty acid, either by itself or in concert with a soluble protein for exogenous cholesterol does not stimulate cholesterol ester formation, but synthesis is stimulated by free oleic acid[73]. This would suggest that the ester's function must, somehow, be related to fatty acid metabolism or re-utilisation, either for the transfer of fatty acids for chain elongation or degradation, or perhaps as a means of detoxifying and storing surplus free fatty acids.

4.3.4 Ganglioside

Biochemists have long been intrigued by the possible role of gangliosides in nervous tissue, for these complex macromolecules have both lipid and hydrophilic character. The discovery of an accumulated ganglioside in the brain of patients with Tay Sachs' disease stimulated research into the enzymic degradation of these glycolipids but until recently rather less interest was taken in their biosynthetic pathways. Thus the structure of the ganglioside was quickly established and the missing catabolic enzyme identified. The study of biosynthesis has not been so easy. Problems in establishing the number and size of precursor pools together with differences in results of *in vitro* and *in vivo* experiments have made the task more difficult. In the following account of ganglioside biosynthesis and degradation the nomenclature used is that of Svennerholm[75, 76].

4.3.4.1 Biosynthesis

Total synthesis of gangliosides is difficult to demonstrate using any given precursor, partly because of the number of steps in formation and what appears to be division of potential precursor materials into different pools[77]. There are individual pools for each particular ganglioside, for example, two separate G_{M1} pools for the formation of disialo gangliosides (G_{D1a} and G_{D1b}) can be distinguished *in vivo*[77, 78]. Since mixing of precursor pools may occur *in vitro*, results from experiments on intact animals are to be preferred for the delineation of ganglioside formation[79, 80]. Several routes of synthesis appear possible and all may be operable to a greater or lesser degree (Figure 4.4). The principal difficulty seems to lie in determining the exact steps between lactosylceramide and G_{D1} gangliosides. It has been proposed that G_{M3} is an intermediate in this pathway as follows: lactosylceramide $\rightarrow G_{M3} \rightarrow G_{M2} \rightarrow G_{M1} \rightarrow G_{D1a}$ or G_{D1b}[81, 82]. However, Kanfer and Ellis[79] suggest that G_{M3} is not an intermediate. Following intracerebral injection of N-acetyl [^{3}H]mannosamine into 15-day-old rats, instead of the expected precursor-product relationship, labelling of G_{M3} was found to

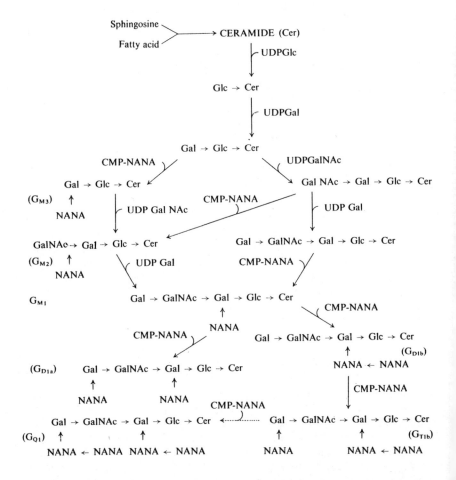

Figure 4.4 Major pathways of ganglioside biosynthesis. The Svennerholm numbering is used. NANA = N-acetylneuraminic acid; Glc = glucose; Gal = galactose; Cer = ceramide

lag behind that of G_{M2}. G_{M3} is also distinguished by a lower stearic acid content (50%) compared to other gangliosides which contain more than 80% of stearate[83]. This again suggests that G_{M3} is not directly on the metabolic pathway for the formation of the more complex gangliosides. The biosynthetic route based on Kanfer and Ellis's research could fit the following sequence:

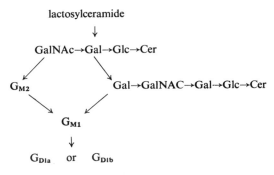

The problem of the multiple pools must again be considered when attempting to rule out G_{M3} as a direct intermediary. It may be that a small metabolically active pool may exist containing G_{M3} with a high level of stearic acid concentration which if measured under the appropriate labelling conditions would be shown to have a higher specific activity than G_{M2}. Further work will be required to clarify these possibilities.

4.3.4.2 Degradation

Steps involved in ganglioside catabolism have come to light as a result of the complete absence or partial depletion of certain enzymes in patients suffering from a number of related neurological syndromes. Identification of the resultant intermediate, pinpoints a number of steps in ganglioside degradation (Figure 4.5). Deficiencies have been described involving each of the catabolic enzymes except for neuraminidase, the enzyme responsible for removing NANA[84, 85]. This particular enzyme would appear to be localised in the synaptosomal fraction,[86, 87], whereas the other degradative enzymes are considered to be lysosomal in origin[87]. Ohman has, however, questioned whether neuraminidase is indeed a synaptosomal enzyme or lysosomal material associated with or trapped by the synaptosomal fraction[87]. Neuraminidase also has been shown to have different levels of activity in various parts of the brain; cerebral cortex was found to have 10–25 times the activity of white matter and cerebellar cortex had greater activity than cerebral cortex[88]. Unlike the other glycosidases involved in ganglioside degradation neuraminidase activity is age dependent[89]. There is little or no activity in the human foetal brain until 15–20 weeks. The NANA-cleaving enzyme activity increases with development[90, 91] while the activity of β-galactosidase, β-glucosidase and β-N-acetylhexosaminidase, all lysosomal enzymes[87], is independent of age[89]. Ohman and Svennerholm have suggested that the levels of neuraminidase

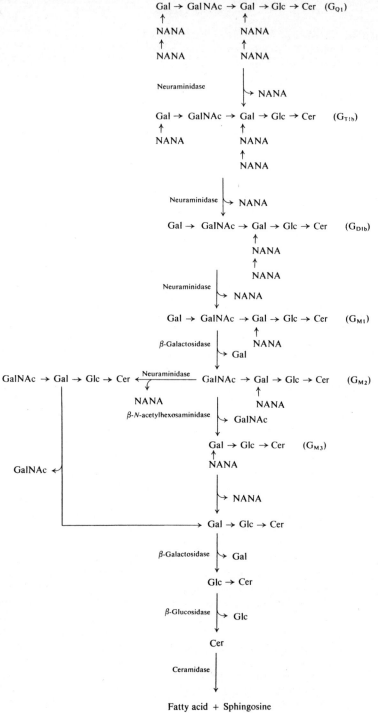

Figure 4.5 Major pathways of ganglioside degradation

may be low in developing brain so that the biosynthetic activities are favoured over the catabolism of gangliosides[89]. Neuraminidase, the key enzyme for removal of NANA, is necessary for the glycosidases to act. A third characteristic that separates neuraminidase from the other glycosides is its lack of substrate specificity. The same enzyme can remove any of the NANA residues from ganglioside[5], or even from non-ganglioside substrates[92,93]. G_{D1b} was found to be hydrolysed more slowly by neuraminidase than other gangliosides, suggesting that a NANA with a $2 \rightarrow 8$ linkage was more resistant to attack than one with $2 \rightarrow 3$ linkage[93]. The remaining glycolytic enzymes are quite substrate-specific. Gatt[94] first demonstrated that there were separate enzymes for cleaving galactose and glucose. Since there is an external and an internal galactose moiety in the G_{M1} molecule separate enzymes may exist for cleavage of each. N-Acetylhexosaminidase is highly specific and is present in at least two isoenzyme forms[95]. Separate enzymes for removal of N-acetylglucosamine and N-acetylgalactosamine have, as yet, not been found[96,97]. The activity of ceramidase, the last step in ganglioside breakdown, like neuraminidase, increases in activity with age[98,99]. However, this reaction is not specific to ganglioside degradation.

4.3.4.3 Turnover

Double labelling experiments carried out with acetate and glucosamine have enabled Holm and Svennerholm[80] to study the biosynthesis and turnover of the sphingosine, sialic acid, stearic acid and hexosamine moieties of the ganglioside. Gangliosides were found to be labelled in proportion to their pool size. After incorporation of isotope Holm and Svennerholm found equal rate of turnover of ganglioside, sphingosine and stearic acid over prolonged time periods. In addition, the drop in specific activities was the same for all four major gangliosides (G_{M1}, G_{D1a}, G_{D1b} and G_{T1}), indicating that the half-lives of the various gangliosides are all of approximately the same duration. The half-lives themselves varied with the age of the animals under study. Adult rat gangliosides had a half-life of ca. 60 days, whereas two-month-old rat gangliosides had a half-life of at most 20 days. Younger animals demonstrated even shorter half-lives.

4.3.5 Glycosphingolipids not containing sialic acid

In brain this group of lipids is characteristically represented by galactosylceramide (cerebroside) and the corresponding sulphatide, galactosylceramide sulphate. Both the synthesis and catabolism of cerebroside and sulphatide have been subjects of exhaustive reviews[100].

Brain also contains cerebroside esters in which the galactose moiety is esterified with intermediate chain-length fatty acids[101]. The subcellular localisation and metabolism of these esters has not been investigated. Another curious compound, psychosine (galactosyl sphingosine), has been claimed to be an intermediate in cerebroside biosynthesis. This material has never been isolated from normal brain possibly because it is highly toxic. However, human brain apparently possesses an enzyme that degrades psychosine; brain tissue from patients with Krabbe's disease is deficient in this enzyme[102]. Similarly, this inborn error is also characterised by a lack of cerebroside galactosidase[103].

4.3.6 Cellular metabolism

4.3.6.1 Whole tissue

In the central nervous system the question of the metabolic relationships between a given neural cell and its cellular neighbours (glia or neurones) has presented formidable difficulties. *In situ* investigations have generally utilised labelled precursor and autoradiography. This approach was used by Torvick and Sidman[104] who studied labelled acetate, serine and cholesterol uptake by brain. They were able to demonstrate the rapid uptake of labelled substances by neurones and the slower label acquisition by the glial cells. By following the distribution of ^{14}C and [7α-^{3}H]-cholesterol over a six month period it was possible to demonstrate the cellular retention of label. Radioactivity present in the oligodendroglial cells, and accompanying myelin, was of a diffuse nature and not at all localised, metabolically stable lipid was absent from nerve cell bodies. This was evidence of the slow turnover of myelin membrane constituents.

The ability of the central nervous system cells to re-utilise and redistribute lipid membrane components has also been described using autoradiography in conjunction with injected precursor–time investigations. By use of intracerebral injection of labelled mevalonic acid into rats Gautheron, Petit and Chevallier[105] beautifully demonstrated first the formation of cholesterol near the point of injection and then the gradual shifting of this labelled cholesterol to other cells surrounding this area. Autoradiography of animals 3 months after injection clearly indicated how dispersed the cholesterol had become— an excellent example of the dynamic interactions of the neural cells with regard to a common membrane component. This transferring and recycling of cholesterol has also been exhibited in peripheral nerve. Rawlins *et al.* have used autoradiography at the light[106] and electron microscopy[107] levels to show how peripheral nerve tissue, degenerated after experimental segmentation of the nerve, contained labelled cholesterol which had previously been present only in the original intact nerve. This work also demonstrated that the lipid used in the generation or repair of cellular membranes was in part *de novo* synthesis and in part contributed by the other cells in the area. As stated earlier, these experiments have a number of limitations. A tool that may be useful for relatively short-term studies of dynamic cellular relationships is that of organ explants. The first to describe a viable system using tissue from normal adult brain were Kiernan and Pettit[108]. Menkes has also introduced a viable brain tissue explant culture system and has described lipid metabolic studies by use of these explants[109-111]. While the explant techniques do work for adult brain, he found that rat brain of a pre-myelination age (three days) was most successful at incorporating [2-^{14}C]mevalonic acid, [1-^{14}C]palmitic acid and [1-^{14}C]stearic acid[109]. The value of such experiments with respect to lipid metabolism can be seen in terms of cholesterol and cerebroside formation. Cholesterol formation from labelled mevalonate was found to proceed at a linear rate for at least 72 h, reinforcing the concept of slow CNS cholesterol synthesis found by others in intact animals[59,112,113]. Cerebroside, with ^{14}C stearic acid as a precursor was not strongly labelled until 24 h after the beginning of the experiment. Detailed examination of the entire pathways of formation of these two lipids which has not been practicable before using an *in vitro* system, may now be possible. However, the profile of incorporation into all the lipids by brain tissue explants using [^{14}C]stearic and [^{14}C]palmitic acids was not exactly what one would believe is necessarily true of the intact brain. Thus when these two fatty acids were added to the system a great deal of labelled triglyceride was formed. The same was found to be true if sodium [1-^{14}C]acetate was added, but not if [1-^{14}C]lignoceric acid was used. Menkes[110,111] Sun and Horrocks[114] found *in vivo* labelled triglyceride formation after intracerebral administration of labelled fatty acid to animals but only of the order of 15% of the label was triglyceride whereas the explants had up to 67% of labelled palmitic acid in triglyceride. Such factors as insufficient phosphorylated nucleotides or lack of other co-factors may be the cause of this accumulation of labelled fatty acid into triglycerides. Of the phospholipids and galactolipids in the brain explants, phosphatidylcholine took up ^{14}C fatty acids faster than other complex lipids. The future use of the brain explant technique looks promising, the method has already been used to culture diseased brain and potential lipid metabolic capabilities have been measured[110,111].

4.3.6.2 *Neural cell-enriched fractions*

The value of biochemical analysis of brain lipids from whole tissue or its subcellular fractions is somewhat limited since no account can then be taken of regional and cellular heterogeneity. Methods have been devised for the analysis of single cells but very special procedures have to be employed. The use of gradient density centrifugation for the separation of neuronal and glial enriched cell fractions offers material for analysis by more conventional methods. At present a variety of isolation techniques are being utilised but refinements must be expected which will bring about a degree of methodological uniformity, as was the case with subcellular fractionation a few years ago.

In Table 4.2, data regarding the lipid composition of neural cell types as assessed by several investigators has been presented. In addition to this, fatty acid analysis has been carried out on the phospholipids[13,115] and the galactosyl ceramides[116] of the glial and neuronal-enriched fractions. No great differences were seen by Hamberger and Svennerholm[13] in the fatty acid compositions of the individual phospholipids of the respective cell types. Raghavan and Kanfer[116], however, found that whereas *ca.* 80% of the myelin galactosyl ceramide fatty acids were hydroxy-fatty acids, 21.5% of the astroglial galactosyl ceramide fatty acid was hydroxy-fatty acids. Since only normal fatty acids were found in the neuronal-enriched galactosyl ceramides, there may be a definite need for hydroxy fatty acids in myelin and not in other neural membranes. It would be of interest to determine the relative turnover rates of these three different galactosyl ceramide 'pools' (myelin, neuronal and glial) and to know if the percentage of hydroxy fatty acid-containing galactosyl ceramide in the glial bears any relationship to the manner in which myelin cerebrosides and sulphatide is being exchanged or metabolised.

Most of the common labelled tracers of lipid biosynthesis have now been used in studies on cell-enriched preparations. With the aid of labelled orthophosphate, phospholipid biosynthesis has been explored by Freysz, Bieth and Mandel[117]. They were able to determine that neuronal phospholipids had a faster turnover rate than glial and that inositol and choline phosphatides turned over faster than other phospholipids. Labelled acetate has been used by Bernsohn and Cohen[118] to assess its uptake into phosphoglyceride fatty acids. When acetate was incubated with 10-day-old rat brain enriched cell preparations, glial cells showed *ca.* 1.5 times the uptake of neuronal cells. At 21 days of age both cells were less active in their uptake of acetate, the neuronal cell preparations being incorporated more than twice as much as the glial cells. The fatty acids labelled were similar in both cell-enriched fractions. Incorporation of [^{14}C]acetate has also been used to show that *in vivo* glial cells are almost entirely responsible for sterol synthesis. The same was found to be true when labelled mevalonic acid was used as a cholesterol precursor *in vitro*[119]. The use of [^{14}C]serine *in vitro* incubations of cell-enriched fractions from rat brain enabled Jones and his colleagues[120] to examine the labelling of gangliosides and other neural lipids. Incorporation of the serine into neuronal lipids was slow, still increasing after 20 h incubation, while uptake into the glial had stabilised after 5 h. Incorporation of labelled serine into ganglioside was most marked in the neuronal preparation while the glial cells preferentially utilised the serine for phospholipid synthesis. When calculated on the basis of uptake per cell the same amount of labelled serine led to the synthesis of *ca.* 13 times as much ganglioside per neurone as per glial cell but about equal amounts of galactolipid were found per cell type; using labelled *N*-acetylneuraminic acid 27 times as much ganglioside was synthesised by nerve cells compared to glial. The reverse was true with regard to mevalonate incorporation; glia incorporated five times as much into sterol as did neurones. This interesting distribution of biosynthetic activity between the two major cell types has also been observed by Radin and his group[121]. Galactosyltransferase was found to be of similar levels of activity in both cell types, but glucosyltransferase needed for ganglioside formation, was concentrated in the neuronal-enriched fraction. Less is known about the distribution of degradative pathways between the cell types.

4.4 EXCHANGE OF BRAIN LIPIDS

4.4.1 Cholesterol exchange

Although it had been demonstrated that labelled lipid remained for long periods in myelin after injection into developing animals it was surprising to find that radioactivity was readily incorporated into what was thought to be inert myelin of adult animals[122]. In addition, it was unexpectedly observed that ^{14}C-cholesterol remained for as long in mitochondrial membranes as in myelin[2, 3] and re-utilisation of labelled sterol was suggested as one of the possible explanations of the results, for other evidence suggested that the mitochondrial membrane undergoes steady metabolism. Smith and Eng[123] also showed that different myelin lipids turned over at different rates (see p. 166) and the polyphosphoinositides of myelin were demonstrated by Eichberg and Dawson[124] to be metabolically quite active. It was next found that desmosterol (24-dehydrocholesterol) could be detected in small amounts in myelin from young animals but unexpectedly the cholesterol precursor did not remain in the myelin of mature animals. After injection of anti-cholesterolaemic drugs (e.g. Triparanol or AY-9944) into developing animals, much of the total sterol appeared in the brain as either 24-dehydro- or 7-dehydrocholesterol, whereas on discontinuing drug treatment, these sterol precursors were soon replaced in myelin and subcellular fractions by cholesterol[125]. Thus when AY-9944 was injected into 5 day-old rats followed by ^{14}C acetate given intraperitoneally, myelin 7-dehydrocholesterol was extensively labelled when the rats were 12 days old. From then onwards the precursor disappeared and an equivalent of radioactive cholesterol was found subsequently in the myelin, indicating a possible localised conversion of 7-dehydrocholesterol into cholesterol. This suggested that contrary to earlier views, myelin sterols in the developing brain are metabolically active and that either the myelin sterols could undergo exchange or that the cholesterol precursors could be reduced to cholesterol within the developing sheath. However, the 7-dehydrocholesterol reductase responsible for the conversion was present in brain microsomes and not in purified myelin[126]. These observations, therefore, suggest that in the developing rat brain the 7-dehydrocholesterol can migrate from myelin, become reduced by microsomal reductase and the resultant cholesterol molecules be re-incorporated into myelin. In elegant autoradiographic studies Hedley-Whyte, Rawlins, Salpeter and Uzman[127] demonstrated probable re-utilisation of [1,2-^{3}H]cholesterol incorporated into developing mouse sciatic nerve. 46 days after injecting radioactive cholesterol, grain counts showed uniform labelling on the outer, middle and inner myelin lamellae but radioactivity was detected in Schwann cell cytoplasm. It was proposed that the myelin sheath and its formative cell formed a closed system in which continuous exchange was possible. Later work demonstrated re-incorporation of labelled cholesterol released after Wallerian degeneration. In extension of this research to adult animals [^{14}C]acetate was injected into 16 day-old animals and the relative cholesterol specific activities of cerebral subcellular fractions were determined[128]. At different times (10 min, 6 and 7 months) after injection, there was uniform specific activity of cholesterol in all fractions. In another series of experiments[128] adult rats were given intravenous injections of [4-^{14}C]cholesterol and killed at intervals. 10 Days later, the cholesterol specific activity of myelin had increased to that of the sterol in all other fractions, a finding completely unexpected if adult myelin were inert. Similar results have been obtained in adult animals after labelling with 7-dehydrocholesterol.

4.4.1.1 Experiments in vitro

In order to test further the possibility of exchange of sterol molecules, myelin labelled with [^{14}C]lipid or with dehydrocholesterol was incubated at 37 °C for

30 min with whole of rat brain homogenate or with isolated microsomes labelled with [^{14}C]protein. After incubation myelin and microsomes were re-isolated. Contamination was checked for by determining [^{14}C]protein and cyclic 2′ 3′-nucleotide 3′-phosphohydrolase in the separate fractions. Labelled cholesterol and 7-dehydrocholesterol were found to undergo up to 10% exchange between myelin and microsomes. Phospholipid and cerebroside also underwent limited exchange from one membrane to another. Graham and Green[129] had already demonstrated a 57% exchange of cholesterol from adult myelin with plasma lipoproteins after 24 h incubation at 37°C. It is also well established that cholesterol of erythrocyte stroma undergoes ready exchange with plasma sterols. In our experiments, enhanced exchange occurred with whole-brain suspensions rather than myelin plus isolated microsomes. It has been demonstrated that exchange *in vitro* requires a heat-labile supernatant factor, identified as a lipoprotein carrier[29, 130] for the exchange of phospholipid molecules between rat liver mitochondria and microsomes[131]; and a similar protein carrier may be important in the exchange of cerebral lipid (p. 153).

4.4.2 Phospholipid exchange

Despite the fact that synthesis of phospholipids occurs mainly in brain microsomes, extensive labelling of phospholipid is found in mitochondrial and synaptosomal membranes soon after injection of radioactive phosphorus into intact animals[132-134]. It seems likely that as for cholesterol it is possible that phospholipid molecules readily exchange between the various cerebral membranes leading to eventual isotopic equilibrium as occurs in liver[22, 131, 135]. Jungalwala and Dawson[131] have examined incorporation of [^{32}P]phosphate U-[^{14}C]glycerol and [2-^{14}C]ethanolamine into microsomal and a highly purified myelin fraction of adult rat brain. Initially, the specific activity of microsomal phospholipids was higher than that of myelin but, by 20 days after intracerebral injection, the reduced specific activities were similar. Persistence of labelling occurred after 10-day-old rats had been injected, thus supporting the previous concept that two metabolic pools exist in myelin, one a rapidly-exchanging pool and the other a more persistent and slowly-turning-over part which can be significantly labelled in the developing animal. This recent work indicates that the pool of exchangeable myelin lipid is larger and more accessible than had been previously assumed. In the brain *in vitro* studies[134] show [^{32}P]phospholipid can exchange between isolated microsomal and mitochondrial fractions of guinea-pig brain when incubated under suitable conditions at pH 7.4. Miller and Dawson[134] found no evidence of exchange between [^{32}P]microsomes and myelin and only slow transfer into nerve endings. All of the isolated individual synaptosomal membranes were capable of acquiring phospholipid on incubation with a [^{32}P]labelled brain supernatant fraction although a greater percentage was exchanged by the nerve ending mitochondria. In the exchange reaction between [^{32}P]brain microsome and unlabelled mitochondria similar rates (*ca.* 50% in 1 h) were found for all phospholipids. Exchange *in vitro* was inhibited by lysolecithin or myelin and transfer from microsomes was greatly stimulated by a non-dialysable heat labile macromolecular component of the brain soluble fraction. In the experiments of Miller and Dawson possible contamination by labelled microsomes was assessed by measuring NADPH- cytochrome C reductase activity and corrections were made. Evidence for slight exchange between myelin and microsomal phospholipid reported by Banik and Davison[126] may perhaps be due to impurities in the myelin fractions for similar corrections were not made in this study.

Exchange of myelin lipid will also depend on the binding of lipid to myelin protein. Braun and Radin[137] have demonstrated special interaction between sulphatide, phosphatidylserine and phosphatidylinositol with proteolipid protein which may be compared to the weaker binding of the non-ionic lipids cholesterol and cerebrosides. It has also been shown that polyphosphoinositides and lecithin readily combine with the myelin basic protein[138]. These myelin components are all known to undergo rapid turnover and one may, therefore, be combined in some form in the myelin sheath. Despite the ready exchange of lipid within the nervous system it appears to be difficult to modify the brain lipids by altering whole brain metabolism. Thus no change in brain lipids was noted in severely-starved rats by Dobbing[139] and only minor changes were seen in brain fatty acids in chronic deficiency of essential polyunsaturated fatty acids[140].

The possibility of exchange is one which has therefore to be considered in interpreting experiments attempting to localise specific changes in phospholipid to areas and organelles within the nervous system[141]. Turnover experiments are also obviously complicated by exchange and since released phospholipid may be catabolised by cytoplasmic phospholipase but since cholesterol is not easily degraded, this may account for the apparently different turnover rates found for these lipids in myelin.

4.5 LIPIDS AND TRANSMISSION IN THE NERVOUS SYSTEM

Study of lipid metabolism in relation to transmission processes has been concentrated on examining the possible involvement of lipids at the molecular level in ion transport. There is good reason to believe that release of transmitter acetylcholine induces conformational changes of the post-synaptic cholinergic receptor protein. Possibly as many as 40 000 ions move in each direction as a result of liberation of one molecule of acetylcholine[142]. One way in which acetylcholine may act was discovered by Hokin and Hokin[143] who found that the neurotransmitter stimulated incorporation of ^{32}P into avian salt gland and brain lipids *in vitro*. Further work demonstrated that these effects could be attributed to the enhanced exchange of [^{32}P]orthophosphate with phosphatidic acid and phosphatidylinositol (Table 4.6). Larrabee and his colleagues[144] also found that electrical stimulation of sympathetic ganglia causes an increased incorporation of [^{32}P]orthophosphate into phosphatidylinositol. In this preparation acetylcholine stimulated incorporation into both phosphatidic acid and phosphatidylinositol. Since stimulation is abolished by freezing it was suggested that intact subcellular structures, such as the nerve endings, were implicated[145]. The effect of phosphatidylinositol metabolism appears to be post-synaptic and associated with cellular components of the neurone other than synaptic membranes, for Hokin[146] showed (using radioautography of acetylcholine treated sympathetic ganglia) that the inositol phospholipid labelled after stimulation was distributed throughout the postsynaptic neurone. No significant effect of conducted impulses on labelling of inositol levels was found in vagus, sympathetic or phrenic nerve trunks or non-synaptic ganglia of the vagus[144, 147] although possibly some change of triphosphoinositide metabolism may occur as a result of electrical stimulation[148]. Certain experiments[149] indicated that enhanced incorporation into phosphatidylinositol following stimulation is localised at the level of the acetylcholine-receptor proteolipid. Lapetina and Michell[141] measured acetylcholine-stimulated incorporation of ^{32}P into phosphatidylinositol of rat cerebral cortical subcellular fractions. Incorporation into phosphatidylinositol of the nuclear, microsomal and synaptic-vesicular fraction was greatest, whereas myelin was least labelled. Stimulation by acetylcholine had a uniform effect on uptake of ^{32}P of all subcellular fractions. This agrees with observations of Hokin[146] who showed a lack of specific autoradiographic localisation

Table 4.6 Labelling of phosphatidylinositol in rat cerebral cortex slices[141]

Addition of acetylcholine (10 mM) and eserine (0.1 mM)	Radioactivity of phosphatidylinositol	
	Control	Experimental
10 min stimulation (8 expts)	100 ± 19	138 ± 14
120 min stimulation (6 expts)	100 ± 14	206 ± 31

of labelled inositol phospholipid in the post-synaptic neurone after stimulation. In contrast, using guinea-pig cortical subcellular fractions[150, 151] it was found that added acetylcholine preferentially stimulated [^{32}P]phosphate uptake into cholinergic nerve ending phosphatidic acid and phosphatidylinositol. The effect was not significant in other subcellular fractions, including a denser, possibly 'non-cholinergic', nerve ending fraction. Labelling of phosphatidylinositol phosphate and the diphosphate was inhibited by acetylcholine suggesting possible degradation of polyphosphoinositide following acetylcholine stimulation. It has been suggested[152] that probably Pi is incorporated directly by diglyceride kinase into pre-existing diglyceride (formed by the action of phospholipase C on phosphoinositides). This process may be stimulated by acetylcholine releasing endogenous diglyceride. Acetylcholine does not, however, stimulate diglyceride kinase activity[153]. There is a great deal of evidence to show that calcium ions, besides, specific phospholipids, are important components of the receptor macromolecules in the cerebral cortex and other tissues when synaptic transmission is affected by drug action. The demonstration of relatively rapid metabolism of di- and tri-phosphoinositide in the myelin fraction isolatable from brain and nerve, suggests that these polyphosphoinositides may be associated in some way with ion transport, perhaps at the axolemma. It has been suggested that enzyme hydrolysis of calcium phosphatidylinositol phosphates could be a sensitive mechanism for the local transient release of calcium ions from the lipid, so inducing a conformational change of membrane protein. Hawthorne postulated that this process could be initiated by depolarisation allowing axoplasmic triphosphoinositide phosphomonoesterase to attack the membrane polyphosphoinositide[154]. The phosphoinositides may thus function by regulating the flow of sodium ions through the sodium channels of neuronal membranes, thus controlling action potential[155-157]. These various experiments strongly suggest an association between nervous transmission, inositol phospholipid metabolism and ionic flux. The relation of the system with the acetylcholine-receptor proteolipid[149] is another fascinating area for study.

4.6 DEVELOPING AND AGEING

4.6.1 Development

In discussing the development of the central nervous system the one overriding event is myelination. Myelin contains the bulk of the lipid of the adult brain and its formation is associated with the maximal rate of lipid synthesis. Although it was thought that myelin had a constant composition throughout ontogeny, this view has recently been modified. Horrocks and his colleagues[158] first noted that crude myelin isolated from developing mouse brain was different in composition from that of adult mice. Similarly in 10–20-day-old rat brain, Cuzner and Davison[159] found that isolated myelin contained a higher molar proportion of phospholipid to cholesterol and there was a relative deficiency in cerebroside. Examination of myelin from rabbit brain and spinal cord at a number of different ages during development has added additional evidence to the lipid transitions of myelin during its maturation[160]. Such data could be ascribed to extensive contamination of myelin fractions, for example, by microsomal fragments present in an unusually high proportion in developing brain, or the changes in lipid composition could be due to alterations in the composition of the developing myelin sheath. Evidence for the latter possibility came from observations by Banik, Blunt and Davison[161]. It was found that although myelin rings could be seen by light microscopy in 10-day-old kitten optic nerve stained with sudan black or luxol fast blue, the myelin could not be stained with osmium tetroxide until optic nerves from animals at least 25 days old were examined. Wolman[162] had also observed that osmiophilia of rat brain, spinal cord and sciatic nerve myelin sheath does not coincide with the first appearance of the sudanophilic myelin rings. Electron microscopy confirmed that myelin was present in the optic nerves of 10-day-old kittens and that myelination was well established by 16 days postnatally. Lipid analysis revealed that the proportion of the typical myelin-lipid cerebroside was much less in the developing optic tract than was expected from the accumulation of cholesterol and from the amount of histologically stainable myelin. Only traces of galactocerebrosides were found to be present in the 10–16-day-old kitten optic nerve and even in the 28-day-old animal the molar proportion of cerebroside to cholesterol was less than a fifth of that found in the adult cat. As development proceeds, the high proportion of myelin phospholipid lessens and phosphoglyceride fatty acid chain length increases[161]. Thus, it appeared that the lipid composition of 'early' myelin (i.e. that formed during the first period of myelinogenesis) was more similar to that of the plasma and other cytomembranes than to the unique lipid composition of myelin. These data on the changing lipid ratios during development together with present knowledge of alterations in the protein

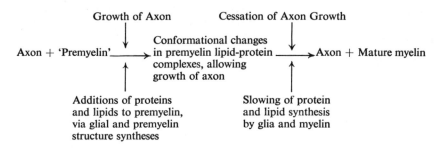

Figure 4.6 Accommodation of axon growth during deposition of myelin[7]

components of a developing myelin form the basis of a workable mechanism of myelination. It has been suggested by Braun and Radin[137] that as the developing myelin takes on additional lipid components, a conformational change of the myelin proteins will also occur. In developing a scheme, however, one must also take into account the influence of the axon as well as changes in the oligodendroglial membrane prior to compact myelin synthesis.

The scheme of Johnston and Roots[7] (Figure 4.6) accounts for these chemical changes occurring during myelination.

4.6.2 Ageing

Well defined degenerative changes are known to occur in the ageing brain. For example, neurofibrillary tangles and some accumulation of insoluble lipofuscin particles may be seen in intellectually well-preserved individuals. In dementia these changes may be exaggerated and additional pathological alterations also appear. The cause of memory loss and mental deterioration is unknown but is presumably associated with the accumulation of lipofuscin, senile plaques, neurofibrillary tangles and the characteristic granulo-vacuolar degeneration. The granulo-vacuolar degeneration is almost diagnostic of senile dementia[163]. This characteristic change affects selected hippocampal neurones which appear to contain intracytoplasmic vacuoles with a dense core. They are particularly prominent in neurones containing neurofibrillary tangles and bear some resemblance to lipofuscin[164]. It has been suggested that some of these degenerative changes are linked with impaired autodigestion of abnormal microtubules[164]. A failure of cerebral auto-oxidation may be another of the causes of the pathological changes[165, 166].

4.6.2.1 Lipofuscin

Lipofuscin is a dense granule that characteristically possesses an autofluorescent pigment. The granule, isolated from heart, contains a relatively high proportion of lipid (25%); the lipid fraction is *ca.* 75% phospholipid[167]. Lipofuscin has also been isolated from human brain[168].

The origin of lipofuscin has for long been debated. As judged by the complement of typical lysosomal hydrolases a portion of neural lipofuscin is derived from lysosomes[169]. Since lipofuscin is rich in phospholipids, this transformation of lysosomes may be brought about, in part, by peroxidation reactions that oxidise the polyunsaturated fatty acids, characteristically found in phospholipids. Labilisation of lysosomal membranes[170, 171] and the increase in lipofuscin-like pigment in tocopherol deficiency[172-174], leads us to suggest that a release of acid hydrolases (or an increase in the permeability of lysosomes) occurs during the transformation of lysosomes to lipofuscin. Another source of lipofuscin may be lipoperoxidised microsomes[175]; this material may be engulfed but not dispersed by neural lysosomes. The resultant particle may be identical to lipofuscin.

4.6.2.2 Lysosomes and lipid peroxidation

In the ageing brain there is some evidence for increase of soluble lysosomal enzyme activity[99] and lysosomes of adult neurones appear to be more permeable than those from young animals[176]. Cerebral ischaemia also results in an increase in the cytosol activity of certain acid hydrolases[177]. Ten weeks after transient anoxia, the level of brain ganglioside acetylneuraminic acid but not cholesterol was reduced; by subjective assessment, the animals displayed behavioural deficiencies[178]. This unsubstantial report is of interest because a decrease in ganglioside is one of the only consistent biochemical alterations in Alzheimer's disease[179] and senile dementia[180]. Since gangliosides are characteristically found in synaptosomes and as it has been claimed that neuronal loss progressively increases with ageing in brain[181], the decrease in gangliosides in the dementias can be viewed as a reflection of neuronal loss. It is pertinent that like microsomes synaptosomes and synaptic plasma membranes contain a high proportion of long-chain unsaturated fatty acids[182, 183] and an NADPH-dependent enzymic system for lipid peroxidation[184]. Therefore, the decrease in ganglioside seen in the dementias might be, in part, due either to cerebral anoxia or lipoperoxidation causing the release of sialidase (neuraminidase). This acid hydrolase has been identified in brain lysosomes[185] and in the membranes of synaptosomes[86]. In vitamin E deficiency one might expect the levels of brain gangliosides to fall. However, in nutritional encephalomalacia, produced in chicks fed a diet deficient in a-tocopherol, higher brain contents of gangliosides have been observed[185]. The data discussed above attempt to establish a link, albeit tenuous, between brain lipids, the histological changes seen in aged human brain, and at least one underlying degenerative process (auto-oxidation) that might be important in ageing brain. Since most animal species have a shorter life span than man most of the research on the neurochemistry of ageing has been carried out on laboratory animals. This is inapplicable because the human brain appears to be unique in accumulating both senile plaques and neurofibrillary tangles. Furthermore, human brain differs from rodent brain in loss of brain weight and some constituents with advancing age[187]. It, therefore, seems likely that studies with animals may be of limited importance in establishing the nature of the underlying process in ageing human brain.

4.7 BRAIN DISEASE

The primary lipid abnormalities that affect the brain fall generally into the categories of lipid accumulation (lipidoses) or degeneration (for example as seen in the aminoacidurias and leucodystrophies). The lipidoses and other metabolic diseases that affect the nervous system have recently been reviewed. In addition, alterations in the lipid composition of the central nervous system may result from a primary pathological process, a good example being the loss of ganglioside in Jacob–Creuzfeld disease or the degenerative changes in infarction of brain tissue.

4.7.1 Lipidoses

It has become increasingly difficult to distinguish one lipidosis from another on purely clinical grounds. Moreover, differentiation on purely pathological grounds is equally difficult due to the variety of histological changes variously involving nervous system and viscera. With these difficulties in mind we have classified the lipidoses according to their lipid composition (Table 4.7) rather than using a clinical or even enzyme deficiency basis. The detailed chemistry and pathology of the individual lipidoses is not included. Since most of these lipid storage diseases have been described in domestic animals and since their usefulness is enhanced because of their potential in selective breeding and laboratory housing some examples are also shown. Because of species variation, as well as general lack of knowledge concerning veterinary neurology, it is necessary to approach classification of these lipid storage diseases cautiously, many which apparently resemble their human counterparts may not be identical diseases. Their value is therefore in providing a model of a general type of human disease.

The discovery of specific enzyme deficiencies in many of the lipidoses has facilitated detection of homozygotes *in utero* following cultivation of amniotic foetal cells. Changes in enzyme activity have also been detected in skin fibroblasts and leucocytes. Current work is directed to attempting to supply the missing enzyme as a pure protein or by organ transplant (as for example in Fabry's disease).

4.7.2 Aminoacidurias

Defects in amino acid metabolism may also lead to abnormalities in brain lipid composition such as lack of myelin lipids[191]. In some cases, there is in

Table 4.7 Chemical classification of lipidoses (diseases affecting man and some cases in domestic animals)

General lipid class:	Sphingolipid						
	Ganglioside				Cerebroside		
	Gm_1	Gm_2					
Lipid stored	Gm_1	Tay Sachs	Sandhoff's	Juvenile gangliosidosis	Sulphatide	Galacto cerebroside	Gluco cerebroside
Disease	generalised and juvenile gangliosidosis		disease		Metachromatic leucodystrophy	Globoid cell leucodystrophy (Krabbe's)	Gaucher's disease
Enzyme deficiency	β-Galactosidase	Hexosaminidase A	B + A	A (partial)	Aryl sulphatase A	β-galactosidase	Glucocerebrosidase
Also affects	Cat	Dog			Mink	Dog, cat	Sheep (viscera only)

General lipid class:	Sphingolipid			Neutral Lipids	
	Ceramides			Cholesterol	
Lipid stored	Lactosyl ceramide	Ceramide trihexoside	Sphingomyelin	Cholesteryl ester	Cholesterol
Disease	Lactosyl ceramide	Fabry's disease	Niemann-Pick	Sudanophilic leucodystrophy	Xanthomatosis
Enzyme deficiency	galactosyl hydrolase	Ceramide trihexoside α-galactosidase	Sphingomyelinase	Phytanic acid Refsum's disease	Phytanic acid α-hydroxylase
Also affects			Cat		

addition accumulation of neutral lipid. In sheep, copper deficiency is associated with ataxia of the newborn lamb (Swayback) and it has been proposed by Howell and Davison[192] that changes in brain cytochrome oxidase activity may account for the amyelination of the central nervous system in this condition. Several mutants of mice have shown abnormalities of normal lipid accumulation. Both the 'Jimpy' and 'Quaking' mice show a notable lack of brain galactolipid[193, 194]. It would appear, however, that in these mutants the primary defect is not in lipid metabolism but in protein synthesis[195]. Another animal brain disorder quite similar to Quaking and Jimpy mouse has been described in Landrace pigs[196].

4.7.3 Demyelinating diseases

It has previously been suggested that the brain of patients with multiple sclerosis is abnormal in lipid composition before the onset of the disease. There are reports that the apparently normal areas of white matter from such patients are deficient not only in basic protein[197] but in galactolipids and that there are alterations in lipid fatty acid composition. However, these changes in apparently normal tissue are probably due to microscopic areas of demyelination for enzymic changes and appearance of cholesterol esters have now been found to accompany other differences in lipid composition[198].

Various hypotheses involving lipids have been advanced to explain the appearance of plaques in the demyelination diseases. It may be more than fortuitous that 80% of plaques can be traced to veins and that lymphocytes often associate with plaques. In some patients with multiple sclerosis (M.S.), just before the exacerbation of the disease, there is a considerable increase in plasma cholesterol: lecithin acyltransferase activity leading to the release of lysolecithin[199]. It is proposed (see Figure 4.7) that formation of the surface-

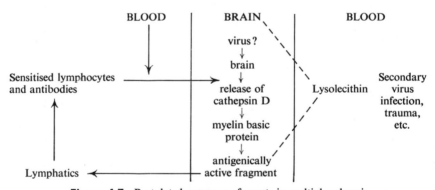

Figure 4.7 Postulated sequence of events in multiple sclerosis

actant lysolecithin, possibly as a result of a secondary virus infection, could account for the attack on the myelin membrane, but there is little evidence for raised phospholipase activity and lysolecithin formation in areas of active demyelination. The elevation in acid proteinase or cathepsin D at the edge of

M.S. plaques, is of special significance because the same enzyme degrades basic protein to an antigenically active fragment.

4.8 CONCLUSION

Although much of the chemistry and the main pathways of lipid metabolism in the brain have been well established, some quite unexpected discoveries have revitalised interest in this field. For example, the demonstration that sphingosine could act as a precursor of plasmlogen was completely unanticipated; the possibility, suggested by Popják and his colleagues, of another pathway in the catabolism of cholesterol intermediates opens up a new and exciting field of enquiry. Our complacency was also shaken by experimental evidence for exchange of whole lipid molecules between one membrane and another, making it necessary to completely re-evaluate past concepts of membrane structure and metabolism. There is still, however, much room for progress. Thus, we need to learn more about the role in the central nervous system of long-chain fatty acids and their relation to the biochemistry of prostaglandins. Moreover, the suggestion that polyunsaturated fatty acids are essential for neural development and function is certainly worthy of further study. Since a high proportion of enzymes are membrane bound, lipid molecules may serve to provide a suitable milieu for physio-chemical interaction or even act as participants in catalytic reactions and little is known about these possibilties in the brain. Finally, the application of biochemistry to neuropathology is a new and intriguing area for future work. This, in our view, deserves a great deal more attention for the combination of detailed lipid analyses together with studies on enzymes and gel electrophoresis of proteins from diseased tissue is likely to lead to very significant advances in the next decade.

Acknowledgements

We are grateful to the National Institutes of Health (Grant No. 1F02 NS-51982-01), Bethesda, Maryland, U.S.A., and the Wellcome Foundation for support.

References

1. Folch, J. P., Lees, M. and Sloane-Stanley, G. H. (1957). *J. Biol. Chem.*, **226**, 497
2. Cuzner, M. L., Davison, A. N. and Gregson, N. A. (1966). *Biochem. J.*, **101**, 618
3. Khan, A. A. and Folch, J. P. (1967). *J. Neurochem.*, **14**, 1099
4. Fried, R. and Menzies, W. (1971). *Methods in Neuro chemistry*, Vol. 1, 286 (New York: Plenum Press)
5. Johnson, A. R. and Davenport, J. B. (1971). *Biochemistry and Methodology of Lipids* (New York: Wiley Interscience)
6. Mokrasch, L. C. (1971). *Methods in Neurochemistry*, Vol. 1, 1 (New York: Plenum Press)
7. Johnston, P. V. and Roots, B. I. (1972). *Nerve Membranes* (Oxford: Pergamon Press)
8. Rouser, G. and Yamamoto, A. (1969). *Handbook of Neurochemistry*, 121 (A. Lajtha, editor) (New York: Plenum Press)

9. Norton, W. T., Poduslo, S. E. and Suzuki, K. (1966). *J. Neuropath. Exp. Neurol.*, **25**, 582
10. Patterson, D. S. P., Swensey, D. and Hebert, C. N. (1971). *J. Neurochem.*, **18**, 2027
11. Howell, J. McC., Davison, A. N. and Oxberry, J. M. (1964). *Res. Vet. Science*, **5**, 376
12. Norton, W. T. and Poduslo, S. E. (1971). *J. Lipid Res.*, **12**, 84
13. Hamberger, A. and Svennerholm, L. (1971). *J. Neurochem.*, **18**, 1821
14. Poduslo, S. E. and Norton, W. T. (1972). *J. Neurochem.*, **19**, 727
15. Norton, W. T. and Autilio, L. A. (1966). *J. Neurochem.*, **13**, 213
16. Lapetina, E. G., Soto, E. Y. and De Robertis, E. (1968). *J. Neurochem.*, **15**, 437
17. Sun, G. Y. and Sun, A. Y. (1972). *Biochim. Biophys. Acta*, **280**, 306
18. McIlwain, H. and Bachelard, H. S. (1971). *Biochemistry and the Central Nervous System*, 323 (Edinburgh: Churchill Livingstone)
19. Porcellati, G., Biasion, M. G. and Pirotta, M. (1970). *Lipids*, **5**, 734
20. Freysz, L., Lastennet, Z. and Mandel, P. (1972). *J. Neurochem.*, **19**, 2599
21. Miller, E. K. and Dawson, R. M. C. (1972). *Biochem. J.*, **126**, 805
22. McMurray, W. C. and Dawson, R. M. C. (1969). *Biochem. J.*, **112**, 91
23. Lunt, G. G. and Lapetina, E. G. (1970). *Brain Res.*, **18**, 451
24. Abdel-Lalif, A. A. (1971). *Abstr. Commun., Third Int. Meet., Int. Soc. Neurochem., Budapest*, 247
25. Miani, N. (1964). *Progr. Brain Res.*, **13**, 115
26. Ansell, G. B. (1971). *Chemistry and Brain Development*, 63 (R. Paoletti and A. N. Davison, editors) (New York-London: Plenum Press)
27. Bell, O. E., Blank, M. L. and Snyder, F. (1971). *Biochim. Biophys. Acta*, **231**, 579
28. Schmid, H. H. O., Muramatsu, T. and Su, K. L. (1972). *Biochim. Biophys. Acta*, **270**, 317
29. Stoffel, W., Le Kim, D. and Heyn, G. (1968). *HoppeSeylers Z. Physiol. Chem.*, **35**, 875
30. Kishimoto, Y. and Radin, N. S. (1966). *Lipids*, **1**, 47
31. Green, E. and Almann, D. W. (1968). *Metabolic Pathways*, **2**, 1
32. Bowen, D. H. and Radin, N. S. (1968). *Advan. Lipid Res.*, **6**, 255
33. D'Adamo, Jr. A. F. (1970). *Handbook of Neurochemistry*, Vol. 3, 525 (A. Lajtha, editor) (New York: Plenum Press)
34. Su, K. L. and Schmid, H. H. O. (1972). *J. Lipid. Res.*, **13**, 452
35. Su, K. L. (1972). *Biochem. Biophys. Res. Commun.*, **48**, 94
36. Takahashi, T. and Schmid, H. H. O. (1970). *Chem. Phys. Lipids*, **4**, 243
37. Blank, M. L. and Snyder, F. (1970). *Lipids*, **5**, 337
38. Aeberhard, E. and Menkes, J. H. (1968). *J. Biol. Chem.*, **243**, 3834
39. Boone, S. C. and Wakil, S. J. (1971). *Biochemistry*, **9**, 1470
40. Baumann, N. A., Jacque, C. M., Pollet, S. A. and Harpin, M. L. (1968). *Europ. J. Biochem.*, **4**, 340
41. Gerstl, B., Eng, L. F. and Tavaststjerna, M. (1970). *J. Neurochem.*, **17**, 677
42. Lane, M. D. and Moss, J. (1971). *Metabolic Pathways*, **5**, 23
43. MacDonald, R. C. (1966). *Ph.D. Thesis*, University of California
44. Morell, P. and Radin, N. S. (1970). *J. Biol. Chem.*, **245**, 342
45. Rolleston, F. S. and Newsholme, E. A. (1967). *Biochem. J.*, **104**, 519
46. Williamson, D. H. and Buckley, B. M. (1973). *Inborn Errors of Metabolism, Symposium of Developmental Biochemistry*. In press
47. Levis, G. M. and Mead, J. F. (1964). *J. Biol. Chem.*, **239**, 77
48. Mead, J. F. and Hare, R. S. (1971). *Biochim. Biophys. Res. Commun.*, **45**, 1451
49. Poole, A. R. (1970). *Biochem. J.*, **118**, 543
50. Bazan, N. G. and Joel, C. D. (1970). *J. Lipid. Res.*, **11**, 42
51. Rouser, G., Galli, C. and Kritchevsky, G. (1965). *J. Amer. Oil Chem. Soc.*, **42**, 404
52. Bazan, N. G. (1970). *Biochim. Biophys. Acta*, **218**, 1
53. Bazán, Jr. N. G., de Bazán, Haydée, E. P., Kennedy, W. G. and Joel, C. D. (1971). *J. Neurochem.*, **18**, 1387
54. Ramsey, R. B., Jones, J. P., Nagvi, S. H. M. and Nicholas, H. J. (1971). *Lipids*, **6**, 154
55. Linn, T. C. (1967). *J. Biol. Chem.*, **242**, 990
56. Kandutsch, A. A. and Saucier, S. E. (1969). *Arch. Biochem. Biophys.*, **135**, 201
57. Paoletti, R., Grossi-Paoletti, E. and Fumagalli, R. (1969). *Handbook of Neurochemistry*, Vol. 1, 195 (A. Lajtha, editor) (New York-London: Plenum Press)
58. Popják, G. (1971). *Harvey Lect.*, **65**, 127

59. Ramsey, R. B., Jones, J. P. and Nicholas, H. J. (1971). *J. Neurochem.*, **18**, 1485
60. Chesterton, C. J. (1968). *J. Biol. Chem.*, **243**, 1147
61. Shah, S. N. (1972). *FEBS Letters*, **20**, 75
62. Scallen, T. J., Srikantaiah, M. V., Skidlant, H. B. and Hansbury, E. (1972). *FEBS Letters*, **25**, 227
63. Ramsey, R. B., Aexel, R. T., Jones, J. P. and Nicholas, H. J. (1972). *J. Biol. Chem.*, **247**, 3471
64. Weiss, J. F., Galli, C. and Paoletti, E. G. (1968). *J. Neurochem.*, **15**, 563
65. Ramsey, R. B., Aexel, R. T. and Nicholas, H. J. (1971). *J. Biol. Chem.*, **246**, 6393
66. Holstein, J. J., Fish, W. A. and Stokes, W. M. (1966). *J. Lipid. Res.*, **7**, 634
67. Paoletti, R., Galli, C., Paoletti, E. G., Fiecchi, A. and Scala, A. (1971). *Lipids*, **6**, 134
68. Scallen, T. J., Candie, R. M. and Schroepfer, G. J., Jr. (1962). *J. Neurochem.*, **9**, 99
69. Hinse, C. M., McKean, C. M. and Shah, S. N. (1970). *Fed. Proc.*, **29**, 887
70. Doncher, N. Ch. (1971). *Compt. Rend. Acad. Bulg. Sci.*, **24**, 977
71. Alling, C. and Svennerholm, L. (1969). *J. Neurochem.*, **16**, 751
72. Werb, Z. and Cohn, Z. A. (1972). *J. Exp. Med.*, **135**, 21
73. Eto, Y. and Suzuki, K. (1971). *Biochim. Biophys. Acta*, **176**, 146
74. Eto, Y. and Suzuki, K. (1972). *J. Neurochem.*, **19**, 117
75. Svennerholm, L. (1970). *Comp. Biochem.*, **18**, 201
76. Svennerholm, L. (1970). *Handbook of Neurochemistry*, Vol. 3, 425 (A. Lajtha, editor) (New York-London: Plenum Press)
77. Arce, A., Maccioni, H. J. and Caputto, R. (1971). *Biochem. J.*, **121**, 483
78. Suzuki, K. and Korey, S. R. (1964). *J. Neurochem.*, **11**, 647
79. Kanfer, J. and Ellis, D. (1971). *Lipids*, **6**, 959
80. Holm, M. and Svennerholm, L. (1972). *J. Neurochem.*, **19**, 609
81. Arce, A., Maccioni, H. J. and Caputto, R. (1966). *Arch. Biochem. Biophys.*, **116**, 152
82. Kaufman, B., Basu, S. and Roseman, S. (1966). *Proc. 3rd Int. Symp., Cerebral Sphingolipidosis*, 193 (S. M. Aronson and B. N. Volk, editors) (New York: Pergamon Press Inc.)
83. Suzuki, K. and Chen, G. C. (1967). *J. Lipid. Res.*, **8**, 105
84. Hultberg, B., Ockerman, P.A. and Sjoblad, S. (1972). *Europ. Neurol.*, **7**, 101
85. Brady, R. O. (1972). *Handbook of Neurochemistry*, Vol. 7, 33 (A. Lajtha, editor) (New York-London: Plenum Press)
86. Schengrund, C. L. and Rosenberg, A. (1970). *J. Biol. Chem.*, **245**, 6191
87. Ohman, R. (1971). *J. Neurochem.*, **18**, 89
88. Ohman, R. (1971). *J. Neurochem.*, **18**, 531
89. Ohman, R. and Svennerholm, L. (1971). *J. Neurochem.*, **18**, 79
90. Roukema, P. A., van Den Eijnden, D. H., Heijlman, D. N. and van der Berg, G. (1970) *FEBS Letters*, **9**, 267
91. Schengrund, C. L. and Rosenberg, A. (1971). *Biochemistry*, **10**, 2424
92. Tettamanti, G. and Zambotti, V. (1968). *Enzymologia*, **31**, 61
93. Ohman, R., Rosenberg, A. and Svennerholm, L. (1970). *Biochemistry*, **9**, 3774
94. Gatt, S. (1967). *Inborn Disorders of Sphingolipid Metabolism*, 261 (S. M. Aronson and B. W. Volk, editors) (New York: Pergamon Press)
95. Robinson, D. and Stirling, J. L. (1968) *Biochem. J.*, **107**, 321
96. Frohivein, Y. Z. and Gatt, S. (1967). *Biochemistry*, **6**, 2775
97. Robinson, D., Jordan, T. W. and Horsburgh, T. (1972). *J. Neurochem.*, **19**, 1975
98. Radin, N. S., Hof, L., Bradley, R. M. and Brady, R. O. (1969). *Brain Res.*, **14**, 497
99. Bowen, D. M. and Radin, N. S. (1969). *J. Neurochem.*, **16**, 501
100. Morell, P. and Braun, P. (1972). *J. Lipid. Res.*, **13**, 293
101. Kishimoto, Y., Wajda, M. snd Radin, N. S. (1968). *J. Lipid. Res.*, **9**, 27
102. Miyutake, T. and Suzuki, K. (1972). *Biochim. Biophys. Acta*, **48**, 538
103. Suzuki, K., Schneider, E. L. and Epstein, C. J. (1971). *Biochim. Biophys. Res. Commun.*, **45**, 1363
104. Torvik, A. and Sidman, R. L. (1965). *J. Neurochem.*, **12**, 555
105. Gautheron, C., Petit, L. and Chevaleier, J. (1969). *Exp. Neurol.*, **25**, 18
106. Rawlins, F. A., Hedley-Whyte, E. T., Villegas, G. and Uzman, B. C. (1970). *Lab. Invest.*, **22**, 237
107. Rawlins, F. A., Villegas, G. M., Hedley-Whyte, E. T. and Uzman, B. C. (1972). *J. Cell Biol.*, **52**, 615

108. Kiernan, J. A. and Pettit, D. R. (1971). *Exp. Neurol.* **32,** 111
109. Menkes, J. H. (1971). *J. Neurochem.*, **18,** 1433
110. Menkes, J. H. (1972). *Lipids, Malnutrition and the Developing Brain*, A CIBA Foundation Symposium, 179 (K. E. Elliott and J. Knight, editors) (Amsterdam: Associated Scientific Publishers)
111. Menkes, J. H. (1972). *Lipids,* **7,** 135
112. Ramsey, R. B., Jones, J. P., Rios, A. and Nicholas, H. J. (1972). *J. Neurochem.*, **19,** 101
113. Ramsey, R. B., Jones, J. P. and Nicholas, H. J. (1972). *J. Neurochem.*, **19,** 931
114. Sun, G. Y. and Horrocks, L. A. (1969). *J. Neurochem.*, **16,** 181
115. Tamai, Y., Matsukawa, S. and Sataki, M. (1971). *Brain Res.*, **26,** 149
116. Raghavan, S. and Kanfer, J. N. (1972). *J. Biol. Chem.*, **247** 1055
117. Freysz, L., Bieth, R. and Mandel, P. (1969). *J. Neurochem.*, **16,** 1417
118. Bernsohn, J. and Cohen, S. R. (1972). *Lipids, Malnutrition and the Developing Brain*, A CIBA Foundation Symposium, 159 (K. E. Elliott and J. Knight, editors). (Amsterdam: Associated Scientific Publishers)
119. Jones, J. P., Ramsey, R. B. and Nicholas, H. J. (1971). *Life Sci.*, **10,** 997
120. Jones, J. P., Ramsey, R. B., Aexel, R. J. and Nicholas, H. J. (1972). *Life Sci.*, **11,** 309
121. Radin, N. S., Arora, R. C., Ullman, M. D., Brenkert, A. L. and Austin, J. (1972). *Res. Commun. Chem. Path. Pharmacol.*, **3,** 637
122. August, C., Davison, A. N. and Williams, F. M. (1961). *Biochem. J.*, **81,** 8
123. Smith, M. E. and Eng, L. F. (1965). *J. Amer. Oil Chem. Soc.*, **42,** 1013
124. Eichberg, J. and Dawson, R. M. C. (1965). *Biochem. J.*, **96,** 644
125. Fumagalli, R., Smith, M. E., Urna, G. and Paoletti, R. (1969). *J. Neurochem.*, **16,** 1329
126. Banik, N. L. and Davison, A. N. (1971). *Biochem. J.*, **122,** 751
127. Hedley-Whyte, E. T., Rawlins, F. A., Salpeter, M. M. and Uzman, B. C. (1969). *Lab. Invest.*, **21,** 536
128. Spohn, M. and Davison, A. N. (1972). *J. Lipid Res.*, **13,** 563
129. Graham, J. M. and Green, C. (1967). *Biochem J.*, **103,** 16C
130. Wirtz, K. W. A., Kamp, H. H. and Van Deenen, L. L. M. (1972). *Biochim. Biophys. Acta,* **274,** 606
131. Jungalwala, F. B. and Dawson, R. M. C. (1970). *Biochem. J.* **117,** 481
132. Mandel, P. and Nussbaum, J. L. (1966). *J. Neurochem.*, **13,** 629
133. Lapetina, E. G., Arnaiz, G. R. L. and De Robertis, E. (1969). *J. Neurochem.*, **16,** 101
134. Miller, E. K. and Dawson, R. M. C. (1972). *Biochem. J.*, **126,** 823
135. Wirtz, K. W. A. and Zilversmit, D. B. (1968). *J. Biol. Chem.*, **243,** 3596
136. Jungalwala, F. B. and Dawson, R. M. C. (1971). *Biochem. J.*, **123,** 683
137. Braun, P. E. and Radin, N. S. (1969). *Biochemistry,* **8,** 4310
138. Palmer, F. B. and Dawson, R. M. C. (1969). *Biochem. J.*, **111,** 627
139. Dobbing, J. (1968). *Applied Neurochemistry,* **287,** (A. N. Davison and J. Dobbing, editors) (Oxford: Blackwell)
140. Galli, C., White, H. B., and Paoletti, R. (1970). *J. Neurochem.*, **17,** 347
141. Lapetina, E. G. and Michell, R. H. (1972). *Biochem. J.* **126,** 1141
142. Nachmansohn, D. (1971). *Proc. Nat. Acad. Sci.*, **68,** 3170
143. Hokin, L. E. and Hokin, M. R. (1953). *J. Biol. Chem.*, **203,** 967
144. Larrabee, M. G. and Leicht, W. S. (1965). *J. Neurochem.*, **12,** 1
145. Durell, J. and Sodd, M. A. (1966). *J. Neurochem.*, **13,** 487
146. Hokin, L. E. (1969). *Structure and Function of Nervous Tissue*, Vol. 3, 161 (G. Bourne, editors) (New York-London: Academic Press)
147. Salway, J. G. and Hughes, I. E. (1972). *J. Neurochem.*, **19,** 1233
148. Birnberger, A. C., Birnberger, K. L., Eliasson, S. G. and Simpson, P. C. (1971). *J. Neurochem.*, **18,** 1291
149. Lunt, G. G., Canessa, O. and De Robertis, E. (1971). *Nature (London),* **230,** 187
150. Schacht, J. and Agranoff, B. W. (1972). *J. Biol. Chem.*, **247,** 771
151. Yagihara, Y. and Hawthorne, J. N. (1972). *J. Neurochem.*, **19,** 355
152. Durell, J., Garland, J. T. and Friedel, R. O. (1969). *Science,* **165,** 862
153. Lapetina, E. G. and Hawthorne, J. N. (1971). *Biochem. J.*, **122,** 171
154. Hawthorne, J. N. and White, G. L. (1973). *Biochem. Soc. Trans.*, **1,** 359
155. Kai, M., Salway, G. and Hawthorne, J. N. (1968). *Biochem. J.*, **106,** 791
156. Durell, J. and Garland, J. T. (1969). *Ann. N.Y. Acad. Sci.*, **165,** 743

157. Hawthorne, J. N. and Kai, M. (1970). *Handbook of Neurochemistry*, Vol. 3, 491. (A. Lajtha, editor). (New York-London: Plenum Press)
158. Horrocks, L. A., Meckler, R. J. and Collins, R. L. (1966). *Variations in Chemical Composition of the Nervous System*, 46 (G. B. Ansell, editor) (Oxford: Pergamon Press)
159. Cuzner, M. L. and Davison, A. N. (1968). *Biochem. J.*, **106**, 29
160. Dalal, K. B. and Einstein, E. R. (1969). *Brain Res.*, **16**, 441
161. Banik, N. L., Blunt, M. J. and Davison, A. N.. (1968). *J. Neurochem.*, **15**, 471
162. Wolman, M. (1957). *Bull. Res. Coun. Israel E.*, **6**, 163
163. Tomlinson, B. E. and Kitchener, D. (1972). *J. Path.*, **106**, 165
164. Hirano, A., Dembitzer, H. M., Kurland, L. T. and Zimmerman, H. M. (1968). *J. Neuropath. Exp. Neurol.*, **27**, 167
165. Terry, R. D. (1971). *J. Neuropath. Exp. Neurol.*, **30**, 8
166. Terry, R. D. and Wisniewski, H. M. (1972). *Ageing and the Brain*, 94 (C. M. Gaitz, editor) (New York: Plenum Press)
167. Samorajski, T., Ordy, J. M. and Keefe, J. R. (1965). *J. Cell Biol.*, **26**, 779
168. Siakotos, A. N., Watanabe, I., Saito, B. and Fleischer, S. (1970). *Biochem. Med.*, **4**, 361
169. Barden, H. (1970). *J. Neuropath. Exp. Neurol.*, **25**, 76
170. Jibril, B. O. and McCay, P. B. (1965). *Nature (London)* **205**, 1714
171. Roels, O. A. (1969). *Lysosomes in Biology and Pathology*. Vol. 1, 274 (J. T. Dingle and Honor B. Fell, editors) (Amsterdam: North-Holland Publ. Co.)
172. Voskneseroskii, O. M. and Lovitskii, A. P. (1970). *Voprosy Med. Khimii*, **16**, 563
173. Einarson, L. (1953). *J. Neurol. Neurosurg. Psychiat.*, **16**, 98
174. Nishioka, N., Takahata, N. and Iizuka, R. (1972). *Histochemie*, **30**, 315
175. Arstila, A. U., Smith, M. B. and Trump, B. F. (1972). *Science*, **175**, 530
176. Brunk, U. and Brun, A. (1972). *Histochemie*, **30**, 315
177. Clendenon, N. R., Komatsu, T., Allen, N. and Gordon, W. A. (1971). *Neurology, Minneap.*, **21**, 103
178. Rastogi, R. N., Prichrard, J. S. and Lowden, J. A. (1968). *Pediat. Res.*, **2**, 125
179. Suzuki, K. and Chen G. (1966). *J. Neuropath. Exp. Neurol.*, **25**, 396
180. Cherayil, G. D. (1969). *J. Neurochem.*, **16**, 913
181. Wahal, H. M. and Riggs, H. H. (1960). *Arch. Neurol. Psychiat.*, **2**, 151
182. Cotman, C., Blank, M. L., Moehl, A. and Snyder, F. (1969). *Biochemistry*, **8**, 4606
183. Kishimoto, Y., Agranoff, B. W., Radin, M. S. and Burton, R. M. (1969). *J. Neurochem.* **16**, 397
184. Bishayee, S. and Balashbramanian, A. S. (1971). *J. Neurochem.*, **18**, 909
185. Koenig, H. (1969). *Lysosomes in Biology and Pathology*, Vol. 1, 111 (J. T. Dingle and Honor B. Fell, editors) (Amsterdam: North-Holland Publ. Co.)
186. Witting, L. A. and Sakr, A. H. (1969). *Biochim. Biophys. Acta*, **176**, 889
187. Samorajski, T. and Ordy, J. M. (1972). *Ageing and the Brain*, 41 (C. M. Gaitz, editor) (New York: Plenum Press)
188. Raivio, K. O. and Seegmiller, J. E. (1972). *Ann. Rev. Biochem.*, **41**, 543
189. Hultberg, B., Ockerman, P. A. and Sjoblad, S. (1972). *Europ. Neurol.*, **7**, 101
190. Cumings, J. N. (1970). *Handbook of Neurology*, Vol. 10, 325
191. Martin, J. J. and Schlote, W. (1972). *J. Neurol. Sci.*, **15**, 49
192. Howell, J. McC. and Davison, A. N. (1959). *Biochem. J.*, **72**, 365
193. Hogan, E. L. and Joseph, K. C. (1970). *J. Neurochem.*, **17**, 1290
194. Hogan, E. L., Joseph, K. C. and Schmidt, C. (1970). *J. Neurochem.*, **17**, 75
195. Costantino-Ceccarini, E. and Morell, P. (1971). *Brain Res.*, **29**, 75
196. Patterson, D. S. P., Sweasey, D. and Harding, J. D. (1972). *J. Neurochem.*, **19**, 2791
197. Einstein, E. R., Csejtey, J., Dalal, K. B., Adams, C. W. M., Bayliss, O. B. and Hallpike, J. F. (1972). *J. Neurochem.*, **19**, 653
198. Cuzner, M. L. and Davison, A. N. (1973). *J. Neurol. Sci.*, **19**, 29
199. Paoletti, R. (1972). *Lipids, Malnutrition and the Developing Brain. A CIBA Foundation Symposium*, 177, (K. E. Elliott and J. Knight, editors). (Amsterdam: Associated Scientific Publishers)

5
The Biosynthesis of Unsaturated Fatty Acids

M. I. GURR
Unilever Limited, Sharnbrook, Bedford

5.1	INTRODUCTION	182
	5.1.1 *Aims*	182
	5.1.2 *Terminology*	183
	5.1.3 *Scope of the review*	184
	5.1.4 *Historical*	184
5.2	THE ANAEROBIC PATHWAY OF MONOENOIC ACID BIOSYNTHESIS	187
	5.2.1 *Reviews*	187
	5.2.2 *The pathway*	187
	5.2.3 *Properties and role of the dehydratase*	189
5.3	THE AEROBIC MECHANISM	190
	5.3.1 *Reviews*	190
	5.3.2 *Introduction: the pathway*	191
	5.3.3 *Form of the substrate*	193
	5.3.3.1 Acyl-S-CoA	193
	5.3.3.2 Acyl-S-ACP	194
	5.3.3.3 *Complex lipids as substrates for the desaturase*	195
	5.3.4 *Alternatives to the direct desaturation pathway*	199
	5.3.4.1 *Evidence for an alternative pathway*	199
	5.3.4.2 *Evidence for direct desaturation*	200
	5.3.4.3 *Connexion between different metabolic sequences*	202
	5.3.4.4 *Critical assessment of the alternative pathway*	203
	5.3.5 *Evidence for acyl-enzyme intermediates in desaturation*	204
	5.3.6 *Products of the desaturase reaction*	204
	5.3.6.1 *The nature of the products*	204
	5.3.6.2 *Inhibition by the products of desaturation and its relief*	206
	5.3.6.3 *Reactions competing for substrate and product*	208

5.3.7	Co-factors for the aerobic desaturation reaction	209
	5.3.7.1 Electron acceptors: oxygen: effect of temperature	209
	5.3.7.2 Electron donors	210
5.3.8	Inhibitors of desaturation	211
	5.3.8.1 SH Blocking and protecting agents	211
	5.3.8.2 Metal chelators	211
	5.3.8.3 Cyanide	211
	5.3.8.4 Cyclopropene fatty acids	212
	(a) Background	212
	(b) Mechanism of the inhibition	212
	(c) The alternative pathway for oleic acid biosynthesis	214
5.3.9	Assay of desaturases	215
	5.3.9.1 Methods depending on measurement of conversion of [^{14}C] substrate into [^{14}C] product	215
	5.3.9.2 Methods depending on NAD(P)H utilisation	216
	5.3.9.3 Methods depending on labelled hydrogen release	216
5.3.10	The organisation of the desaturase complex	216
	5.3.10.1 The requirement for lipids	216
	5.3.10.2 The purification of desaturases	217
	5.3.10.3 Indirect methods for studying electron transport components	219
	5.3.10.4 Control of fatty acid desaturation	220
5.3.11	The stereochemistry and mechanism of hydrogen removal	221
	5.3.11.1 Stereochemistry of hydrogen removal	221
	5.3.11.2 Mechanism of hydrogen removal	223
5.3.12	Specificity of desaturases with regard to double-bond position	225
	5.3.12.1 Monoenoic acid biosynthesis	225
	5.3.12.2 Dienoic acid biosynthesis	228
	5.3.12.3 Classification of desaturases	230
5.3.13	Conclusions	230
ACKNOWLEDGEMENTS		231

5.1 INTRODUCTION

5.1.1 Aims

It is not my intention in this article to give an exhaustive review of the literature concerning the biosynthesis of unsaturated fatty acids. Many excellent reviews exist which very thoroughly document this literature, certainly up to 1970 and these will be pointed out as sources for further detailed reading. The literature of the past 2–3 years will be covered a little more thoroughly but even so will not be reviewed exhaustively.

The chief difference between this and former reviews will be one of emphasis. Most reviews in the past have covered a particular aspect of unsaturated fatty-acid biochemistry such as 'higher plants', 'micro-organisms' or 'essential fatty acids', or have simply been confined to the work of the particular author's laboratory. My aim will be to span the barriers of biological classification and examine the process of desaturation at a more molecular level, concentrating on unifying concepts rather than highlighting differences which distract attention from the basic mechanism.

I shall begin by attempting to summarise our existing knowledge of unsaturated fatty acid biosynthesis to give the general reader an idea of the 'state of the art'. I shall try to point out the significant major advances of the past few years and discuss techniques which are helping to make such advances possible—a feature which is often omitted from general reviews of biosynthesis. It is important for the reader who has not been engaged in lipid research to get a 'feel' for the problems associated with handling water-insoluble compounds, particularly in the assay of enzymic reactions *in vitro* where the interpretation of kinetic data is not straightforward. These difficulties are often underestimated, or ignored, by the lipid and non-lipid biochemist alike.

This will lead on naturally to a discussion of the major outstanding problems and those areas which are ripe for investigation.

Such a treatment will necessarily involve a larger degree of personal assessment of the field than would be the case for a straightforward literature review; for which fact I make no apology.

5.1.2 Terminology

I shall deal mainly with a relatively few commonly occurring fatty acids, namely: palmitic (hexadecanoic, 16:0); stearic (octadecanoic, 18:0); oleic (*cis*-9-octadecenoic, 9-18:1); vaccenic (*cis*-11-octadecenoic, 11-18:1); linoleic (*cis,cis*-9,12-octadecadienoic, 9,12-18:2); α-linolenic (*cis,cis,cis*-9,12,15-octadecatrienoic, 9,12,15-18:3); γ-linolenic (*cis,cis,cis*-6,19,12-octadecatrienoic, 6,9,12-18:3); and arachidonic acid (*cis,cis,cis,cis*-5,8,11,14-eicosatetraenoic, 5,8,11,14-20:4).

The shorthand system is extremely useful and will be employed extensively, particularly in diagrams. The number before the colon represents the carbon chain length; the number after the colon refers to the number of double bonds and the position of these double bonds is indicated by a prefix followed by a hyphen.

The prefix Δ, followed by a number, is often used to denote the position of the first of the pair of carbon atoms of the double bond measured from the carboxyl end; since it is common practice to number from the carboxyl end, the omission of Δ assumes this system of numbering. It is sometimes useful in comparing the structures of some unsaturated acids to define the double-bond position with respect to the terminal methyl group, as acids derived from one another by chain elongation can then be clearly identified. I shall use the Greek ω for this purpose, which was in common use in older literature, but is now frowned upon by the I.U.P.A.C.–I.U.B. Nomenclature Commission, who prefer the more cumbersome system, ($n-x$ where n is the total number of carbon atoms in the chain, x the position of the substituent or double bond from the carboxyl terminal.

The terms monoenoic or monounsaturated, which are interchangeable, refer to fatty acids with a single double bond; similarly, polyenoic or polyunsaturated refer to acids with more than one double bond. Specific types of polyenoic acids are dienoic, trienoic tetraenoic..... etc. with 2, 3, 4..... double bonds.

A more detailed treatment of fatty acid nomenclature can be found in Ref. 1.

5.1.3 Scope of the review

The emphasis here will be rather on the unity of mechanism and the economy of enzymic species by which the vast array of naturally occurring unsaturated fatty acid molecules are produced, rather than the diversity of the processes which is the impression one might get from a superficial glance at the literature. In order to make the most of this approach, I shall concentrate on a quite limited number of very commonly occurring acids and forego a detailed treatment of 'special' or 'unusual' unsaturated fatty acids such as cyclic acids, hydroxy acids, conjugated acids and prostaglandins. Knowledge of these is either very scant (conjugated acids) or they now form a specialised subject of their own, as is the case now for the prostaglandins.

The types of desaturation reactions I shall discuss and the positions at which double bonds are introduced in natural fatty acids are summarised in Table 5.1.

5.1.4 Historical

The extent of the literature on unsaturated fatty acid biosynthesis today readily conceals the short life of the subject. Few references date before 1961. An oft-quoted cornerstone of knowledge on naturally occurring fatty acids, *'The Chemical Constitution of Natural Fats'*,[1] whose 4th Edition was published in 1964, has this to say about unsaturated fatty acid biosynthesis: '. . . . Millions of tons of unsaturated C_{18} acids (mainly oleic and linoleic) must therefore be produced annually in the vegetable and animal kingdom, and an adequate explanation of their synthesis is the most pressing aspect of the general problem of fat biosynthesis.

Yet almost the only approach to it so far has been hypothetical and rather naïve: perhaps taking the easiest way out, it is not infrequently taken for granted that all these unsaturated acids arise by 'desaturation' (dehydrogenation in a peculiarly selective way) of stearic acid. . . . . Let us take by way of example a tobacco seed oil with component acids, palmitic 7%, stearic 3%, oleic 15% and linoleic 75% (by weight): we are asked to believe that in the seed, stearic acid is continuously synthesised and concurrently selectively desaturated, first to oleic acid and then to linoleic, with the final result that 75% of linoleic acid is produced and 15% of oleic, leaving a minute proportion (3%) of the stearic intermediate in the finished article.

It seems a remarkably inefficient manner of synthesis for a natural process, especially from the standpoint of the energy changes involved. . . . .'

Table 5.1 Positions at which double bonds occur in natural unsaturated fatty acids

Position of double bond numbering from the carboxyl group (Δ)	Occurrence—Remarks
trans-2	Occur only as intermediates in the biosynthetic pathway to saturated fatty acids or in the β-oxidation of fatty acids (Section 5.2.2)
cis-3	Intermediates in the anaerobic pathway to unsaturated fatty acids (Section 5.2.2)
trans-3	Present in photosynthetic tissue and in some seed oils as trans-3-16:1 specifically as component of phosphatidyl glycerol
cis-4	Animals (Section 5.3.2) and algae (Sections 5.3.2 and 5.3.3.3) can insert a double bond into an already polyunsaturated acid, e.g. the formation of cis,cis,cis,cis-4,7,10,13-16:4
cis-5	Animals (Section 5.3.2 and 5.3.6.2) desaturate already polyunsaturated acids at the $\Delta 5$ position. The major acid is arachidonic, cis,cis,cis,cis-5,8,11,14,-20:4. Some species of higher plants and some bacteria form $\Delta 5$- monoenes
cis-6	Plants form $\Delta 6$-monoenoic acids, e.g. Petroselenic, cis-6-18:1. Animals insert a double bond at position 6 into an already unsaturated chain, e.g. conversion of cis-9-18:1 into cis,cis-6,9-18:2 (Sections 5.3.2 and 5.3.6.2). May also arise from chain elongation after $\Delta 4$ introduction
cis-7	Plants and algae form cis-7-monoenoic acids, e.g. cis-7-16:1 (Sections 5.3.2, 5.3.3.3 and 5.3.12.1). May also arise from chain elongation after $\Delta 5$ introduction
cis-8	No direct desaturase. Only arise from chain elongation after $\Delta 6$ introduction
cis-9	The major desaturase; certainly the one for which most data exist (Sections 5.3.2, 5.3.4–5.3.12). May also arise from chain elongation after $\Delta 7$ introduction.
cis-10	Direct desaturation exists only in bacteria. May arise from chain elongation of compounds with $\Delta 8$ unsaturation
cis-11	No direct desaturase. Arises in animals and plants by chain extension of acids with $\Delta 9$ unsaturation. In Eubacteria it is an end-product of the anaerobic pathway (Section 5.2.2)
cis-12	Direct desaturase only exists in plants. Main example is the formation of linoleic acid (cis,cis-9,12-18:2) from oleic acid (cis-9-18:1) (Sections 5.3.2, 5.3.3, 5.3.11, 5.3.12.2)
cis-13	Arises only by chain elongation, e.g. in formation of erucic acid (cis,-13-22:1) from oleic acid
cis-14	Present in long chain polyunsaturated acids particularly in animals after a series of desaturations and chain elongations (e.g. arachidonic cis,cis,cis,cis-5,8,11,14-20:4, Section 5.3.2, 5.3.6.2)
cis-15	Direct desaturase only in plants, e.g. in formation of α-linolenic acid (cis,cis,cis-9,12,15-18:3) from linoleic acid
higher double-bonds	A number of higher double bonds occur in polyunsaturated fatty acids

* This is not a full list of all fatty acid types but merely gives examples of the types of double bond whose formation will be discussed in this article.

Hilditch[1] firmly believed that unsaturated acids are built up by an entirely different mechanism from that which operates in the synthesis of saturated acids and went on to say that . . . 'the order of synthesis of *unsaturated* acids may be presumed, in the light of present knowledge, to be: trienoic⟶ dienoic⟶monoenoic'.

Few researchers writing on this subject today would reject the idea of desaturation in this way or even deny that it provided the major pathway for unsaturated acid biosynthesis. This should not blind us to the fact that there are controversies on detail arising from the fact that in higher plants it is difficult to demonstrate desaturation of exogenous stearic acid *in vitro* and similarly difficult to demonstrate, in many plants, a precursor–product relationship between stearic and oleic acids derived from labelled acetate *in vivo*. Equally, 'saturases' or 'biohydrogenases' have been characterised in rumen micro-organisms which effectively catalyse the sequence envisaged by Hilditch and it is possible, though not probable, that such mechanisms could be encountered in higher plants.

What then have been the major steps that have led us to the present day belief that, in general, unsaturated fatty acids are produced by sequential desaturation of more saturated precursors?

That unsaturated acids could be directly formed from saturated ones by 'desaturation' was first brought to light by the, then, novel tracer methods of Schoenheimer and Rittenberg[2], who, in 1936, fed mice with deuterium-labelled saturated acids. Fatty acids with different degrees of unsaturation were separated by fractional crystallisation of lead salts. These studies were extended to show direct conversion of palmitic into palmitoleic acid as well as chain elongation and shortening by Stetten and Schoenheimer[3].

It was not until 1958 that Bloomfield and Bloch[4] first published a short note on the formation of Δ9-unsaturated acids from saturated acids in *cell-free extracts* of yeast. The necessary co-factors were molecular oxygen and a reducing coenzyme, NADPH. The authors converted unsaturated fatty acids into dihydroxy acids which could then be separated from labelled precursor saturated fatty acids by silicic acid chromatography.

During the late 1950s and early 1960s these observations were extended to rat liver[5-7], bacteria[8,9], protozoa[10,11], algae[12-14], higher plant leaves[13,15,16] and seeds[17-19]. By this time chromatographic techniques such as radiochemical gas–liquid chromatography and argentation thin-layer chromatography were becoming available for direct separations of labelled product and precursor fatty acids and this in itself speeded up progress in our knowledge of desaturation reactions.

During the 1960s, too, the general mechanism for the dehydrogenation was clarified. It was found to involve a stereospecific removal of two hydrogen atoms from specific positions in the saturated hydrocarbon chain, resulting in a *cis*-unsaturated fatty acid with the same carbon chain length as the precursor fatty acid[20,21]. The same mechanism was also shown to be responsible for the introduction of subsequent double bonds into the chain to give rise to the so-called polyunsaturated fatty acids in both plants and animals[13,21-23]. The desaturation in every case required molecular oxygen and no artificial electron acceptor that could substitute for oxygen was found, nor indeed has been found to this day.

Yet Bloch was well aware that many *obligately anaerobic* bacteria contained monounsaturated fatty acids in their lipids and he became interested in testing the ability of these anaerobic micro-organisms to synthesise their own fatty acids. The result of these studies, which Bloch[24] has reviewed, was the discovery of the so-called 'anaerobic pathway' of monounsaturated fatty acid biosynthesis.

These two pathways probably represent the only routes for unsaturated fatty acid formation. Bloch himself[12] has said that the two mechanisms appear to be mutually exclusive in the sense that no organism has yet been found to possess both pathways.

Other pathways have been invoked to explain some apparent anomalies in the aerobic pathways of some animals and plants. As I shall discuss later (Sections 5.3.4 and 5.3.10.4), I believe these to be red herrings and unnecessary complications in a basically simple biosynthetic pattern, while acknowledging this to be a controversial viewpoint.

The anaerobic pathway is relatively unimportant in the sense that it is restricted to a limited number of bacterial species, the *Eubacteriales*. My discussion of the anaerobic pathway will be brief in comparison with my treatment of aerobic desaturation.

5.2 THE ANAEROBIC PATHWAY OF MONOENOIC FATTY ACID BIOSYNTHESIS

5.2.1 Reviews

Aspects of the anaerobic pathway have been covered in the Refs. 24–30.

5.2.2 The pathway

While the first tentative experiments to look at the biosynthesis of unsaturated acids in cell-free extracts were taking place, research into the mode of biosynthesis of saturated fatty acids was already gaining momentum. Fatty acid synthetases were being isolated and one of the features that could be discerned was that, whereas the mammalian, avian and yeast synthetases produced only saturated acids, the eubacterial synthetases, of which the most extensively studied was that of *E. coli*, produced saturated *and* unsaturated (monoenoic) acids.

The first clue as to how this might arise came from the studies of Goldfine and Bloch[31] on *Clostridium butyricum*. This organism, like almost all other bacteria, produces only monoenoic fatty acids. Two pairs of monoene isomers are produced. $\Delta 7$ and $\Delta 9$ hexadecenoic and $\Delta 9$ and $\Delta 11$ octadecenoic acids. Cultures of *C. butyricum* extended added radioactive C_8 and C_{10} saturated acids to labelled saturated *and* monounsaturated acids, whereas no unsaturated acids arose from substrates longer than C_{10}[31,32].

The occurrence of the monoene isomer pairs can be explained if there are 'branch points' in fatty acid synthesis at C_8 and C_{10} according to the scheme in Figure 5.1.

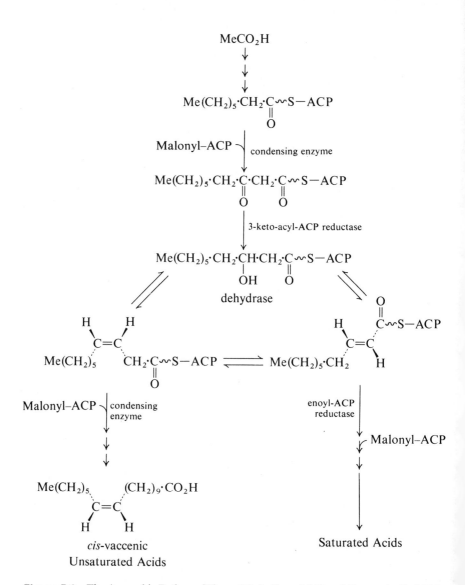

Figure 5.1 The Anaerobic Pathway [Figure 2.8. in Gurr, M. I. and James, A. T. (1971) *Lipid Biochemistry: An Introduction*. (London: Chapman and Hall) reproduced by kind permission of the publisher]

Fatty acid synthesis in anaerobic bacteria occurs by stepwise condensation of a 2-carbon with a 3-carbon unit, and simultaneous elimination of CO_2. All the intermediates on this sequence are bound to an acyl carrier protein (ACP). The pathway is described by Vagelos[33] in this volume (Chapter 3) and elsewhere.

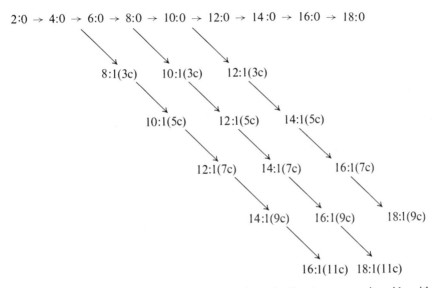

Figure 5.2 Branch points on the anaerobic pathway leading to monoenoic acids with different double bond positions. [Figure 5.14. from Hitchcock, C. H. S. and Nichols, B. W. (1971). Plant Lipid Biochemistry (London: Academic Press) reproduced by kind permission of authors and publisher]

At the branch points, the normal *trans*-2-enoyl-ACP is isomerised to *cis*-3-enoyl-ACP, which is not a substrate for the enoyl-ACP-reductase, but *is* capable of being elongated. Thus the *trans*-isomer formed at the branch point gives rise to a long-chain *saturated* acid while the *cis*-isomer yields a long-chain *cis*-unsaturated acid (Figure 5.1).

The positions of the double bonds in the final products, therefore, depend on the positions of the branch-points, as illustrated in Figure 5.2.

5.2.3 Properties and role of 3-hydroxydecanoyl thiolester dehydratase

Subsequent studies on the mechanisms of the branching process have been done using *E. coli*[25,34-48]. The crude fatty acid synthetase contains an enzyme that catalyses the dehydration of 3-hydroxydecanoyl-ACP to a mixture of *trans*-2- and *cis*-3-decenoates and also the interconversion of the two monoene isomers[37,42]. The enzyme can also accept the *N*-acetylcysteamine (NAC)[41,42] derivatives of the fatty acid intermediates as substrates.

The use of 'model' thiolesters has the advantages of more ready availability, ease of preparation and purification and cheapness when compared with the ACP derivatives, even though the rates are only one-sixth of those of the natural substrates.

The enzyme is called 3-hydroxydecanoyl thiolester dehydratase and has been purified to homogeneity[41]. It appears to be a truly multifunctional enzyme in the sense that at all stages of the 1200-fold purification it catalyses the dehydration of 3-hydroxy thiolesters into a mixture of 85% *trans*-2- and 15% *cis*-3-decenoates when assayed under initial velocity conditions and also the interconversion of the two monoene isomers. Studies[45] in which the path of labelled hydrogen atoms has been followed from different specifically-labelled substrates during the reaction, have indicated that *trans*-2-decenoyl-NAC must be formed directly by dehydration of 3-hydroxydecanoyl-NAC, whereas *cis*-3-decenoyl-NAC arises only by isomerisation of the *trans*-2-monoene. The interconversion of the three substrates involves a common intermediate, namely, enzyme-bound *trans*-2-decenoate.

Further confirmation that the enzyme is truly multifunctional comes from the fact that a substrate analogue, 3-decynoyl-NAC specifically and irreversibly inhibits *all* the transformations catalysed by the dehydratase[42,45,48]. The inhibitor is converted by the enzyme itself into the corresponding allene that then reacts with an active site histidine.[48]

These studies threw up two apparent paradoxes. First, although the major products of fatty acid synthesis in *E. coli in vivo* are unsaturated fatty acids[25,42], the precursor of these acids, *cis*-3-decenoate is a minor component (15%) of the equilibrium mixture produced by the dehydratase *in vitro*. Secondly, although 3-decynoyl-NAC inhibits unsaturated fatty acid biosynthesis *in vivo* and *in vitro*, it does not affect saturated fatty acid production[25]. It seems as if it may be the function of another dehydratase to furnish the *trans*-2-monoenoic acid precursor for saturated acid biosynthesis. This idea was confirmed by the finding in Vagelos'[49] laboratory that an *E. coli* mutant, lacking the specific dehydratase, can synthesise saturated acids normally, but not unsaturated ones. Altogether three dehydratases have now been found to be involved in fatty acid synthesis in *E. coli*[50,51].

It seems, therefore, that although the specific dehydratase is essential for the *formation* of long chain unsaturated acids, it is not responsible for the *proportion* of these acids. This is probably determined *in vivo* by the relative rates of the two reactions (reduction and chain elongation) that remove isomeric decenoates.

5.3 THE AEROBIC MECHANISM

5.3.1 Reviews

General reviews can be found in Refs. 27, 28, 52. More specific reviews deal with unsaturated fatty acid biosynthesis in animals[53,54], plants[29,55-58] and micro-organisms[24,30]. Of these, Refs. 24, 54, 56, 57 concentrate on research in the particular author's laboratory; the remainder are general reviews.

5.3.2 Introduction: the pathway

All organisms, with the exception of anaerobic bacteria, form unsaturated fatty acids by direct introduction of a double bond into the hydrocarbon chain of a preformed fatty acid, whether it be saturated, monounsaturated or polyunsaturated.

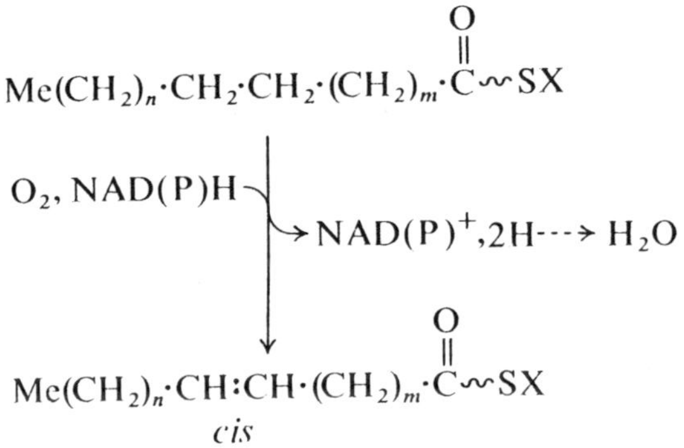

$$Me(CH_2)_n \cdot CH_2 \cdot CH_2 \cdot (CH_2)_m \cdot \overset{O}{\overset{\|}{C}} \sim SX$$

$$O_2, NAD(P)H \downarrow$$
$$\rightarrow NAD(P)^+, 2H \cdots \rightarrow H_2O$$

$$Me(CH_2)_n \cdot CH{:}CH \cdot (CH_2)_m \cdot \overset{O}{\overset{\|}{C}} \sim SX$$
$$cis$$

Figure 5.3 Aerobic desaturation pathway

Other types of fatty acids may arise by subsequent chain elongation, but most of the evidence to date suggests that the mechanism of the dehydrogenation itself is identical whether the organism be a higher or lower animal, higher or lower plant (but see Section 5.3.11).

Differences exist only in co-factor requirements, specificity of double bond position and possibly, though this is less certain, differences in the form of the substrate.

Except for bacteria, all organisms are capable of introducing more than one double bond into the chain. These double bonds are always separated by a methylene group and are known as methylene-interrupted polyunsaturated fatty acids. Some bacteria do have polyunsaturated acids[59-61] but these are not of the methylene-interrupted type and except in one special case[61] the pathways have not yet been worked out. In animals, the second and subsequent double bonds are introduced between the first double bond and the carboxyl group (Figure 5.4 and Refs. 22, 23, 28, 53, 54), whereas in plants, the new double bonds are introduced towards the methyl end (Figure 5.4 and Refs. 13, 18, 55–57). Some primitive organisms, intermediate in the evolutionary scale such as the phytoflagellate *Euglena*, have the ability to desaturate in either direction[12,29].

In some cases, the formation of a particular unsaturated fatty acid appears not to be by direct desaturation and a number of alternative mechanisms have been discussed (see Sections 5.3.4 and 5.3.8.4). I shall argue that these processes all involve direct desaturation, the discrepancies occurring mainly because of differences in availability of particular substrates to the enzyme.

There is also no evidence that the basic mechanism for the introduction of the 2nd, 3rd, ...nth double bonds in polyunsaturated acids is different from that for monoenoic acid formation.

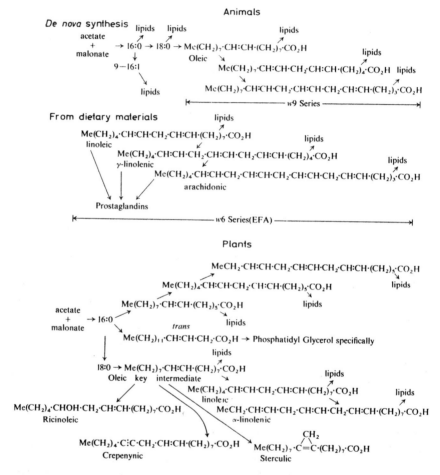

Figure 5.4 Some important pathways of fatty acid biosynthesis in animals and plants. [Figure 2.10 in Gurr, M. I. and James A. T. (1971). *Lipid Biochemistry: An Introduction.* (London: Chapman and Hall) reproduced by kind permission of the publisher.]

5.3.3 The form of the substrate

No case is known where the unesterified carboxylic acid is desaturated without prior 'activation'. Methods are available for the preparation of acyl-S-CoA[62-64] acyl-S-ACP[40,65] and O-ester lipids[66-68] as substrates.

5.3.3.1 Acyl-S-CoA

In the first demonstration of the direct desaturation of palmitic acid by cell-free extracts of yeast, Bloomfield and Bloch[4,69] showed that when palmitic acid was the substrate, both a particulate and a soluble fraction were required and that the essential co-factors were coenzyme A, ATP, Mg^{2+}, NADPH and molecular oxygen. When palmitoyl-S-CoA was added to the incubation mixture, CoASH, ATP, Mg^{2+} and the soluble enzyme could be dispensed with. They deduced correctly that 'activation' of the free acid to the acyl-S-CoA thiol ester was a necessary preliminary to desaturation.

$$\text{Palmitic acid} \xrightarrow[\text{Acid: CoA ligase}]{\text{CoASH, ATP, } Mg^{2+}} \text{palmitoyl-S-CoA} \qquad (5.1)$$

$$\text{Palmitoyl-S-CoA} \xrightarrow[O^2, \text{ NADPH}]{\text{desaturase}} \text{palmitoleoyl-S-CoA} \qquad (5.2)$$

Mammalian tissues[7,70], seed tissues[65] and bacteria[8] have similar requirements. As far as photosynthetic tissue is concerned, the desaturase that converts oleic acid into linoleic acid was shown by Harris et al.[13] to have similar requirements, but many attempts to demonstrate the conversion of stearoyl-S-CoA into oleoyl-S-CoA apparently failed. This question will be discussed in Section 5.3.4.

In animal tissues, the acid: CoA ligases are widely distributed[71,72]. Since the microsmal fraction is claimed to be a major source of these enzymes[71,72] and because the microsmal fraction is the major site of desaturases in all animal tissues so far examined[7,54,70,73] the soluble fraction should not strictly be required for the desaturation of unesterified fatty acid. Nevertheless, in my experience, the rate of desaturation is always greater in hen liver microsmal fractions with acyl-S-CoA as the substrate, than with non-esterified fatty acid supplemented with all the co-factors whereas in the presence of soluble fraction there is little difference between the rates of desaturation of the two forms of the substrate[74]. Marsh and James also found that in rat liver the presence of supernatant was necessary for optimum desaturation of non-esterified fatty acid. In the alga, *Chlorella vulgaris*, the acid CoA: ligase activity is unequivocally in the soluble fraction[13,68,75] whereas the oleoyl-S-CoA desaturase activity, catalysing the reactions:

$$\text{Oleoyl-S-CoA} \longrightarrow \text{Linoleoyl-S-CoA} \qquad (5.3)$$

$$\text{Linoleoyl-S-CoA} \longrightarrow \text{Linolenoyl-S-CoA} \qquad (5.4)$$

is confined to a particulate fraction. Both fractions are, therefore, necessary for the formation of linoleic acid from non-esterified oleic acid. In *Chlorella*, too, the supernatant fraction has a stimulating effect over and above the acid CoA ligase activity[13]. One interpretation may be that the acyl-S-CoA derivative is not the direct substrate and that a further conversion must occur (possibly to acyl-S-ACP, see below) before the substrate is in the correct form for acceptance by the desaturase. A second possibility is that the soluble fraction supplies a factor which is limiting in particulate fractions and which is involved in the removal of product. Glycerol -3(*sn*)-phosphate, a necessary precursor for complex lipid synthesis is one such possibility which is discussed in Section 5.3.6.2. Recently, a detailed account of the properties of acid:CoA ligases was published[76].

5.3.3.2 Acyl-S-ACP

Among the first attempts to demonstrate direct desaturation of stearic acid in higher plant leaves were those of Stumpf and James[15]. Their failure to do so has been a common experience throughout the years and has given rise to the idea of a separate 'plant pathway' for monoenoic acid synthesis which I shall discuss in Section 5.3.4.

Instead of invoking an alternative pathway one might postulate that the substrate is required in a particular form, into which exogenous stearic acid cannot be converted. One such form could be the acyl-S-ACP derivative.

ACP, originally discovered by Vagelos and his group[26] in *E. coli* was shown to be present in higher plants by Stumpf's Laboratory[77,78] and to be obligatory for fatty acid biosynthesis from acetate in lettuce chloroplasts[79].

Nagai and Bloch[14,80] obtained direct desaturation of stearic to oleic acid in a soluble preparation from *Euglena gracilis*. When the organism was grown under photoautotrophic conditions, the required substrate was stearoyl-S-ACP; the CoA derivative was not a substrate; when the organism was grown in the dark on an organic medium, the CoA thiolester but not the ACP derivative was a substrate. In a similar way, *Chlorella vulgaris* fails to desaturate added stearic or palmitic acids if grown on a medium containing carbohydrate; if, on the other hand, the organism is denied access to any carbon source other than CO_2, direct aerobic desaturation does take place[75,81]. However, unlike *Euglena*, *Chlorella* desaturates both stearoyl-S-ACP and stearoyl-S-CoA at an equal rate.[82]

Only one report has appeared on the ability of a higher plant preparation to utilise ACP-derivatives for desaturation. Nagai and Bloch reported that spinach chloroplasts directly desaturate stearoyl-S-ACP[83] and elongate short chain acyl-S-ACP derivatives[84].

On the negative side, Vijay and Stumpf[85] showed that a particulate fraction of maturing safflower seeds would convert oleoyl-S-CoA *but not oleoyl-S-ACP* into linoleate. No other direct studies have been reported of the ability of higher plants or algae other than *Euglena* or *Chlorella* to utilise ACP derivatives. One can speculate that this has been largely due to both the tedious procedures for obtaining adequate quantities of purified ACP and the additional difficulties in synthesising ACP thiolesters.

It should be pointed out that the ACP used to make the substrates in Nagai and Bloch's experiments was isolated from *E. coli*. However, there seems to

5.3.3.3 Complex lipids as substrates for desaturases

The alga, *Euglena gracilis*, provides a link between the animal and plant kingdoms and this is illustrated vividly by its lipid metabolism.

When the organism is grown on a strictly inorganic medium in the presence of light and a supply of CO_2, its predominant lipids are galactosyl diglycerides and its fatty acid spectrum and metabolism are predominantly of the plant type[12,86-89] (See Section 5.3.2). That is to say it produces linoleic acid (9,12–18:2), and α-linolenic acid (9,12,15–18:3) presumably by the sequential desaturation from oleic acid as in *Chlorella*[13]. A similar sequence of polyunsaturated acids occurs in the C_{16} series, namely 7–16:1; 7,10–16:2; 7,10,13–16:3 and 4,7,10,13-16:4[86,87,89]. This last acid is *specifically* a product of photoautotrophic growth and *specifically* a component of galactosyl glycerides. The organism can, however, be grown readily on an organic medium in complete absence of light. No galactosyl glycerides are synthesised under these conditions and the major fatty acids are typical of those found in animals, namely, those derived from oleic acid by desaturation towards the carboxyl group such as arachidonic and other C_{20} and C_{22} polyunsaturated acids[12,86-89].

In studies designed to show that the C_{16} unsaturated acids were indeed related to one another by sequential desaturations, Gurr and Bloch synthesised [7,8-$^3H_2$]16:1 and fed it to growing photo-autotrophic cultures of *E. gracilis*[90]. No radioactivity appeared in any of the more highly unsaturated C_{16} acids but instead, an elongation step to oleic acid was followed by subsequent desaturations *of the animal type*. The radioactive products were in all cases located in phospholipids, *not* in galactosyl glycerides.

```
16:1 (7)─╫→16:2 (7, 10)─╫→16:3 (7, 10, 13)─╫→16:4 (4, 7, 10, 13)
  │         phospholipids    phospholipids
  ↓       ↗                ↗
18:1 (9)──→18:2 (9, 12)
  │                        phospholipid      phospholipid
  ↓                      ↗                ↗
  20:2 (11, 14)──→20:3 (8, 11, 14)──→20:4 (5, 8, 11, 14)
  ↘                                     │                       phospholipid
    phospholipid                        ↓                     ↗
                                        22:4 (7, 10, 13, 16)──→22:5 (4, 7, 10, 13,16)
                                        ↘
                                          phospholipid
```

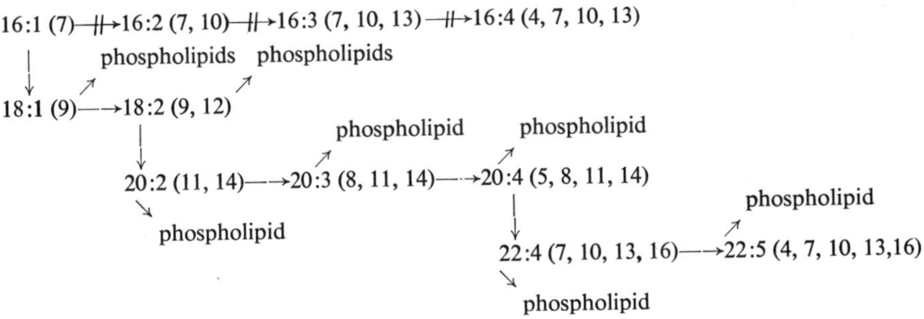

Figure 5.5 Pathways of metabolism of exogenous *cis*-7- hexadecenoic acid in photoautotrophic *Euglena gracilis*. [Gurr, M. I. and Bloch, K. unpublished results.]

This experiment raised the possibility that the formation of certain polyunsaturated fatty acids could be strongly associated with certain types of complex lipids. Looking at it conservatively, one could envisage compartments in which precursor fatty acids were available only to be esterified to specific lipids. Before a particular desaturation could take place (for example, 7,10,13–16:3⟶4,7,10,13–16:4) the substrate would be donated by the specific lipid to the enzymic site for desaturation either directly (Figure 5.6a) or through the intermediacy of a thiolester (Figures 5.6b,c).

A more unorthodox model would envisage a series of double bond introductions into the hydrocarbon chain of the fatty acid while it was still attached to the glycerol back-bone through its ester linkage (Figure 5.6d).

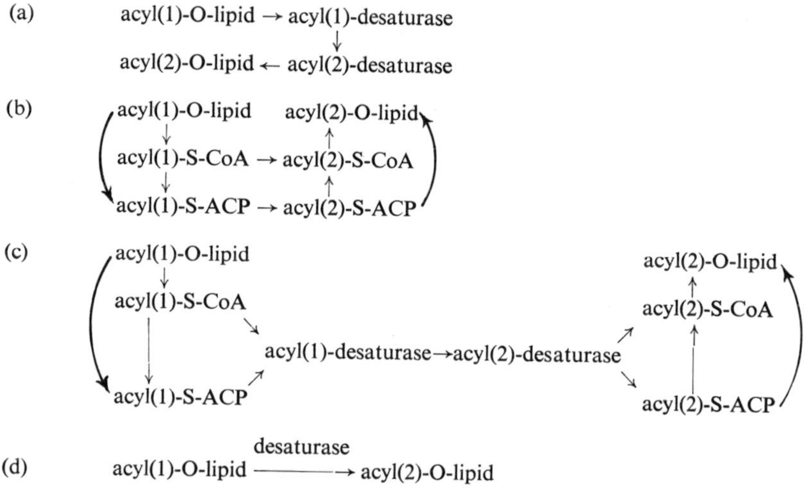

(a) acyl(1)-O-lipid → acyl(1)-desaturase
     ↓
     acyl(2)-O-lipid ← acyl(2)-desaturase

(b) ⎛ acyl(1)-O-lipid   acyl(2)-O-lipid ⎞
     ⎜   ↓   ↑   ⎟
     ⎜ acyl(1)-S-CoA → acyl(2)-S-CoA   ⎟
     ⎜   ↓   ↑   ⎟
     ⎝ acyl(1)-S-ACP → acyl(2)-S-ACP   ⎠

(c) ⎛ acyl(1)-O-lipid   acyl(2)-O-lipid ⎞
     ⎜   ↓   ↑   ⎟
     ⎜ acyl(1)-S-CoA   acyl(2)-S-CoA ⎟
     ⎜   │   acyl(1)-desaturase → acyl(2)-desaturase   ↑   ⎟
     ⎜   ↓   ⎟
     ⎝ acyl(1)-S-ACP   acyl(2)-S-ACP ⎠

(d) acyl(1)-O-lipid ──desaturase──→ acyl(2)-O-lipid

Figure 5.6 Possible pathways of desaturation utilising different substrate forms

At about the same time, Nichols, James and co-workers[91] were experimenting with *Chlorella vulgaris* in respect to the relationships between esterification of fatty acids into lipids and the sequential desaturation of those fatty acids.

Nichols[92] found a group of three lipids that became particularly strongly labelled when *Chlorella* was grown in the presence of [^{14}C]acetate namely phosphatidyl glycerol (PG), phosphatidyl choline (PC) and monogalactosyl diglyceride (MGDG).

A discussion of plant lipid structures can be found in Ref. 29. Once the maximum uptake of label had occurred, the total amount of radioactivity in these lipids remained constant for a considerable time. Initially most of this label was present in the saturated acids, 16:0 and 18:0. As the incubation proceeded, even though the total amount of radioactivity in a particular lipid remained constant, the label in the saturated acids fell, whilst the label in the corresponding monoenes first increased, then declined and so on, through to the trienoic acids (Figure 5.7).

These experiments confirmed the idea of 'compartments' and in 1968–1969 we were able to furnish more direct proof that the lipid itself could provide the acyl group for desaturation without the participation of thiolesters as obligatory intermediates[68]. The experiment was done in two ways. At first we used the more direct approach, that is to synthesise a lipid esterified with [1-^{14}C]oleic acid. The lipid employed was phosphatidyl choline, since this lipid was shown to have the highest specific activity both with respect to [^{14}C]oleic acid and [^{14}C] linoleic acid when [1-^{14}C]oleic acid was added as a precursor to either growing cultures of *Chlorella* or to subcellular preparations from the same organism.

When the labelled lipid was incubated with a particulate fraction from *Chlorella*, the product of the reaction was [^{14}C]linoleoyl-phosphatidyl choline:

$$[^{14}\text{C}]\text{oleoyl-PC} \longrightarrow [^{14}\text{C}]\text{linoleoyl-PC} \tag{5.5}$$

The possibility that small amounts of radioactive fatty acid were released from the lipid by a lipase, converted into thiolester, desaturated and the product re-esterified as illustrated in Figure 5.6b or c was eliminated by making use of the fact that the particle fraction contained *no acid: CoA ligase activity whatsoever*.

Several difficulties were encountered with this method, not the least being that complex lipids are difficult to 'solubilise' and present large micelles in which probably very little of the substrate is available. This results in extremely poor conversions. It turned out, too, that PC micelles had themselves an inhibiting effect on the enzyme, which could be relieved to some extent by making mixed micelles with other (unlabelled) lipids.

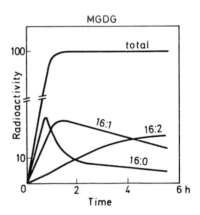

Figure 5.7 Fatty acid transformations in monogalactosyl diglyceride (MGDG) subsequent to *de novo* synthesis from [2-^{14}C]acetate. The ordinate represents the radioactivity of each fatty acid as a percentage of the total radioactivity in MGDG. [Figure 2 in Gurr, M. I. (1971). *Lipids* **6**, 266, reproduced by kind permission of the editors of *Lipids*.]

The second approach[68] was designed to overcome these difficulties and made use of the fact that when an obligatory co-factor for desaturation is withheld, the precursor fatty acid accumulates in the specific lipid. We, therefore, generated labelled oleoyl-PC within the intracellular membranes by growing the alga in the presence of labelled oleic acid but in an atmosphere of nitrogen and in the dark (to prevent 'internal' photosynthetic generation of O_2). When maximum accumulation of label had occurred, the cells were fragmented and a particle fraction isolated, great care being taken to maintain anaerobiosis throughout. Again the object of preparing a particulate fraction was to eliminate fatty acid activation enzymes. When the particle fraction had been carefully washed free of unesterified precursor and soluble protein, the particle preparation was exposed to an aerobic atmosphere and incubated further. About 10% of oleoyl-PC labelled *in situ* was transformed into linoleoyl-PC.

The criticism of this type of experiment is that we cannot be absolutely certain that all last traces of oleoyl-S-CoA generated during the incubation *in vivo* had been scavenged. However, the results of the two different types of experiments *taken in conjunction* strongly suggest that in this particular organism, desaturation of a fatty acid is taking place while the acyl moiety is esterified in the lipid in a manner analogous to the introduction of a methylene group across a double bond in the biosynthesis of bacterial cyclopropane fatty acids.

The results of Nichols[92-94], however, include the possibility that thiolester intermediates are involved within 'compartments' in the manner depicted in Figure 5.6b, or 5.6c. One cannot make the general conclusion that PC is specifically involved in the desaturation of oleic to linoleic acid since blue-green algae, which have no PC at all, are very efficient in synthesising linoleic acid.[95]

Indirect confirmation of these results in higher plants has come from the work of Roughan[96]. In labelling studies of pumpkin leaf lipids from [^{14}C]acetate, he showed that the changes in labelling of the fatty acids within the 'PC compartment' were consistent with the sequence 18:1→18:2→18:3.

Another organism that appears to utilise phospholipids as substrates for Δ12 desaturation is the fungus *Neurospora crassa*. Baker and Lynen[97] used a technique for generating labelled oleoyl-lipid *in situ* by incubating isolated microsmal fraction with oleoyl-S-CoA in the absence of pyridine nucleotide. As in Gurr's[68] experiments, unused precursor was removed by washing techniques and the labelled particle preparation incubated in the presence of NADH. Under these conditions, lipid-bound linoleic acid was formed at the normal rate. It is noteworthy that the same mechanism, with one possible exception, has never been shown to apply for the Δ9 desaturase. In Baker and Lynen's experiments[97] when the same techniques were applied to stearoyl-S-CoA, no desaturation occurred on addition of reduced pyridine nucleotides. Similarly, our own attempts to demonstrate desaturation of lipid-bound stearic acid in hen liver microsmal fractions always failed[98]. However, Japanese workers claim to have isolated a Δ9-desaturase from maturing soya bean cotyledons which acts on stearic acid bound to phosphatidyl choline[99,100].

More direct evidence has recently been supplied for the direct conversion of dioleoyl-PC into dilinoleoyl-PC in Bloch's laboratory[101] with the yeast, *Torulopsis utilis* and in Kates' laboratory[102] with the yeast *Candida lipolytica*.

In some tissues there is evidence, on the contrary, that the pathway involving lipid-bound substrates is definitely not operating. Vijay and Stumpf[85] attempted to eliminate the participation of acyl-*O*-lipids in desaturation in safflower microsomes by an ingenious procedure due to Barron and Mooney[103]. This method utilises the fact that reduction of the contents of the incubation with NaBH$_4$ converts thiolesters into alcohols. These are subsequently converted into trimethylsilyl-ethers. At the same time, *O*-esters in the incubation mixture are saponified, converted into fatty acid methyl esters and separated from the trimethylsilyl-ethers by g.l.c. This method should, therefore, be a sensitive monitor of the amounts of *O*- and *S*-ester in the incubation tube at any particular instant. Vijay and Stumpf[85] observed a very rapid transfer of acyl groups, both of precursor oleic acid and newly formed linoleic acid to phosphatidyl choline. Using the same technique as Baker and Lynen[97] of withholding NADH from the reaction, they were able to build up oleoyl-PC with no desaturation to linoleate. To eliminate all traces of oleoyl-S-CoA, that had not been incorporated into endogenous lipids, they added GPC as a 'scavenger'. The Barron and Mooney[103] technique indicated that all thiolester had been removed. When NADH was added, no further desaturation occurred indicating that oleoyl-lipid was not an intermediate and that oleoyl-S-CoA was an obligatory substrate.

In summary, there is no evidence to suggest that the pathway for desaturation which utilises lipids as direct substrates is a universal one and indeed it may represent a very minor and localised pathway or one which has been eliminated during evolution. In a later section on the substrate specificity of desaturases (Section 5.3.12.2) I shall discuss evidence that two separate $\Delta 12$ desaturases exist in *Chlorella* and leaf tissue and that there is a sound mechanistic reason for one of them to act specifically on lipid-bound substrates.

5.3.4 Alternatives to the direct desaturation pathway

5.3.4.1 Evidence for an alternative pathway

It has often been written that higher plants cannot form monoenoic acids by direct desaturation of palmitic or stearic acids and that there is, therefore, a pathway to oleic acid which is quite independent of stearoyl-S-CoA desaturase. The whole subject has been very thoroughly and critically reviewed by Stearns[58].

Three types of observations support this idea. First, the inability to demonstrate direct desaturation of added palmitic or stearic acids exemplified by the experiments of Stumpf and James[104]. Although all labelled short- and medium-chain saturated carboxylic acids up to $C_{14:0}$ were directly incorporated into C_{16} and C_{18} monounsaturated acids by lettuce chloroplasts, palmitic and stearic acids themselves were not utilised. Similarly, slices of ice-plant[105] do not desaturate added stearic acid but metabolise [1-^{14}C]myristate to 9-14:1 and oleic acid.

Secondly, on the basis of studies with cell-free preparations of avocado mesocarp, in which acetyl-S-CoA was a precursor of oleic acid whereas malonyl-S-CoA was a precursor only of saturated fatty acids, Barron and Stumpf[106] postulated separate pathways for saturated and unsaturated fatty acid biosynthesis in this tissue.

Thirdly, there is the evidence from many experiments in which the time course of incorporation of radioactivity from a pulse of [^{14}C]acetate into long-chain fatty acids has been followed[58,107,108]. This shows that: (a) the first acid to be labelled is oleic acid; (b) the specific activity of this acid rises very rapidly to high values; (c) that the labelling of stearic and palmitic acids is very slow and the specific activity of stearic acid does not relate to that of oleic acid as precursor to product according to the classical Zilversmit scheme[109]. In general, the relationship between the specific activities of oleic, linoleic and linolenic acids is in reasonable agreement with a sequential desaturation starting from oleic acid.

These observations, taken together, have led to the suggestion that oleic acid is the precursor of the polyunsaturated acids (which is consistent with studies at the enzyme level[13]) and that monounsaturated acids are formed by a separate pathway. This pathway[29] is usually conceived as involving the introduction of a double bond into an acid of relatively short chain length such that subsequent elongation would produce the long-chain monoene with the double bond in the appropriate position (Figure 5.8).

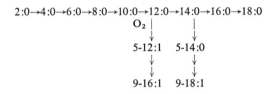

Figure 5.8 'The plant pathway' for monoenoic acid synthesis. [Redrawn from Figure 5.15 in Hitchcock, C.H.S. and Nichols, B. W. (1971). *Plant lipid Biochemistry* (London: Academic Press) by kind permission of the authors and publisher].

This pathway has features in common with the bacterial anaerobic pathway (Section 5.2.2) but differs from it in being unequivocally oxygen-dependent. A similar pathway has also been proposed for animal tissues to explain (a) the appearance of label from [^{14}C]acetate in oleic acid before its appearance in stearic acid and (b) the apparent ability of sterculic acid to inhibit oleate formation from stearic acid but not from acetate. This will be discussed in detail in Section 5.3.8.4.

A consequence of the 'plant pathway' is that the biosynthesis of two homologous Δ9 acids such as palmitoleic (9-16:1) and oleic (9-18:1) acids would require two distinct desaturases since elongation of palmitoleic acid would produce *cis*-vaccenic (11-18:1) *not* oleic acid.

5.3.4.2 Evidence for direct desaturation

Before discussing critically the evidence for an alternative pathway, it is worthwhile quoting a few examples of experiments in which direct desaturation *has* been demonstrated.

As early as 1957 there were reports in a Japanese journal that extracts from maturing soy bean cotyledons could directly desaturate myristic, palmitic, stearic and 12-hydroxystearic acids. Stearic acid was desaturated at the 9-position and palmitic at position 7[100]. These authors also partially purified the extracts and claimed to have demonstrated two distinct Δ9 desaturases, one acting on stearic acid, the other on stearic acid bound to phosphatidyl choline[99]. Inkpen and Quackenbush[110] have also obtained direct desaturation of added [1-^{14}C]stearic acid in subcellular fractions of maturing soy bean cotyledons.

Recently, my colleagues and I have been studying fatty acid and triglyceride synthesis in developing seeds of *Crambe abyssinica* and have obtained very small direct conversions of [1-^{14}C]stearic into [1-^{14}C]oleic acid in both whole seeds and subcellular fractions[111].

In photosynthetic tissue evidence is sparser. Nagai and Bloch reported that spinach chloroplasts can directly desaturate stearic acid only in the form of stearoyl-S-ACP[80,83] and can elongate short-chain acyl-S-ACPs[84].

Another approach that has been used to demonstrate the ability of higher plants to desaturate stearate is due to Harris and James[75]. Although chopped leaf preparations could not convert added labelled palmitic or stearic acids into the corresponding monoenoic acids, they were taken up by the leaf tissue and incorporated into complex lipids. Thus the problem of inability to desaturate stearic acid was not due to impermeability of the cell wall or to the absence of 'activation' enzymes, since 'activation' is a prerequisite for esterification of fatty acids into glycerides. Acetate, however, was readily incorporated into monoenoic acids by the preparations as long as the atmosphere in the incubation vessel was aerobic. Under anaerobic conditions no unsaturated acids were formed and labelled stearate accumulated. After performing the experiment under anaerobic conditions[75], Harris and James washed out the excess labelled acetate and exhausted the unused pool of endogenous acetate by continued anaerobic incubation. At this stage the only significant label present was in endogenous saturated acids. When the incubation was then exposed to oxygen, the radioactivity in stearic acid declined and was replaced by an approximately equal amount of oleic acid.

Combining these results with Bloch's data[80,83] on the need for ACP, one could make the interpretation that the immediate substrate for the desaturase is acyl-S-ACP and that in order to achieve desaturation of *added* stearic acid, the appropriate series of activations and transfers of fatty acid must occur. James' ideas on the subject are illustrated in Figure 5.9.

This scheme provides a framework for understanding the differences between monounsaturated fatty acid formation in plants on the one hand and in

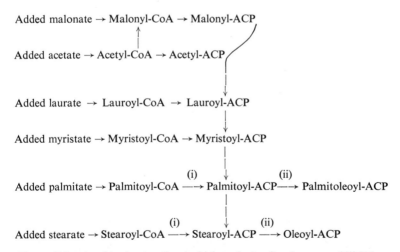

Figure 5.9 Aerobic desaturation in higher plants after James *et al.*[163] [Figure 5.17 from Hitchcock, C.H.S. and Nichols, B. W. (1971). *Plant Lipid Biochemistry* (London: Academic Press) by kind permission of the authors and publisher.]

animals and micro-organisms on the other. One can postulate that stearoyl-S-CoA:ACP acyl transferase is lacking in higher plants; in algae, such as *Chlorella*, it may be inducible[81] by transferring heterotrophically grown cells to autotrophic conditions. Animals, whose tissues are constantly in contact with circulating long chain fatty acids, have developed the mechanism for transferring exogenous stearic acid to the site at which it is to be desaturated. We can picture at least two metabolically different states of stearic acid. One is synthesised *in situ* from acetyl-CoA (derived from carbohydrate breakdown) and is desaturated to oleic acid without becoming mixed with the other pool which is available to exogenously added stearate. In plants, apparently, although palmitic and stearic acids are not accessible to the multi-enzyme complex, shorter chain acids such as myristate are accessible[104].

5.3.4.3 Connexion between different metabolic sequences

This raises the whole question of 'linking processes' between multi-enzyme complexes, a subject that has great significance for the compartmentalisation of biosynthetic processes within cells, but which has received very little attention. In the case of lipid synthesis we have to consider the interlinking of: (a) saturated fatty acid synthesis; (b) fatty acid desaturation; (c) fatty acid 'activation' to thiolesters; (d) further elongation of long chain fatty acids; (e) chain shortening (retroconversion); (f) esterification of fatty acids into lipids and finally (g) transfers of substrates between these complexes (Figure 5.10).

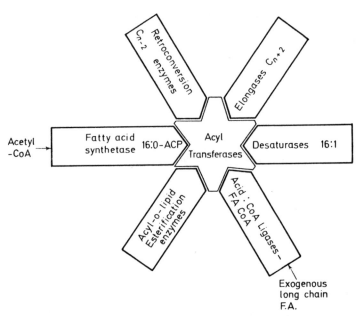

Figure 5.10 Diagrammatic representation of possible close interrelationships between the different multi-enzyme complexes of lipid biosynthesis

When fatty acids are synthesised *de novo* from acetate, the intermediates are bound to a protein that is probably an ACP-like component of the synthetase[26,77,112,113]. Therefore the end-product of the fatty acid synthetase (depending on the organism) must initially be in the form of palmitoyl or stearoyl-S-ACP. When we know more about what happens to this substrate in between its formation as the end-product of the fatty acid synthetase and its utilisation as starting product for desaturation, esterification or simply release from the complex as NEFA (Figure 5.10), we shall be better able to decide whether there are real mechanistic differences between monounsaturated fatty acid formation in plants and animals, or as I predict, the differences are traceable to differences in cellular and enzymic architecture. One way in which the integration of such enzyme complexes may be investigated is by the use of specific inhibitors; this subject is discussed in Section 5.3.8.4.

5.3.4.4 Critical assessment of the alternative pathway

In kinetic experiments in which the labelling of oleic acid is observed to precede that of stearic acid at 'short-labelling periods', the shortest period studied appears to be 2 min. It may be that such time periods are actually rather long in terms of the kinetics of the labelling of that particular stearate pool that may be giving rise to oleic acid. The practical difficulties of working at shorter times have inhibited studies in this direction. Also, we know virtually nothing about the relative rates of the competing processes depicted in Figure 5.10 and the problems are unlikely to be resolved until we can separate the various multi-enzyme complexes involved.

Mudd[55] has cited an example of extremely rapid incorporation of stearate into lipids by cell-free preparations. If esterification is rapid compared with desaturation, the formation of oleate will be slow or non-existent owing to effective removal of the substrate. If the rates of 'activation' and elongation of shorter chain acids are faster than the rates of esterification, their superiority as precursors for unsaturated acids could be explained. I shall discuss the controlling effects of competing reactions of lipid biosynthesis on desaturation further in Section 5.3.6.3.

In summary, failure to demonstrate the complete universality of a well established mechanism in general leads to two sorts of postulates. First, an alternative pathway may be envisaged. In this case, the proponents of such a pathway are under obligation to demonstrate the existence of and identify all postulated intermediates on such a pathway. Secondly, one may resort to the idea of separate compartments or metabolic pools of precursors that may not be available to the exogenous substrate, at least in the form in which it is presented by the experimenter. This is the current situation with regard to oleic acid formation in plants (and in animals under certain conditions, see Section 5.3.8.4.)

The question under dispute is whether oleic acid *always* arises by direct introduction of a double bond into stearic acid, or whether, in some cases, the double bond may be introduced at a position close to the carboxyl group allowing chain elongation to produce the desired end-product. Nobody doubts that oxygen is required for either process or that the actual double bond introduction, wherever it may occur in the chain, is by essentially the same mechanism of direct oxidative dehydrogenation.

The adherents of the alternative pathway have not produced evidence in the form of correctly identified intermediates, and until this is achieved, the specific activity data are not sufficient proof of an alternative pathway. Harwood et al.[114] have recently shown that avocado supernatant fraction converts labelled dodecanoic acid into 3-hydroxydodecanoic acid and a *very* small quantity of what *may* be dodecenoic acid. This might suggest a pathway similar to that depicted in Figure 5.8, but in which 3-hydroxydodecanoic acid is dehydrated to *cis*-3-dodecenoic acid followed by chain elongation to oleic acid (9-18:1). However no oleic acid was formed in their system and until the double-bond position of the C_{12} monoene is established, this pathway remains highly speculative. Arsenite, which inhibits elongation of 16:0 to 18:0[115], does not inhibit oleate formation providing circumstantial evidence for this alternative pathway.

With the techniques available at present, it will be even more difficult to prove specific compartmentalisation and many people are going to regard this hypothesis as a convenient synonym for lack of knowledge.

5.3.5 Evidence for acyl-enzyme intermediates in desaturation

In my discussion in Section 5.3.3., I referred frequently to acyl-enzyme intermediates in desaturation (e.g. See Figure 5.6). No evidence exists to date that such intermediates occur and all references to them are thus highly speculative. Some evidence has been presented that in certain tissues, covalently-bound acyl-enzyme intermediates are *not* involved in desaturation. This evidence is based on the fact that radioactive coenzyme A does not exchange with substrate (or product) as it should if acyl-enzyme intermediates are formed according to the following scheme:

$$\text{Acyl(1)-S-CoA} + \text{Enzyme} \rightleftharpoons \text{Acyl(1)-Enzyme} + \text{CoASH} \quad (5.6)$$

$$\text{Acyl(1)-Enzyme} \rightarrow \text{Acyl(2)-Enzyme} \quad (5.7)$$

$$\text{Acyl(2)-Enzyme} + \text{CoASH} \rightleftharpoons \text{Enzyme} + \text{Acyl(2)-S-CoA} \quad (5.8)$$

Schulz and Lynen[116] found no evidence of such exchange when they incubated [^{14}C] CoASH with an equimolar proportion of (unlabelled) palmitoyl-S-CoA and a yeast (*Saccharomyces cerevisiae*) microsmal fraction. In a similar experiment, Vijay and Stumpf[85] demonstrated less than 1 % exchange of [^{3}H]CoASH into oleoyl-S-CoA when these substrates were incubated with a microsmal fraction from maturing safflower seeds.

5.3.6 Products of the desaturation reaction

5.3.6.1 Nature of the products

The ultimate fate of the end-products of fatty acid synthesis and desaturation is esterification in complex lipids. Highly purified desaturase preparations are not yet available (See Section 5.3.10.2) and virtually all the preparations in use for desaturation studies also contain active acyl transferases that transfer the product of the desaturase reaction, whether it be as the acyl-S-CoA or acyl-S-ACP, onto the glycerol backbone of a complex lipid.

Thus the products of the reaction are located in phospholipids (or glycolipids) and to a lesser extent triglycerides at a quite early stage in the incubation in *Neurospora* microsomes[97], *Chlorella* particle fraction[68], safflower seed microsomes[85], rat liver microsomes[117], cow[118] and goat[119] lactating mammary gland microsomes, and hen liver microsomes[98]. In hen liver microsomes, after a 30 min incubation with [^{14}C]stearoyl-S-CoA, 90% of the radioactivity was in lipids; 40% was newly-formed oleic acid. Phosphatidyl choline accounted for 51% of the label, phosphatidyl ethanolamine 20%, triglycerides 22%; 4% of radioactivity was 'protein-bound'. Baker and Lynen[97] observed strong binding of acyl-S-CoA to yeast microsomes and a similar phenomenon occurs in rat liver[120]. Care must be taken to work in the region where product formation is proportional to protein concentration, since high amounts of protein bind substrate effectively reducing its concentration.

In hen liver microsomes[98] non-esterified fatty acid accounted for 7% of the activity distribution at the end of the incubation. This can be accounted for by an active thiolester hydrolase in microsomal fractions[116,121,122] that releases fatty acid from the substrate before it can be desaturated, and also from the product before it can be incorporated into lipids. The thiolester hydrolase in rat liver microsomes is extremely active and has the effect of reducing the substrate concentration so that the time-course of the reaction is linear for only a very short time[121,123]. This is especially marked if the microsomal fraction is preincubated with acyl-S-CoA in the absence of NADH[121]. In yeast, although Bloomfield and Bloch[69] identified palmitoleoyl-S-CoA as the end-product of palmitoyl-S-CoA desaturation, Schulz and Lynen[116] were able to demonstrate a very active thiolester hydrolase and described its kinetic properties. In contrast, safflower seed microsomes have virtually no hydrolase activity[85].

The distribution of the product of desaturation among the different lipid classes is not dissimilar from the actual composition of unsaturated fatty acids in the endogenous lipids of rat and hen liver microsomes[98,117]. Quite a different picture emerges in plant preparations, however, where the predominant lipids are galactosyl glycerides but the most highly labelled lipid is almost always phosphatidyl choline[68,85,96]. This is true for seed tissue just as much as for photosynthetic tissue[85]. When isolated spinach leaf chloroplasts are incubated with [^{14}C]acetate, thiolesters are the major products at early stages of the incubation, but at later stages a greater proportion of the labelled fatty acids was observed in predominantly phospholipids, but also in monogalactosyl glycerides and diglycerides[124].

It seems likely that in these studies, a somewhat unphysiological situation is being studied. Because of the presence of an acyl transferase which has a strong preference for transfer of exogenous long-chain fatty acid to an endogenous acceptor to form phosphatidyl choline, the labelling pattern does not reflect the true fatty acid composition of the tissue. Thus, when growing *Chlorella* cells are supplied with acetate, the labelling pattern reflects the true distribution of lipids in the cell; when isolated particles are incubated with a pre-formed long-chain substrate, the label is almost exclusively in PC[68,92,95].

5.3.6.2 Inhibition by the products of desaturation and its relief

In their original full paper on palmitoyl-S-CoA desaturation in yeast, Bloomfield and Bloch[69] found that palmitoleic acid added as the potassium salt to the incubation medium markedly inhibited the reaction, whereas oleic and palmitic acids, at the concentrations tested, were not inhibitory. There was 10% inhibition of desaturation when the palmitoleate was in tenfold excess of substrate and 80% when in 100-fold excess. In contrast, Schulz and Lynen[116] could not inhibit the *Saccharomyces cerevisiae* desaturase with non-esterified fatty acid, but demonstrated by kinetic experiments that palmitoleoyl-S-CoA competitively inhibited the desaturation of palmitoyl-S-CoA while oleoyl-S-CoA competitively inhibited the desaturation of stearoyl-S-CoA. Similarly, in rat liver microsomes, oleic acid had no inhibitory effect on stearoyl-S-CoA desaturation when in a concentration four times greater than the substrate, whereas oleoyl-S-CoA, at the same concentration, inhibited 70%[121].

In view of what was said in Section 5.3.6.1., about the effect of protein in binding acyl-S-CoA, the protein concentration may be expected to have an influence on the degree of product inhibition. The substrate itself is inhibitory at very high concentration ($>100~\mu M$) and this inhibition can be relieved by increasing the concentration of microsomal protein[120]. In fact, Pande and Mead[120] regard the inhibition by acyl-S-CoA thiolesters and non-esterified fatty acids as examples of inactivation by surface active agents not related to specific effects of reaction products. In their experiments, oleic and linoleic acids were equally effective at inhibiting desaturation (40% at 12-fold molar excess of inhibitor over substrate in presence of 0.3 mg protein). In the hands of Raju and Reiser[125], however, a 16-fold molar excess of oleic acid did not inhibit stearic acid desaturation by rat liver homogenate. In this case the protein concentration was not stated, but since the incubation contained '1 ml of crude rat liver extract', the protein concentration was probably extremely large and this may have relieved any product inhibition.

Rather more careful studies have been made of the regulatory effects which the different fatty acids of the liver microsmal desaturation sequence exert upon each other. These sequences are depicted in Figure 5.11.

The fact that animals and plants form different types of polyunsaturated fatty acids by inserting the second and subsequent double bonds between the first double bond and the carboxyl group (animals) or between the first double bond and the methyl end (plants; see Figure 5.4) gives rise to distinct families of polyunsaturated fatty acids that are *not interconvertible in animal tissues*[28,54]. These can be more easily recognisable by numbering the double bonds from the methyl end of the molecule. Hence the three most important families of polyunsaturated acids are $\omega 9$, $\omega 6$ and $\omega 3$ and the first members of each family are oleic, linoleic and *a*-linolenic acids respectively (Figure 5.11).

Each of these acids is converted by a series of desaturations and elongations into a series of polyunsaturated acids belonging to the same ω-family. The preferred routes appear to be by alternate desaturations and elongations rather than a sequence of desaturations followed by elongation[54,126].

Studies on the way in which fatty acids of these families regulate each other's metabolism have been done both *in vivo* and *in vitro*. *In vivo*, for example, increasing amounts of dietary *a*-linolenate (18:3) suppressed the levels of arachidonate (20:4) in tissue lipids, indicating that the conversion of 18:2 to 20:4 is inhibited by dietary 18:3[127].

In a series of studies to examine the regulatory effects arising from competition between the different acids of a series *in vitro*, Brenner and his colleagues have shown that the product of each desaturation step competitively inhibits the desaturation of its precursor[128-133]. Brenner has summarised his and others' results in a recent article[54]. Only unsaturated fatty acids of similar chain length significantly depressed the conversion of oleic, linoleic and linolenic acids into more unsaturated acids. The depressive effect increased with

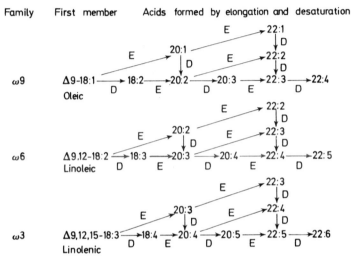

Figure 5.11 The three important families of polyunsaturated acids in animals showing chain elongation (E) and desaturation (D) [Figure 2.11 in Gurr, M. I. and James, A. T. (1971). Lipid Biochemistry: An introduction (London: Chapman and Hall) by kind permission of the publisher]

unsaturation in the order 18:3>18:2>18:1. These results correlate well with results *in vivo* in respect to the inhibition of 5,8,11-20:3 synthesis by fat deficient rats after supplementation of the diet with linoleic or linolenic acid[127].

In addition to this direct product inhibition, Brenner[54] also observes 'feedback inhibition' on the desaturation of the first member of each family by the very long-chain highly unsaturated end-products of the sequence (22:4, 22:5, 22:6).

It is important to point out that in Brenner's experiments relatively small amounts of the end-product acids were added to the incubation mixtures. This is highly suggestive of specific inhibitory effects whereby fatty acids can regulate the metabolism of others within the same family rather than non-specific and non-physiological detergent effects *in vitro*. In assessing the inhibitory effects of fatty acids *in vitro*, therefore, the concentration of both substrate and inhibitor fatty acids and the concentration of protein should be carefully considered.

5.3.6.3 Reactions competing for substrate and product

Brenner's results also show[54,134] that the desaturation of certain acids of lesser degree of unsaturation is *stimulated* by other more highly-unsaturated acids. This is particularly noticeable in regard to the effect of 20:4 on the desaturation of saturated acids and oleic acid. Brenner's data suggest that arachidonoyl-S-CoA competes successfully with palmitoyl, oleoyl, or linoleoyl-S-CoA for esterification into endogenous lipids, thus 'sparing' more of these substrates for the desaturase and increasing their effective substrate concentrations. As a consequence, increasing the level of endogenous acceptors, lysophospholipids, eliminates the activating effect of 20:4 because there is then an excess of acceptor and the substrates need no longer compete. It follows that the rate of desaturation of any substrate *in vitro* will be critically influenced by the level of endogenous acceptors. Vijay and Stumpf[85], explain the rapid fall-off in rate of oleoyl-S-CoA desaturation by safflower microsmal fraction by the rapid removal of substrate by transacylases that esterify acyl groups into endogenous lipids. Since both precursor and product fatty acids are presumably available to these transacylases, their relative affinities for the transacylase will determine whether desaturation is stimulated by preferential removal of inhibitory product or retarded by preferential removal of substrate. This probably explains the wide discrepancies reported in the literature concerning the effects of added 'acceptors'. For example glycerol-3-phosphate stimulated desaturation of stearic acid by rat adipose tissue[70], by goat and sow mammary gland[119] cow mammary gland[118], and rat liver microsomes[135]. However, other authors have found an inhibitory effect on stearate desaturation in rat liver microsomes[121], and on linoleate desaturation[117], or no effect at all on stearate desaturation[117].

Other reactions, apart from esterification, that compete with desaturases for substrates are those that involve either chain elongation[136-138] or chain shortening[139].

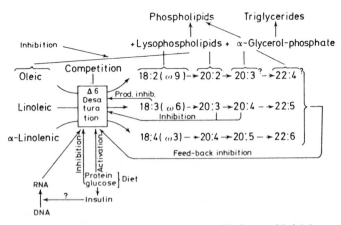

Figure 5.12 Scheme of factors that may modify fatty acid Δ6-desaturation in microsomes. [Figure 4 in Brenner, R. R. (1971) *Lipids*, **6**, 567 by kind permission of the author and editors of *Lipids*.]

In endoplasmic reticulum, acyl-S-CoA derivatives are the substrates for chain elongating enzymes and the additional carbon atoms are derived from malonyl-S-CoA[140]. An acetyl-S-CoA dependent chain elongating system is present in mitochondria[141] but its importance for the biosynthesis of C_{20} and C_{22} polyenoic fatty acids, compared with the very active microsmal malonyl-S-CoA requiring system is questionable. Chain elongation of acyl-S-ACP derivatives by spinach chloroplasts has been demonstrated by Nagai and Bloch[84].

Chain shortening can occur by β-oxidation, which is a mitochondrial process and generally results in complete degradation of the fatty chain to acetyl-S-CoA. However, chain shortening by only two carbon atoms, a process that has come to be known as *retroconversion* is also catalysed by mitochondrial enzymes[139]. This chain shortening reaction is not due to a reversal of chain elongation reactions, but to a partial β-oxidation.

The various reactions that compete for fatty acids are depicted in Figure 5.12.

Much more needs to be known about the relative rates of these competing reactions and much more care taken in monitoring concentrations of components in incubations *in vitro*.

5.3.7 Co-factors for the aerobic desaturation reaction

5.3.7.1 *Electron acceptors: oxygen*

In all organisms except those bacteria that possess the anaerobic pathway (Section 5.2.2.), molecular oxygen is obligatory for the formation of the double bond whether the product is a monounsaturated or a polyunsaturated acid. So far, no artificial electron acceptors that will replace oxygen, such as methylene blue, ferricyanide or phenazine methosulphate[69] have been found.

Difficulties may be encountered in removing all last traces of oxygen, but absolute dependence can be demonstrated by flushing out with nitrogen and arranging to trap expelled oxygen in alkaline pyrogallol. If this is insufficient, small amounts of oxygen can be removed enzymically by including a glucose oxidase–catalase system in the assay[69]. The difficulties are even greater in photosynthetic tissue, where oxygen is being generated internally, and the additional precautions of excluding light can be used to effect.

In photosynthetic tissue the concentration of oxygen is not rate-limiting for desaturation, but in those tissues where oxygen must reach the desaturase by diffusion from the atmosphere or fluid surrounding the tissue, the production of unsaturated acids may be limited by oxygen supply[142,143]. This can explain why the production of unsaturated fatty acids in the seed oils of some plants is greater when they are grown at lower temperature[1,144]. This effect seems to depend on the particular species because, for example, Canvin[144] found that it occurred in seeds of rape, sunflower and flax, but not of safflower or castor.

Two effects could be involved: (a) an increase in level of the desaturase enzymes after growth of the organism at a lower temperature and (b) an effect of temperature on reaction velocity, possibly by regulating oxygen availability.

Harris and James[142,143] investigated this point by measuring unsaturated fatty acid production from acetate in a number of plant tissues at different temperatures and at different oxygen tensions. In non-photosynthetic tissue such as seed[142,143] or bulb[142] supply of oxygen is the major rate-limiting factor in desaturation. Since the effect of decreasing the temperature at a fixed gaseous oxygen concentration is to increase the oxygen concentration in solution, there is therefore an increase in rate of desaturation. In active photosynthetic tissues, oxygen cannot be shown to be rate-limiting in the light because excess of oxygen is produced within the cell by photosynthesis.

This phenomenon probably has physiological significance in that an organism in a cool environment may require lipids of lower melting point, i.e. higher degree of unsaturation, and these are automatically supplied by increased oxygen availability. No elaborate controls are required and the system is self-regulated.

The yeast, *Torulopsis utilis*, yielded cell-free preparations with a more active stearoyl-S-CoA desaturase when grown at 19°C than when grown at 30°C[145,146].

Frogs are among the animals whose body lipid composition changes under different temperature conditions; liver and adipose tissue from cold-adapted frogs contained a higher unsaturated to saturated fatty acid ratio than the corresponding tissues of animals maintained at higher temperatures[147].

In Bacilli, temperature may influence the production of unsaturated fatty acids by mechanisms such as induction of enzyme synthesis that do not involve oxygen supply[148].

5.3.7.2 Electron donors

The reduced pyridine nucleotides NADH or NADPH also seem to be a universal requirement of all aerobic desaturases so far examined. The specificity for these nucleotides, however, does not seem to follow any recognisable pattern. For example, in most desaturations described in the literature NADH is a more effective electron donor than NADPH. This is true in *animals* for the $\Delta 9$ desaturase (rat liver microsomes[121,123]; hen liver microsomes[98,122]; rat adipose tissue[70]; goat mammary gland[119]); the requirement of the $\Delta 6$ desaturase is less clear[23,54,149]. It is also true in fungi for the $\Delta 9$ desaturase (*N. crassa*[97]) and in higher plants for the $\Delta 12$ desaturase (safflower[60]). Thus one cannot discern a distinct difference in nucleotide requirement for desaturases that insert double bonds at different positions, nor can one detect any evolutionary change in nucleotide requirement. Yet, in spite of the foregoing 'rule' the first cell-free desaturase system to be described, the $\Delta 9$ desaturase of *Saccharomyces cerevisiae*[69], had a marked preference for NADPH and in *Mycobacterium phlei*[8,150] the $\Delta 9$ desaturase has an absolute requirement for NADPH. In this latter case, NADH is inhibitory in the presence of NADPH. The $\Delta 12$ desaturases of *N. crassa*[97] and *C. vulgaris*[13] have no preference for either nucleotide. In *E. gracilis*[83] the $\Delta 9$ desaturase in the crude homogenate will utilise either nucleotide with a preference for NADPH, but the purified preparation from this organism, which is incidentally the most highly-purified desaturase yet achieved, has a requirement for NADPH which is absolute.

Desaturases have many features in common with mixed function oxygenases which normally have an absolute specificity for NADPH[121]. We now know that the desaturases and hydroxylases are each associated with an electron-transport chain and it is possible that differences in specificity for the nucleotide reflect the degree of interaction between these chains and the positions of cross-over points between the two chains. I shall discuss this subject at greater length in Section 5.3.10.

In rat liver microsomes, ascorbate supports desaturation of stearoyl-S-CoA, although ten times the molar concentration is required to produce half the rate supported by NADPH[121]. In contrast, ascorbate had no activity in the purified desaturase of *E. gracilis*[83] even when included at 50 times the concentration of NADPH.

In cases of severe vitamin C deficiency, the serum lipid levels of some animals rise and there is a depression of monounsaturated fatty acid levels[151]. Could it be possible that in certain physiological conditions in which the supply of reduced pyridine nucleotide is limited, ascorbate takes on a role as a physiological electron donor and that in these cases, changes in ascorbate levels are reflected in changes in unsaturated fatty acid production?

5.3.8 Inhibitors of desaturation

5.3.8.1 SH Blocking and protecting reagents

It has often been assumed that desaturases in general are, like fatty acid synthetases, *SH*-enzymes, but in fact a great deal of confusion exists over this issue. Quite probably the final answer must await further purification of the enzyme.

p-Hydroxy and *p*-chloromercuribenzoate (PCMB) are invariably strongly inhibitory at low concentration[65,125,149]. *N*-Ethylmaleimide and iodoacetamide are often more weakly inhibitory[125] or do not inhibit at all[65,116,152]. The PCMB inhibition is generally reversible by SH-reagents such as mercaptoethanol[125,152] but this reagent itself is often strongly inhibitory[68]. The *M. phlei* desaturase is strongly inhibited by both classes of compounds.

5.3.8.2 Metal chelators

Only one desaturase (*M. phlei*) has so far been found to have an unequivocal metal requirement, namely for ferrous ion[150]. *o*-Phenanthroline, which forms strong complexes with Fe^{2+} ions, was a more effective inhibitor than EDTA, which is a less specific complexing agent. *o*-Phenanthroline inhibits other desaturases to varying degrees ranging from mild[65,121] to strong[73]. EDTA inhibition also varies enormously[65,73,150].

5.3.8.3 Cyanide

Cyanide strongly inhibits stearoyl-S-CoA desaturation in rat liver microsomes (50% at 10^{-4}M; 95–100% at 10^{-3}M)[121]. Mixed function oxygenation of drugs and steroids, however, is unaffected[121]. From these data, Sato[121,153] and his co-workers have developed the hypothesis that the terminal step in desaturation, at which electrons are transferred to O_2 to form water, involves a cyanide-sensitive protein. Evidence for the existence of this compound and its role in the desaturase complex will be discussed in Sections 5.3.10.2–5.3.10.4.

Most other desaturases are also inhibited by cyanide[65,73,83] but strangely, the yeast palmitoyl-CoA desaturase[69] does not seem to be affected. In spinach chloroplasts the formation of 18:2 was completely blocked by CN^- whereas 18:3 formation was unaffected[124].

5.3.8.4 Cyclopropene fatty acids

(a) *Background*—Cyclopropene fatty acids have the general structure:

$$Me(CH_2)_n \cdot C \overset{\overset{\displaystyle CH_2}{\diagup \ \diagdown}}{=} C \cdot (CH_2)_m \cdot CO_2H$$

The commonest natural cyclopropene acids are malvalic (m = 6; n = 7) and sterculic (m = n = 7) which occur in plants of the order *Malvales*[154]. Feeding animals diets rich in cyclopropenes results in hardening of their body fats due to an accumulation of stearic acid[154-156]. The reason for this is that cyclopropene acids inhibit the formation of oleic acid from stearic acid, as has been shown by studies *in vivo* and *in vitro* with rodents[120,125,157,158], hens[155,159-162] and algae[163].

(b) *Mechanism of the inhibition*—First, what do we mean by inhibition? Sterculic acid is a very specific inhibitor that acts at very low concentrations (90% inhibition at inhibitor:substrate ratio of 1:10)[159]. It is important to take note of concentrations when assessing the effects of inhibitors which are also carboxylic acids since inhibitions at ratios greater than 1:1 may be due to non-specific detergent effects[120,158]. What are the structural features of the molecule that are essential for inhibitor activity?

First a cyclopropene ring; modification of the ring eliminates activity[125,159]. Secondly, a carboxylic function; sterculyl alcohol is less effective than sterculic acid[160]. Reports of inhibition from the alcohol[163,164] or the hydrocarbon sterculene[164] may be due to oxidation of these compounds in the tissues but this remains uncertain. In one report[163], 1,2-dihydroxysterculene was as effective as sterculic acid. It is not yet known whether the CoA thiolester is the true inhibitor. Recent claims that sterculic acid is not a specific inhibitor, could be due to the fact that the free acid was not being efficiently converted into the thiolester[120,161]. Thirdly, inhibitor activity has been shown to decrease as the ring is moved away from the 9,10-position[161] as illustrated in Figure 5.13.

Arguing mainly from the above evidence it has been proposed that sterculic acid inhibits by blocking the active site due to interaction with an essential SH-group[125,160,161]. The interaction of cyclopropenes with SH-groups is well documented[125] though fatty acid synthetase, a well-known *SH*-enzyme, is not inhibited[125].

$$RC\overset{*}{\underset{9}{=}}\overset{\triangle}{\underset{8}{C}}\cdot(CH_2)_6\cdot COCoA \qquad C18$$

$$R\overset{*}{\underset{10}{C}}\overset{\triangle}{=}\overset{*}{\underset{9}{C}}\cdot CH_2\cdot(CH_2)_6\cdot COCoA \qquad C19$$

$$R\underset{11}{C}\overset{\triangle}{=}\overset{*}{\underset{10}{C}}\cdot CH_2\cdot CH_2\cdot(CH_2)_6\cdot COCoA \qquad C20$$

$$R\underset{12}{C}\overset{\triangle}{=}\underset{11\;10}{C}\cdot CH_2\cdot CH_2\cdot CH_2\cdot(CH_2)_6\cdot COCoA \qquad C21$$

$$R\underset{10}{CH_2}\cdot\underset{9}{CH_2}\cdot\underset{8}{CH_2}\cdot(CH_2)_6\cdot COCoA$$

|———— Enzyme ————| Stearic Acid
b   a

Figure 5.13 Positions of the cyclopropene fatty acids and stearic acid in relation to the desaturase enzyme. C_{18}, C_{19}, C_{20}, and C_{21} represent the CoA derivatives of the cyclopropane fatty acids; R is $CH_9(CH_1)_1$—. The site of initial attachment of the substrate or inhibitor to the enzyme is indicated by point a, the site of desaturation of substrate by point b. The configuration of the enzyme surface between a and b is such that desaturation only occurs at the 9, 10-position of the substrate. [Figure 1 in Fogerty, A. C., Johnson, A. R. and Pearson, J. A. (1972) *Lipids*, 7, 335 by kind permission of the authors and editor of *Lipids*]

Table 5.2 The effect of sterculate on tobacco leaf discs incubated anaerobically with acetate and transferred to aerobic conditions (See Ref. 163. Reproduced by kind permission of the authors and publishers of *European Journal of Biochemistry*).

Tobacco leaf discs were incubated with $6\mu C$ [2-^{14}C]acetate under argon for 5 h, washed and then divided into three samples. One sample was extracted (1) and the other two were incubated aerobically for 5 h, one with 3 mM sterculate added (3) and the other with no added sterculate (2).

Incubation	Distribution of activity in fatty acids			
	Palmitic	Palmitoleic	Stearic	Oleic
		Percentage		
(1) 5 h anaerobic	47	0	53	0
(2) 5 h anaerobic + 5 h aerobic	46.5	6.0	37.5	10.0
(3) 5 h anaerobic + sterculate + 5 h aerobic	49.5	traces	42.0	8.5

Apart from the doubts expressed above of whether desaturase is an *SH*-enzyme, one of the strongest pieces of evidence against the idea that the desaturase itself is a direct target for inhibition, is the finding by James et al.[163] that the conversion of 'internally-generated' stearic acid into oleic acid is not inhibited by sterculate. The stearate was generated from labelled acetate by anaerobic incubation; subsequently the incubations were exposed to O_2 in the presence or in the absence of sterculate (Table 5.2).

James[163] has therefore proposed that sterculic acid inhibits not the desaturase but an acyl transferase involved in transferring exogenous stearic acid to the site where it can be desaturated (Figure 5.14).

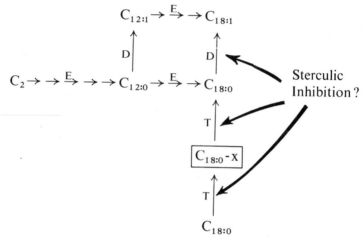

Figure 5.14 Possible sites of inhibition by sterculic acid. (E) represents elongation, (D) desaturation, (T) acyl transfer. [Figure 9 in Gurr, M. I. (1971), *Lipids*, **6**, 266, reproduced by kind permission of the editors of *Lipids*]

This hypothesis could account for the fact that the presence of cyclopropenes in certain plants has no adverse effects on the plant's ability to synthesise unsaturated acids. Higher plants must synthesise all their fatty acids *de novo*; in animals, tissue fatty acids (including both stearic and oleic) may arise by synthesis *de novo* or be supplied unchanged by the diet. In addition, tissue oleic acid may arise by direct desaturation of dietary stearate. If James' hypothesis is correct, only oleate arising from ingested stearate would be susceptible to inhibition by dietary sterculate.

(c) *The alternative pathway for oleic acid biosynthesis*—One of the key experimental findings that any hypothesis about the mechanism of sterculic acid inhibition has to be able to explain, is that the inhibition of oleic acid synthesis is much less when the precursor is [^{14}C]acetic acid than when it is [^{14}C]stearic acid. This has been demonstrated in Reiser's laboratory for the rat[157] and in James' laboratory for *Chlorella vulgaris*[163].

Reiser[157] has proposed that this provides evidence for an alternative pathway to oleic acid in the rat. The features of this suggested pathway (which had been proposed by Favarger[165] on the grounds that oleic acid was labelled much more rapidly from [^{14}C]acetate than was stearic acid) are: (i) that the double bond is introduced into a fatty acid of shorter chain length than stearic acid at such a position that subsequent elongation provides oleic acid and (ii) that unlike the bacterial process with which it bears some resemblance, the desaturation is completely oxygen-dependent. As you can see from Figure 5.14, this pathway resembles the 'plant' pathway discussed in Section 5.3.4.

Evidence has been presented that this pathway can be induced by continuous feeding of sterculate in chicks[162] but this is not the case in the goat[156]. Johnson[166] has pointed out that Reiser's[157] method of expressing the results of inhibition studies in terms of specific activity of unsaturated acid/specific activity of saturated acid may be misleading. After sterculate feeding, the pool size of stearic acid is increased and this will affect the value for *specific activity*.

So far, the evidence for this pathway is indirect; the final proof must rest on the demonstration that the postulated monoenoic intermediates exist *in the inhibited state*.

James[163] claims that the necessity to postulate an alternative pathway only arises if it is held that the stearoyl-S-CoA desaturase itself is the target for inhibition. If one accepts his view that an acyl transferase is the target site, then the difference between the ability of acetate and stearate to act as precursors of oleate in the inhibited state is readily explained since acetate will give rise to 'endogenous stearate' not requiring the transferase for its access to the active site, whereas exogenous stearate cannot gain access to the active site without the involvement of the transferase.

5.3.9 Assay of desaturases and characterisation of the product

5.3.9.1 *Methods depending on the measurement of conversion of* [^{14}C]*substrate into* [^{14}C]*product*

In this category, the usefulness of the method hinges on the ability to separate product from precursor. Several methods have been used: (a) Fractional crystallisation of lead salts (e.g. Ref. 2). (b) Formation of mercuric adducts of the unsaturated products and separation of these from unchanged saturated acids by silicic acid chromatography (e.g. Refs. 83, 150). (c) Oxidative cleavage of the double bond of the product and measurement of radioactivity in the resulting labelled dicarboxylic acid after chromatography (e.g. Ref. 120). (d) Separation of methyl esters by radiochemical g.l.c. (e.g. Refs. 13, 68, 97, 117). (e) Separation of methyl esters by argentation t.l.c. followed by measurement of radioactivity in the fractions by scintillation spectrometry (e.g. Refs. 123, 167).

Only (d) and (e) have extensive use now, but these methods have the disadvantage that the procedure is lengthy, tedious, and very time consuming if large numbers of samples have to be handled. The unsaturated product may be characterised by oxidative cleavage of the double bond and identification of the resulting mono- and di-carboxylic acids (e.g. Refs. 167, 168). The position of the label in the product may be important in order to distinguish between direct desaturation and synthesis *de novo*. This is usually done by oxidative degradation, one carbon at a time beginning at the carboxyl group.

Radioactive carboxyl carbon can be measured by trapping CO_2; each homologous lower fatty acid produced by degradation can be identified by radiochemical g.l.c. (e.g. Refs. 81, 104).

5.3.9.2 *Methods depending on NAD(P)H utilisation*

In practice, the opacity of crude microsomal preparations and the presence of competing reactions makes this method difficult to use with conventional spectrophotometers. Japanese workers have been able to measure reliably NAD(P)H oxidation by use of a triple-beam instrument[169].

5.3.9.3 *Methods depending on labelled hydrogen release*

During the formation of a double bond, hydrogen is removed from the hydrocarbon chain of the fatty acid and eventually appears in cellular water.

If the substrate has tritium atoms at the positions where the double bond is introduced, the release of tritium into water can be used as a rapid and sensitive assay for the desaturase[170-172]. After the incubation, high-molecular-weight material (including tritiated fatty acids) are precipitated with trichloroacetic acid, and after removal of the precipitate an aliquot of the acid-soluble fraction is assayed for tritiated water by scintillation spectrometry.

If tritium release is to be used to calculate *absolute* desaturation rate, three correction factors must be included in the calculation to take into account the following: (a) an isotope-effect discriminating against tritium removal; (b) the stereospecificity of hydrogen removal; only one pair of the four possible hydrogen atoms are removed (See Section 5.3.11.1); (c) the proportion of total tritium in the substrate molecule that is at the positions where the double bond is introduced. These questions are discussed in Refs. 171, 172.

5.3.10 Molecular organisation of the desaturase complex

5.3.10.1 *The requirements for lipids*

It is now recognised that several enzymes in both mitochondria and endoplasmic reticulum require the presence of lipids for optimum enzymic activity. Two main approaches have been used to study lipid-dependence and the controversies that have arisen are, I think, mainly due to the inadequacies of individual techniques for lipid depletion and reconstitution. One approach has been to extract lipids from microsmal membranes with acetone–water mixtures and to demonstrate loss of enzymic activity in the lipid-depleted membrane protein. Ultimate proof that the loss of activity was due to loss of lipid rather than denaturation of the membrane protein was obtained by reconstituting the lipoprotein by adding back lipids to the lipid-free protein and demonstrating restoration of activity.

In the other method, membrane lipid has been degraded by the use of specific hydrolytic enzymes, phospholipases. The advantage here is that the lipid can be removed in a controlled, specific manner, in contrast to the unspecific solvent system that may also cause some denaturation. Of the enzymes that have been used, phospholipase A has the serious disadvantage that the degradation products, lysophospholipids and NEFA are detergents that strongly inhibit most enzymes. Phospholipase C does not suffer from the disadvantage and is proving very useful as a tool for the controlled removal of membrane lipid[173].

Jones et al.[122] found that hen liver microsomes that had been extracted with acetone–water (90:10 v./v.) completely lost the ability to desaturate stearoyl-S-CoA. The activity was partially restored by adding back a micellar dispersion of lipids prepared by ultrasonic treatment. No single lipid could restore activity; a mixture of phospholipid, triglyceride and NEFA was needed for maximum activation, which was generally ca. 60% of the activity of intact microsomes.

The properties of the reconstituted system, its inhibition by p-hydroxymercuribenzoate and cyanide, the product of the reaction, its substrate and co-factor requirements were the same as those of the original microsomal system, suggesting that the removal of lipids had not altered the protein significantly.

In the author's laboratory, attempts to demonstrate lipid dependence by delipidation in a similar way resulted in preparations that still retained 20–50% of their original activity which could not, however, be restored to 100% by adding back micellar solutions of individual or mixed lipids prepared by a variety of techniques[98]. This might have been due to the fact that (a) the lipids were not in a suitable form for reconstituting the complex or (b) that part of the protein was denatured during solvent delipidation. More recently, therefore, we have used phospholipase C to degrade microsomal lipids and have been able to demonstrate complete loss of activity coincident with complete removal of microsomal phospholipid[174]. Experiments to try to demonstrate reconstitution of the desaturase are now in progress.

It seems certain that microsomal stearoyl-S-CoA desaturase is a lipid-dependent enzyme. This may well prove to be a factor of practical importance when attempting the purification of desaturases.

5.3.10.2 *Purification of desaturases*

Most of the desaturases that have been studied have been firmly bound to subcellular membranes and few have been released into the 'soluble' fraction after homogenisation of tissue and subsequent cell fractionation. An exception is the stearoyl-S-ACP desaturase of *E. gracilis*[14,80,83]. When photo-autotrophically grown *Euglena* cells were broken in a pressure cell or by homogenisation with glass beads, the desaturase activity was located almost entirely in the 100 000 g × 90 min supernatant.

In subsequent purification steps, involving salt fractionation, DEAE-cellulose chromatography and gel filtration, three protein components were isolated which in combination were required to restore desaturase activity. One was identifiable as the non-haeme iron–sulphur protein, ferredoxin;

another had NADPH oxidase activity. *Euglena* ferredoxin had nearly ten times the activity of spinach ferredoxin. The third component, referred to as 'desaturase', resisted further attempts at purification and decayed with a half-life of less than 10 h at 0 °C. In frozen crude extracts the activity was stable for at least a month[83].

Before such methods can be applied to the microsomal desaturases of animals, techniques for 'solubilising' the activity must first be devised. The methods that have so far been applied have been 'borrowed' from reported work on other proteins.

Gurr et al.[123] attempted to purify the rat liver microsomal desaturase by treatment of freeze-dried microsomal membranes with high-ionic-strength phosphate buffer, a technique by which Scholan and Boyd[175] had attempted to purify a steroid hydroxylase. Most of the desaturase activity appeared in the 100 000 g $\times$ 4 h supernatant and the specific activity increased threefold. The properties and characteristics of the 'soluble' enzyme were exactly the same as those of the firmly-bound enzyme.

Later, the same authors[98] used the non-ionic detergent Triton-X 100 to solubilise hen liver microsomal membranes, a technique previously used by Widnell[176] to purify the 5'-nucleotidase of plasma membrane. Partial purification, leading to a fourfold increase in specific activity, was achieved by a combination of density-gradient centrifugation and agarose gel filtration. The only fraction having desaturase activity was eluted with the void volume of a sepharose 6B column (mol. wt. $> 4 \times 10^6$). No further increase in specific activity was achieved by adding back other column fractions to the V_0 fraction.

A different technique, involving solubilisation with deoxycholate, has yielded a fivefold purified desaturase from hen liver microsomes, but supplied much more information about the components of the desaturase multienzyme complex than the present author's experiments[177,178]. Holloway and Wakil[177] and Holloway[178] used a 'solubilising solution' containing glycerol, KCl and deoxycholate buffered with citrate, that Lu and Coon[179] had developed for the successful purification of the microsomal fatty acid ω-hydroxylase. By treatment of the soluble extract with N-ethylmaleimide they prepared what they termed an 'NEM-inhibited particle' which itself had no desaturase activity but to which activity could be restored by addition of NADH-cytochrome b_5 reductase[177,178]. The NEM-inhibited particles were 'insoluble' in the absence of deoxycholate but could be dissolved in a solution containing the detergent and further resolved by gel filtration into a large-molecular-weight and a small molecular weight fraction. In the presence of NADH-cytochrome b_5 reductase neither of these subfractions alone had desaturase activity but in combination they were fully active. The large molecular weight fraction was a protein, contained very little lipid and had no characteristic absorption spectrum. The small mol. wt. fraction appears to consist of cytochrome b_5 and phospholipid and this fraction could be inactivated by removal of the lipid or destruction of the cytochrome b_5[178].

Gaylor and co-workers[180a] have recently used the Lu and Coon solubilising solution to isolate a protein that had a high affinity for cyanide and could be assayed by titrating the microsomes with CN^- and measuring the characteristic difference spectrum with λ_{max} at 444 nm and λ_{min} at 405 nm. During

DEAE chromatography of the soluble extract, the CN^--binding protein was eluted with the same retention volume as desaturase activity and Gaylor et al.[180a] suggested that this is the CN-sensitive factor of the desaturase complex. The component was extremely labile with a half-life of ca. 3 h. Doubt about the relevance of this protein will be discussed in Section 5.3.10.4.

The most recent attempt at purification of stearoyl-S-CoA desaturase has been reported by Shimakata et al.[180b]. Rat liver microsomes were solubilised by a mixture containing both Triton X-100 and sodium deoxycholate and the components separated by DEAE-Sephadex chromatography. Desaturase activity could only be reconstituted by adding together fractions containing NADH-cytochrome b_5 reductase, cytochrome b_5 and the 'cyanide-sensitive factor'. The isolated cyanide-sensitive fraction was colourless, contained no haem or cytochromes, but was rich in phospholipids.

5.3.10.3 Indirect methods for studying the electron transport components

As we have seen in the last section, the picture is being built up of the desaturase as a terminal cyanide-sensitive component of an electron transport chain that involves the transfer of electrons from NAD(P)H via a specific reductase and a b_5 cytochrome. It is only recently that any progress has been made in separating the components of the chain and most of our knowledge of its organisation derives from indirect spectrophotometric measurements due to Japanese workers in Sato's laboratory[121,153,169,181-183].

All the available evidence suggests that two different electron transport chains function separately in liver microsomes. One is concerned in the hydroxylation of various drugs and steroids and involves electron transfer from NADPH via an NADPH-dependent flavoprotein to the CO-binding pigment P_{450}[121].

Oshino et al.[121] compared the requirements of aniline hydroxylation with stearoyl-S-CoA desaturation in rat liver microsomes. P_{450} was not involved in desaturation since that process was not CO-sensitive. Reducing equivalents could be supplied from three sources, with the order of efficiency: NADH > NADPH ≫ ascorbate. When NADPH was added to liver microsomes, cytochrome b_5 was rapidly reduced as measured by spectrophotometry with a special 'triple-wavelength' instrument[169]. The reduction was not complete but a steady state was established, based on a balance between rate of reduction of b_5 by NADPH and slow autoxidation of the cytochrome. Addition of stearoyl-S-CoA caused an immediate decrease in the reduction level of b_5 to a new steady state that persisted for only a very short time before returning to the old steady state value. Cyanide interfered with this stearoyl-S-CoA induced shift in b_5 reduction level. The same sequence of events occurred with NADH as electron donor.

The Japanese workers suggest that these events can be explained if the electron transport chain is constituted as in Figure 5.15[169].

Manipulations that alter the level of reduction of b_5, such as the partial inhibition of the NADPH-specific flavoprotein with $HgCl_2$, intensify the stearoyl-S-CoA induced shift. Reduction of b_5 by NADH was so fast that

HgCl$_2$ inhibition had no effect; however, after 99% inhibition of NADH-cytochrome b$_5$ reductase with p-chloromercuribenzenesulphonate, a stearoyl-S-CoA induced shift could be demonstrated. The rate of cytochrome b$_5$ reduction by NADH is over 100 times faster than stearoyl-S-CoA desaturation. This suggests that *in vivo* cytochrome b$_5$ can be used for other metabolic reactions, and probably these include the various reactions of polyunsaturated fatty acid biosynthesis described in Sections 5.3.2 and 5.3.6.2.

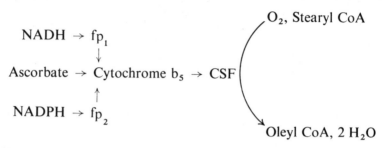

Figure 5.15 Electron carriers in Stearoyl-S-CoA desaturation by rat liver microsomes. f$_{p1}$, NADH-cytochrome b$_5$ reductase; f$_{p2}$, NADPH-specific flavoprotein (NADPH-cytochrome c reductase); CSF = cyanide-sensitive factor. [Figure 1 in Oshino, N., Imai, Y. and Sato, R. (1971). *J. Biochem.,(Tokyo)* **69**, 155, by kind permission of the authors and editors of *J. Biochem. (Tokyo)*.]

Although all the evidence favours the participation of a reduced pyridine nucleotide-linked flavoprotein in desaturation, only one system, namely the Δ9 desaturase of *M. phlei*[150] has been shown to have an absolute requirement for FAD or FMN; on the other hand, the participation of a cytochrome b$_5$-like molecule in the *M. phlei* desaturase has not been investigated.

Whereas electron transport from NADH and NADPH to cytochrome b$_5$ is mediated by their respective flavoproteins, ascorbate reduces the cytochrome directly. Cytochrome b$_5$ can be removed from microsomes to various extents by mild proteolytic digestion with 'Nagarse'. When this is done, the extent of desaturation supported by ascorbate is dependent on the content of b$_5$ remaining in the microsomes. A puzzle still to be solved is that NADH and NADPH-dependent desaturations retained two-thirds of their original activities even when 95% of b$_5$ had been removed[169].

5.3.10.4 Control of fatty acid desaturation

It is now firmly established that the activities of the microsomal desaturases are profoundly affected by different physiological conditions such as age, nutritional and hormonal status[6,54,70,153,183-189]. When animals are starved, desaturase activity decreases to a very low level which can be restored rapidly by re-feeding, especially if the re-fed diet is rich in carbohydrate[183]. Similarly, the stearoyl-S-CoA desaturases of liver and adipose tissue of diabetic animals are markedly depressed *in vitro* when compared with preparations from non-diabetic or diabetic animals treated with insulin[6,54,70,185,186]. The activity of the liver enzyme of rats treated with thyroxine is elevated[184]. Thyroxine was able to substitute to some extent for insulin in repairing the 'desaturase lesion' of diabetic animals[184].

This lesion in diabetes is not, as originally suggested[6] due to an inadequate supply of NADPH since addition of this co-factor to microsomal preparations *in vitro* does not restore desaturase activity[70]. Gelhorn and Benjamin[190] have shown that the insulin-induced 'repair' of desaturase activity involves DNA-directed protein synthesis and that desaturase activity undergoes rapid decay with a half-life of 3–4 h when protein synthesis is inhibited by puromycin. The 'repair' process is preceded by synthesis of an m-RNA template that has a half-life of *ca.* 17 h.

According to the Japanese workers[153] during the time when the desaturase levels are being restored by re-feeding, only the activity of the terminal cyanide-sensitive factor changes rapidly. Other microsomal components either remained constant (e.g. NADPH-cytochrome c reductase), decreased (e.g. P_{450}) or increased only very slowly (e.g. cytochrome b_5 and NADH-b_5 reductase). Cycloheximide, an inhibitor of protein synthesis, completely halted the dietary induction of desaturase and caused a rapid decay of activity with a half-life of 3–4 h. This accords well with the data on the insulin-induced repair of desaturase[190].

The cyanide-sensitive factor (CSF) involved here does not seem to be identical to the CN-binding protein of Gaylor *et al.*[180a] since there was no correlation in Sato's experiments[181] between the content of CN-binding protein, as measured by the difference spectrum, and desaturase activity. In addition, adipose tissue microsomes, which have an active CN-sensitive desaturase, do not show a CN-binding spectrum[169,181].

The rapid turnover of CSF compared with the relative stability of other electron-transport components, can be seen as an extremely economical way of quickly channelling the electron-transport activity into more useful purposes when $\Delta 9$ desaturation is no longer required.

5.3.11 The stereochemistry and mechanism of hydrogen removal

5.3.11.1 Stereochemistry

Although the *R-/S-* nomenclature is for many purposes the method of choice for indicating the absolute stereochemistry of an asymmetric centre, the older D-/L- nomenclature will be used in the following discussion because, in the context of fatty acid desaturation, it is less confusing for the reader. The reason for this will become apparent at the end of this section, when I shall summarise the main conclusions using the *R-/S-* nomenclature.

In the formation of a double bond, two hydrogen atoms are somehow removed. The question arises, which two of the four possible hydrogen atoms are lost; are they of a *cis-* or *trans-* (*erythro-* or *threo-*) configuration relative to each other and is there an absolute specificity in their removal?

These questions were first investigated by Schroepfer and Bloch[20] in the desaturation of stearate to oleate by *Corynebacterium diphtheriae*. They synthesised the four stereospecifically tritium-labelled stearic acids: D- and

L-[9-³H]- and D- and L-[10-³H] stearic acids. During desaturation, tritium from the L-9 and L-10 tritio-substrates was retained in the oleic acid product whereas D-9 and D-10 tritium labels were lost.

Morris and his colleagues[21,191,192] extended these observations to other species and to double bonds introduced in the 12, 13-positions as well as the 9, 10 but their approach was somewhat different since the total synthesis of all the required substrates would have been a large undertaking. They argued that if the relative configuration of the two hydrogen atoms lost on desaturation could be established, then only one pair of stereospecifically labelled substrates would be required for each double bond to establish total stereochemistry. Thus D- and L-[9-³H]stearic acids were prepared for studying oleic acid formation, and D- and L-[12-³H]stearic and oleic acids as precursors for linoleate formation.

Formation of oleic acid from D- and L-[9-³H]stearic acids in *Chlorella*[21,191] hen liver[191] and goat mammary gland[191] was accompanied by retention of L-tritium label in the product and loss of D-tritium. Similarly, formation of linoleic acid from D- and L-[12-³H]stearic acids in *Chlorella*[21,191] and D- and L-[12-³H]oleic acids in castor bean[192] was accompanied by retention of L-tritium and loss of D-tritium. Having thus confirmed that the stereochemistry of hydrogen removal at one position of each double bond was absolute for the D-configuration, they then had to establish the *total* stereochemistry by determining the relative configuration of the two hydrogens removed.

This was done by using racemic but geometrically specific *vic*-dideuterated substrates, either *erythro-* or *threo-* as illustrated in Figure 5.16, which depicts the unsaturated products arising from eliminations of hydrogen atoms in the *cis-* (or *erythro-*) relative configuration and the D-absolute configuration that has already been established. (Loss of hydrogen atoms in the *trans*-relative configuration would give the opposite results to those depicted in Figure 5.16).

The relevant unsaturated products from various incubations with *erythro-* and *threo*-isomers of [9,10-²H$_2$]18:0 and [12,13-²H$_2$]18:1 were isolated and carefully purified by argentation t.l.c. and preparative g.l.c. and their distributions of isotopic species determined by mass spectrometry. In each case, *erythro*-[²H$_2$]labelled precursors gave rise to enrichment of dideutero-species in the products whereas *threo*-[²H$_2$]labelled precursors resulted in enrichment of monodeuterated species in the products.

The two hydrogen atoms removed in all these desaturations were therefore of the *cis* or *erythro* relative configuration.

These results, along with those from stereospecifically tritiated substrates, prove that in the desaturation of stearic acid to oleic acid in *Chlorella*, hen liver, and goat mammary gland, it is the D-9- and D-10-hydrogen atoms that are removed; this is the same stereochemistry as in the bacterial system. Similarly, the D-12- and D-13-hydrogen atoms are the ones removed on desaturation of oleic to linoleic acid in *Chlorella*.

The preceding discussion demonstrates that the stereochemistry of hydrogen removal is identical whether the transformation is from a saturated to a monounsaturated acid or from a monounsaturated to a diunsaturated acid. When the *R-/S-* nomenclature is used, the hydrogen atoms that are removed in going from stearic to oleic acid are the *pro-R-9-* and *pro-R-10*-hydrogen atoms. However, in going from oleic acid to linoleic acid, the presence of the

$$
\begin{array}{c}
\text{R}^2 \\
| \\
\text{D}-\text{C}-\text{H} \\
| \\
\text{D}-\text{C}-\text{H} \\
| \\
\text{R}^1
\end{array}
+
\begin{array}{c}
\text{R}^2 \\
| \\
\text{H}-\text{C}-\text{D} \\
| \\
\text{H}-\text{C}-\text{D} \\
| \\
\text{R}^1
\end{array}
\longrightarrow
\begin{array}{c}
\text{R}^2 \\
| \\
\text{CD} \\
|| \\
\text{CD} \\
| \\
\text{R}^1
\end{array}
+
\begin{array}{c}
\text{R}^2 \\
| \\
\text{CH} \\
|| \\
\text{CH} \\
| \\
\text{R}^1
\end{array}
$$

erythro

$$
\begin{array}{c}
\text{R}^2 \\
| \\
\text{D}-\text{C}-\text{H} \\
| \\
\text{H}-\text{C}-\text{D} \\
| \\
\text{R}^1
\end{array}
+
\begin{array}{c}
\text{R}^2 \\
| \\
\text{H}-\text{C}-\text{D} \\
| \\
\text{D}-\text{C}-\text{H} \\
| \\
\text{R}^1
\end{array}
\longrightarrow
\begin{array}{c}
\text{R}^2 \\
| \\
\text{CD} \\
|| \\
\text{CH} \\
| \\
\text{R}^1
\end{array}
+
\begin{array}{c}
\text{R}^2 \\
| \\
\text{CH} \\
|| \\
\text{CD} \\
| \\
\text{R}^1
\end{array}
$$

threo

Figure 5.16 Summary of olefinic products from racemic *erythro-* and *threo-*[^{2}H$_2$]-labelled precursors on the basis of enzymic desaturation removing a *cis*-pair of hydrogen atoms of the D- configuration. R = Me$_3$(CH$_2$)$_7$ or Me$_3$(CH$_2$)$_1$ or Et; R^1 = HO$_2$C·(CH$_2$)$_7$ or HO$_2$C·(CH$_2$)$_7$·CH: CH.CH$_2$ or HO$_2$C·(CH$_2$)$_7$·CH: CH.(CH$_2$)^4. [Scheme 9 from Morris L. J. (1970) *Biochem J.* **118**, 681, by kind permission of the author and editors of *Biochemical Journal.*]

double bond in the molecule at the 9,10- position results in the hydrogen atoms removed in this desaturation having the *pro-S-12* and *pro-S-13* absolute configurations. This fact makes the discussion of stereochemistry of desaturation somewhat more difficult for the reader to follow when the *R-/S-* system is used and has to be remembered when attempting to translate from the D-/L- nomenclature (which is used in virtually all of the published papers on the stereochemistry of fatty acid desaturation) to the *R-/S-*nomenclature.

5.3.11.2 *Mechanism of hydrogen removal*

The next question that arises is: can the data on release of labelled hydrogen atoms from specifically labelled substrates tell us anything about the mechanism of desaturation?

In their studies of bacterial desaturation, Schroepfer and Bloch[20] compared the release of tritium from all four possible tritium-labelled substrates against the desaturation of a ^{14}C-labelled substrate. The ratio of ^{3}H/^{14}C in unchanged D-[9-^{3}H]stearic acid after the incubations was considerably higher than the original value, indicating a substantial kinetic isotope effect against tritium in the D-9-position.

There was, however, no such effect against tritium in the D-10-position. They suggested, on this evidence, that hydrogen removal in bacterial aerobic desaturation was a stepwise process, and that the first, and rate-limiting, step was removal of the D-9-hydrogen atom.

This would be consistent with the model that envisaged desaturation as a classical mixed-function oxygenase reaction, in which the first step is hydroxylation at either the 9- or 10-position, and the second step dehydration as pictured in Figure 5.17a.

However, although many hydroxyacids do occur naturally with the substituent group in a position normally occupied by a double bond[193], no convincing evidence exists that these are obligatory intermediates in the formation of double bonds, in spite of numerous studies to attempt to demonstrate this[7,194-198].

Morris' results[21,191] do not lead one to the conclusion that a stepwise mechanism is involved. If we re-examine Figure 5.16, it is clear that for each two molecules of racemic *erythro*-[^{2}H$_2$]precursor that are desaturated, only one molecule of [^{2}H$_2$]labelled olefin is formed, whereas from each two molecules of *threo*-[^{2}H$_2$]precursor desaturated, two molecules of [^{2}H$_1$]labelled olefin are formed *if there is no kinetic isotope effect*. Thus, if identical amounts of *erythro*- and *threo*-[^{2}H$_2$]labelled substrates are incubated under identical conditions, there should be twice as much enrichment of [^{2}H$_1$]labelled product from the *threo*-substrates as there is [^{2}H$_2$]labelled product from the *erythro*-substrates. However, in fact the enrichment of [^{2}H$_2$]labelled product from the *erythro*-substrate was equal to (in the 18:2 product) or considerably greater than (in the 18:1 product), the enrichment of [^{2}H$_1$]labelled product from the *threo*-substrate.

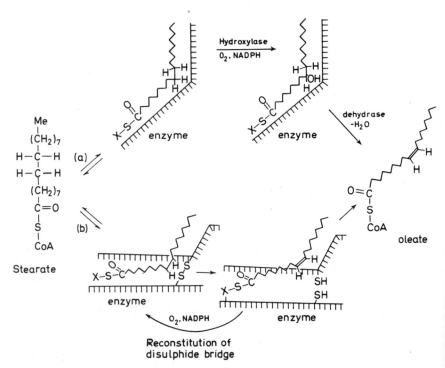

Figure 5.17 Hypothetical schemes for oxidative desaturation. [Figure 2.9 in Gurr, M. I. and James, A. T. (1971). Lipid Biochemistry: An Introduction (London: Chapman and Hall) by kind permission of the publisher.]

Now, if there were a kinetic isotope effect at only one of the two positions, sufficiently pronounced to totally block desaturation of a substrate with a D-[^{2}H] atom at that position, then one enantiomer from each substrate would be completely unreactive, resulting in equal enrichments of [^{2}H$_2$] and [^{2}H$_1$] labelled products from *erythro*- and *threo*-precursors respectively. This could account for Morris' results for linoleate formation but not for oleate formation, where *ca.* 3–4 times as much [^{2}H$_2$]oleic acid was produced from the *erythro*-precursor as [^{2}H$_1$]oleic acid from the *threo*-precursor.

Only one explanation would seem possible to account for these results, namely that there is a substantial kinetic isotope effect exerted by deuterium atoms of the D-configuration at both the 9- and 10-positions, so that neither of the *threo*-enantiomers is as efficient a substrate as the L-9-, L-10-[^{2}H$_2$]*erythro*-enantiomer. Johnson and Gurr[171] recently demonstrated that the isotope effect involved in the removal of the two tritium atoms at the 9- and 10-positions of stearic acid was twice that for the removal of one tritium at either the 9- or 10-positions. This suggests an isotope effect at both positions and is consistent with the results of Morris[191] for deuterium removal.

These considerations imply that desaturation in *Chlorella* and animal microsomes involves concerted removal of a pair of hydrogen atoms which in turn rules out any formal oxygenated compound as an intermediate in fatty acid desaturation. A highly speculative model for the desaturase active site based partly on these results and partly on the results discussed in Section 5.3.12., has been proposed in James' laboratory and is illustrated in Figure 5.17b.

5.3.12 Specificity of desaturases with regard to double-bond position

Table 5.1, although not presenting by any means an exhaustive list of naturally-occurring fatty acids, gives an idea of the variety of structures existing, particularly with regard to double-bond position. Several questions may be posed: How many of these structures can be accounted for by direct desaturation; that is, insertion of the double bond directly into the precursor of the same chain length? Does this variety arise out of a large number of enzymes, each having a rather limited substrate specificity as regards chain length and number of existing double bonds, or can it be explained by a relatively small number of enzymes with a fairly wide range of specificity? What are the structural features of the substrate that the desaturase recognises?

To try to answer these questions, the approach of James, Morris and their colleagues[167,168,199,200] has been to synthesise chemically a wide variety of fatty acid structural types to test as potential substrates in as many different desaturase systems as possible.

5.3.12.1 Monoenoic acid biosynthesis

To begin with, they tested the ability of the desaturases of hen liver and goat mammary gland microsomes, *Torulopsis bombicola* and *Chlorella vulgaris* to

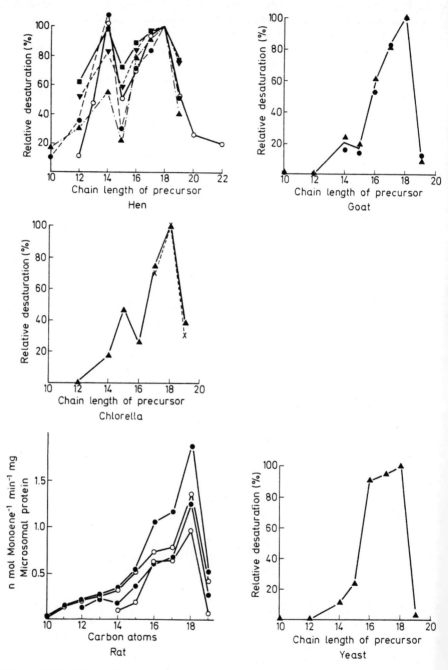

Figure 5.18 Effect of chain length on desaturation in different species [Figure 3 in Gurr, M. I., Robinson, M. P., James, A. T., Morris, L. J. and Howling, D. (1972). *Biochim. Biophys. Acta*, in Press by kind permission of the editors and Elsevier.]

form monenoic acids from saturated substrates of different chain lengths[168,199]. Similar experiments were done by Johnson et al.[160] for hen, Paulsrud et al.[201], Raju and Reiser[202], and Nakagawa et al.[203] for rat and by Schulz and Lynen[116] for *Saccharomyces cerevisiae*. All saturated fatty acids from 10:0 to 22:0 yield a monoenoic acid with a *cis*-double bond at positions 9 and 10. Maximum activity is always at C_{18} but in the hen, goat and *Chlorella* a secondary peak of activity occurs at C_{14} suggesting that another desaturase with specificity towards acids of shorter chain length occurs in some species (Figure 5.18). Hen[168] and yeast[116] are interesting in being able to form the monoenoic acid with a terminal double bond from decanoic acid.

The situation is somewhat more complex in *Chlorella* than in the other species in that there is a separate desaturase that inserts a Δ7 double bond; this enzyme has a rather more limited chain-length specificity and is only effective on C_{14}, C_{15} and C_{16}[168,199].

The specificity of double-bond introduction regardless of chain length argues strongly for a binding site holding the carboxyl end of the chain (which may include a functional group such as a thiol ester) fixed and precisely locating the 9- and 10-positions at the active centre. The only other interactions that can account for substrate binding are cumulative van der Waals forces between the methylene groups of the fatty acid hydrocarbon chain and apolar groups on the enzyme surface. The shorter the chain length, the smaller the van der Waals interactions, a factor that contributes to the poorer activity towards shorter substrates.

Non-polar interactions of this sort might be maximised if the substrate were closely enfolded in a cleft of the protein. In order to investigate the precision of such enfolding, James, Morris and their colleagues[168] synthesised almost all the positionally isomeric methylstearic acids and tested them as substrates for the hen and *Chlorella* desaturases. Only methyl stearates with the methyl group at positions 2, 3, 4, 16 and 17 were substrates. Absolutely no activity was measured when methyl groups were in positions 5–15 inclusive.

Piecing together these data, the picture emerges of a cleft in the enzyme surface into which the substrate molecule, in the extended chain conformation, fits. The length would be about 26 Å (accounting for the sharp fall-off in activity for substrates longer than C_{18}) and 4 Å wide between C_5 and C_{15}. Two very specific features of the dehydrogenation have to be taken into account in the model. First, the absolute specificity for the 9- and 10-positions requires absolutely precise locking of the substrate into the cleft. Second, the absolute stereochemical specificity of the reaction could be most easily visualised if the 9-D- and 10-D-hydrogen atoms are adjacent to each other, i.e. in the eclipsed conformation. In the extended chain configuration, the D-hydrogen atoms at positions 9 and 10 are on opposite sides of the chain because of the zigzag arrangement of carbon atoms. Thus, one could imagine a substantial conformational change occurring in the enzyme when it interacts with the substrate:

'Open enzyme' + substrate → 'closed enzyme'–substrate complex

During this conformational change, it would not be impossible for a rotation to occur about the 9-10 C—C bond to bring the two D-hydrogens together in the eclipsed conformation (and creating a bend in the molecule of almost the same angle as that conferred by the *cis* double bond of the product, oleic acid). The model in Figure 5.17b is an extension of this idea in which the active centre of the enzyme is suggested to be a disulphide bridge which is made in forming the 'closed' enzyme and broken in forming the 'open' enzyme as follows:-

Stearoyl-S-CoA + 'open' reduced enzyme $\xrightarrow{\text{acyl transferase}}$ Stearoyl-S-'open' inactive enzyme

$\downarrow$ O$_2$/NADH

Oleoyl-S-CoA $\xleftarrow{\text{acyl transferase}}$ Oleoyl-S-'open' enzyme $\leftarrow$ Stearoyl-S-'closed' active enzyme

5.3.12.2 Dienoic acid biosynthesis

The question now arises: does the enzyme that introduces a second double bond into a monoenoic acid measure from the carboxyl end of the chain, from the methyl end, or solely with reference to the first double bond? To investigate this, three series of monoenoic acids were synthesised and tested with the *Chlorella* enzyme[200]. The members of series (1) and (2) differed in chain length but all those in series (1) had a $\Delta 9$ double bond and in series (2) the double bond was at a fixed distance of nine carbon atoms from the methyl end of the chain, (i.e. $\omega 9$). Series (3) were 18-carbon acids in which the bond was neither $\Delta 9$ nor ω-9. The results, quoted in Table 5.3 show that the chain length of the monoene is not critical within the range C_{15}–C_{19}. Outside this range, desaturation is severely restricted. Within this range all $\Delta 9$ and $\omega 9$ monoenes were desaturated to yield a methylene-interrupted diene. Monoenes without a double bond in these positions were not substrates (except *cis*-12-octadecenoate, which was converted into linoleic acid).

These results can only be interpreted as indicating two enzymes capable of producing a diene from a monoene, one of which is specific for substrates incorporating structure A; the other for structure B:-

HO$_2$C·(CH$_2$)$_7$·CH:CH·CH$_2$·   ·CH$_2$CH:CH·(CH$_2$)$_7$·Me
(A)   (B)

The relative activities and importance of these two enzymes in plant tissue is not known, but two points are worth making. First, of the substrates tested, only oleic acid incorporates both of the above structural features, and is a more effective substrate than the others tested. Second, an enzyme that recognised a structure based on the methyl end of the chain would provide a mechanism for desaturating lipid-linked acyl groups as discussed in Section 5.3.3.3. An enzyme of this type could account satisfactorily for interconversions of molecular species of galactosyl diglycerides observed by Nichols[94]:

```
 ─18:1(9)[ω9]      ─18:2(9, 12)[ω6, 9]
 ─16:1(7)[ω9] →    ─16:2(7, 10)[ω6, 9]
 ─Gal              ─Gal
```

The active site of the enzyme that recognises the carboxyl end of the chain might have the same features as that proposed for stearoyl-S-CoA desaturase. However, other models have been proposed by Brenner[54] based on a recognition by the enzyme of the existing double bond itself.

An important exception to the rule that only $\Delta 9$ or $\omega 9$ monoenes are desaturated by *Chlorella* was the conversion of *cis*-12-octadecenoate into linoleic acid[200]. It seemed likely that this monoene was in fact a substrate for the $\Delta 9$ desaturase and this was borne out by subsequent experiments in which it was shown that a number of different animal species could convert *cis*-12-octadecenoate into linoleate[167]. Linoleic acid is not normally synthesised by animals tissues[204], and since animals have a specific nutritional requirement for this fatty acid, it must be provided in the diet from vegetable sources. Certain species tested did not have the ability to desaturate *cis*-12-octadecenoate. Among these were the rat[167,205] and the yeast *Candida utilis*[167].

In terms of the model for the $\Delta 9$ desaturase discussed in the last section, it is attractive to envisage the *cis*-12-octadecenoate as a 12- or 13-carbon acid which is a substrate for the short-chain specific enzyme; the conformation of the *cis*-12-double bond has the effect of lifting carbon atoms 14–18 clear of the end of the cleft, ensuring a good fit of the substrate. This proposal certainly

Table 5.3 **The conversion of monoenoic to dienoic fatty acids by *Chlorella vulgaris*** (See Ref. 200. Reproduced by kind permission of the authors and Elsevier Publishing Co.)

Precursor	% Direct conversion*	Double bond position
Group I		
cis-9-Hexadecenoic (*n*-7)‡	21.0	9–10, 12–13
cis-9-Hepadecenoic (*n*-8)	37.4	9–10, 12–13
cis-9-Octadecenoic (*n*-9)	79.0	9–10, 12–13
cis-9-Nonadecenoic (*n*-10)	25.0	9–10, 12–13
Group II		
cis-7-Hexadecenoic (*n*-9)	36.0	7–8, 10–11 (*n*-6, *n*-9)
cis-8-Heptadecenoic (*n*-9)	26.0	8–9, 11–12 (*n*-6, *n*-9)
cis-9-Octadecenoic (*m*-9)	79.0	9–10, 12–13 (*n*-6, *n*-9)
cis-10-Nonadecenoic (*n*-9)	24.1	10–11, 13–14 (*n*-6, *n*-9)
Group III		
cis-7-Octadecenoic (*n*-11)	< 1	—
cis-8-Octadecenoic (*n*-10)	< 1	—
cis-10-Octadecenoic (*n*-8)	< 1	—
cis-11-Octadecenoic (*n*-7)	< 1	—
cis-12-Octadecenoic (*n*-6)	48.0	9–10, 12–13
Group IV		
trans-9-Octadecenoic (*n*-9)	< 1	—
9-Octadecynoic	< 1	—
cis-9,18-Nonadecadienoic (*n*-9)†	72.7	9–10, 12–13, 18–19

* In some cases breakdown and resynthesis *de novo* occurred.
† Presumed precursor from palmitic acid.
‡ Presumed precursor from *cis*-18-nonadecenoic acid. The figure of 72.7% represents radioactivity in the monoene and diene fraction. Individual figures are not available.

fits, qualitatively, the data in Figure 5.18. Other examples can be found of monoenoic acids substituting for saturated acids as substrates for desaturases. In some bacteria (which as a general rule do not synthesise polyenoic acids) dienoic acids can be formed in this manner.

5.3.12.3 Classification of desaturases

The ideas arising out of the chain length specificity studies form the basis for a classification of desaturases. Building on these ideas, two basic 'rules' can be formulated, that can account for virtually all the structures occurring in animal tissues (after taking into account those acids that are formed by chain elongation or retroconversion). *The first 'rule'* is that all structures can be explained on the basis of enzymes 'measuring' from the carboxyl terminal. These we can call Δ-desaturases. (Plants, in addition, have an extra series of ω-desaturases, Section 5.3.12.2.) We can take the $\Delta 6$-desaturase as an example. Evidence that this is a Δ-desaturase comes from the work of Schlenk et al.[206] who showed that $\Delta 9,12,-17:2$ was converted into $\Delta 6,9,12-17:3$ in the rat.

The second 'rule' is that *cis*-monounsaturated or polyunsaturated acids mimic short chain fatty acids, provided that the lowest numbered existing double bond is at least one methylene group towards the methyl group from the reaction centre. Again taking the $\Delta 6$-desaturase, Brenner[54,117] has provided convincing evidence that there is a single enzyme capable of accepting substrates in the order of activity: $9,12,15-18:3 > 9,12-18:2 > 9-18:1$. Applying the foregoing 'rule', each acid fits into the desaturase 'cleft'; as the number of *cis* double bonds increases, so the more is the distal part of the chain lifted away from the enzyme, thus decreasing steric hindrance. These ideas (which readily suggest experiments to test their validity) can be extended to cover the $\Delta 4$ and $\Delta 5$ desaturases, and provide a rational basis for explaining how a wide variety of structures may arise from a limited number of enzymes.

5.3.13 Conclusions

The object of writing this review was twofold: to try to pull together the information on desaturases in such a way that the picture might seem a little less confusing than hitherto; and to point the way to those areas that are ripe for new discoveries.

Outstanding among the questions still to be solved are the following: The question of alternative pathways still has to be resolved; is there a route to oleic acid that does not involve direct desaturation of stearate? How are substrates transferred between the various multi-enzyme complexes of lipid biosynthesis? How does sterculic acid inhibit unsaturated fatty acid formation; at what site?

What are the nature of the electron carriers between NAD(P)H and oxygen; how many are there? Can we stabilise the terminal cyanide-sensitive component sufficiently to enable us to extend the dismantling process and determine its nature? Are SH, S–S or Fe groups involved in this terminal desaturase?

How are the hydrogen atoms removed? By a concerted mechanism? How are electrons passed to O_2 in the terminal step?

New techniques for solubilising microsomal membranes, for studying spectral changes of components of the desaturase complex and for assaying desaturase activity are being explored and extended and should help to answer some of these questions.

Acknowledgements

I wish to thank my colleagues Drs. A. T. James, L. J. Morris, B. W. Nichols and R. Jeffcoate for discussion and advice, Prof. R. Sato and Dr. J. Harwood for making available manuscripts before publication, and Miss M. Phillip for checking the typescript.

References

1. Hilditch, T. P. and Williams, P. N. (1964). *The Chemical Constitution of Natural Fats* (4th ed.) (London: Chapman and Hall)
2. Schoenheimer, R. and Rittenberg, D. (1936). *J. Biol. Chem.*, **113**, 505
3. Stetten, D. and Schoenheimer, R. (1940). *J. Biol. Chem.*, **133**, 329
4. Bloomfield, D. K. and Bloch, K. (1958). *Biochim. Biophys. Acta*, **30**, 220
5. Bernhard, K., von Bulow-Koster, J. and Wagner, H. (1959). *Helv. Chim. Acta*, **42**, 152
6. Imai, Y. (1961). *J. Biochem (Tokyo)*, **49**, 642
7. Marsh, J. B. and James, A. T. (1962). *Biochim. Biophys. Acta*, **60**, 320
8. Fulco, A. J. and Bloch, K. (1962). *Biochim. Biophys. Acta*, **63**, 545
9. Fulco, A. J., Levy, R. and Bloch, K. (1964). *J. Biol. Chem.*, **239**, 998
10. Erwin, J., Hulanicka, D. and Bloch, K. (1964). *Comp. Biochem. Physiol.*, **12**, 191
11. Korn, E. D. (1964). *J. Biol. Chem.*, **239**, 396
12. Erwin, J. and Bloch, K. (1964). *Science*, **143**, 1006
13. Harris, R. V. and James, A. T. (1965). *Biochim. Biophys. Acta.*, **106**, 456
14. Nagai, J. and Bloch, K. (1965). *J. Biol. Chem.*, **240**, pc 3702
15. Stumpf, P. K. and James, A. T. (1962). *Biochim. Biophys. Acta*, **57**, 400
16. James, A. T. (1963). *Biochim. Biophys. Acta*, **70**, 9
17. Simmons, R. O. and Quackenbush, F. W. (1954). *J. Amer. Oil Chem. Soc.*, **31**, 441
18. McMahon, V. and Stumpf, P. K. (1964). *Biochim. Biophys. Acta*, **84**, 359
19. Canvin, D. T. (1965). *Can. J. Bot.*, **43**, 49
20. Schroepfer, G. J. and Bloch, K. (1965). *J. Biol. Chem.*, **240**, 54
21. Morris, L. J., Harris, R. V., Kelly, W. and James, A. T. (1968). *Biochem. J.*, **109**, 673
22. Stoffel, W. (1961). *Biochem. Biophys. Res. Commun.*, **6**, 270
23. Nugteren, D. H. (1962). *Biochim. Biophys. Acta*, **60**, 656
24. Bloch, K. (1969). *Accounts Chem. Res.*, **2**, 193
25. Kass, L. R. and Bloch, K. (1967). *Proc. Nat. Acad. Sci. USA*, **58**, 1168
26. Majerus, P. W. and Vagelos, P. R. (1967). *Advan. lipid. Res.*, **5**, 1
27. Bishop, D. G. and Stumpf, P. K. (1971). *Biochemistry and Methodology of Lipids*, Ch. 19, (A. R. Johnson and J. B. Davenport, editors) (New York: Wiley-Interscience)
28. Gurr, M. I. and James, A. T. (1971). *Lipid Biochemistry: An Introduction* (London: Chapman and Hall)
29. Hitchcock, C. H. S. and Nichols, B. W. (1971). *Plant Lipid Biochemistry* (London: Academic Press)
30. Lennarz, W. J. (1966). *Advan. Lipid Res.*, **4**, 175
31. Goldfine, H. and Bloch, K. (1961). *J. Biol. Chem.*, **236**, 2596
32. Scheuerbrandt, G., Goldfine, H., Baronowsky, P. E. and Bloch, K. (1961). *J. Biol. Chem.*, **236**, pc 70
33. Vagelos, P. R. (1900). CHAPTER 3, this volume

34. Bloch, K., Baronowsky, P. E., Goldfine, H., Lennarz, W. J., Light, R., Norris, A. T. and Scheuerbrandt, G. (1961). *Fed. Proc. Fed. Amer. Soc. Exp. Biol.*, **20,** 921
35. Lennarz, W. J., Light, R. J. and Bloch, K. (1962). *Proc. Nat. Acad. Sci. USA*, **48,** 840
36. Norris, A. T. and Bloch, K. (1963). *J. Biol. Chem.*, **238,** pc 3133
37. Norris, A. T., Matsumura, S. and Bloch, K. (1964). *J. Biol. Chem.*, **239,** 3653
38. Toomey, R. E., Waite, M., Williamson, I. P. and Wakil, S. J. (1965). *Fed Proc. Fed. Amer. Soc. Exp. Biol.*, **24,** 290
39. Weeks, G. and Wakil, S. J. (1967). *Fed. Proc. Fed. Amer. Soc. Exp. Biol.*, **26,** 671
40. Birge, C. H., Silbert, D. F. and Vagelos, P. R. (1967). *Biochem. Biophys. Res. Commun.*, **29,** 808
41. Kass, L. R., Brock, D. J. H. and Bloch, K. (1967). *J. Biol. Chem.*, **242,** 4418
42. Brock, D. J. H., Kass, L. R. and Bloch, K. (1967). *J. Biol. Chem.*, **242,** 4432
43. Kass, L. R. (1968). *J. Biol. Chem.*, **243,** 3223
44. Helmkamp, G. M. Jr., Rando, R. R., Brock, D. J. H. and Bloch, K. (1968). *J. Biol. Chem.*, **243,** 3229
45. Rando, R. R. and Bloch, K. (1968). *J. Biol. Chem.*, **243,** 5627
46. Helmkamp, G. M. Jr. and Bloch, K. (1969). *Fed. Proc. Fed. Amer. Soc. Exp. Biol.*, **28,** 602
47. Helmkamp, G. M. Jr. and Bloch, K. (1969). *J. Biol. Chem.*, **244,** 6014
48. Endo, K., Helmkamp, G. M. Jr. and Bloch, K. (1970). *J. Biol. Chem.*, **245,** 4293
49. Silbert, D. F. and Vagelos, P. R. (1967). *Proc. Nat. Acad. Sci. USA*, **58,** 1579
50. Mizugaki, M., Swindell, A. C. and Wakil, S. J. (1968). *Biochem. Biophys. Res. Commun.*, **33,** 520
51. Mizugaki, M., Weeks, G., Toomey, R. E. and Wakil, S. J. (1968). *J. Biol. Chem.*, **243** 3661
52. Stumpf, P. K. (1969). *Annu. Rev. Biochem.*, **38,** 159
53. Klenk, E. (1965). *Advan. Lipid Res.*, **3,** 1
54. Brenner, R. R. (1971). *Lipids*, **6,** 567
55. Mudd, J. B. (1967). *Annu. Rev. Plant Physiol.*, **18,** 229
56. James, A. T. (1968). *Chem. Brit.*, **4,** 484
57. Gurr, M. I. (1971). *Lipids*, **6,** 266
58. Stearns, E. M. (1970). *Progr. Chem. Fats Other Lipids*, **9,** 455
59. Asselineau, C., Montrozier, H. and Prome, J. C. (1969). *Eur. J. Biochem.*, **10,** 580
60. Fulco, A. J. (1969). *Biochim. Biophys. Acta*, **187,** 169
61. Fulco, A. J. (1970). *J. Biol. Chem.*, **245,** 2985
62. Goldman, P. and Vagelos, P. R. (1961). *J. Biol. Chem.*, **236,** 2620
63. Seubert, W. (1960). *Biochem. Prep.* Vol. 7, 80, (H. A. Lardy, editor) (New York: John Wiley)
63a. Al-Arif, A. and Blecher, M. (1969). *J. Lipid. Res.*, **10,** 344
64. Galliard, T. and Stumpf, P. K. (1968). *Biochem. Prep. Vol.* 12, 66, (W. E. M. Lands, editor) (New York: John Wiley)
65. Vijay, I. K. and Stumpf, P. K. (1972). *J. Biol. Chem.*, **247,** 360
66. van Deenen, L. L. M. and de Haas, G. H. (1964). *Advan. Lipid. Res.*, **2,** 167
67. Lowenstein, J. M. (1969). *Methods Enzymol.*, **14,** 654
68. Gurr, M. I., Robinson, M. P. and James, A. T. (1969). *Eur. J. Biochem.*, **9,** 70
69. Bloomfield, D. K. and Bloch, K. (1960). *J. Biol. Chem.*, **235,** 337
70. Gellhorn, A. and Benjamin, W. (1964). *Biochim. Biophys. Acta*, **84,** 167
71. Pande, S. V. and Mead, J. F. (1968). *J. Biol. Chem.*, **243,** 352
72. Lippel, K., Robinson, J. and Trams, E. G. (1970). *Biochim. Biophys. Acta*, **206,** 173
73. Stoffel, W. and Schiefer, H. G. (1966). *Hoppe-Seyler's Z. Physiol. Chem.*, **345,** 41
74. Gurr, M. I. Unpublished observations
75. Harris, R. V., James, A. T. and Harris, P. (1967). Biochemistry of Chloroplasts, Vol. 2, 241 (T. W. Goodwin, editor) (London: Academic Press)
76. Pande, S. V. (1972). *Biochim. Biophys. Acta*, **270,** 197
77. Overath, P. and Stumpf, P. K. (1964). *Fed. Proc. Fed. Amer. Soc. Exp. Biol.*, **23,** 166
78. Simoni, R. D., Criddle, R. S. and Stumpf, P. K. (1967). *J. Biol. Chem.*, **242,** 573
79. Brooks, J. L. and Stumpf, P. K. (1965). *Biochim. Biophys. Acta*, **98,** 213
80. Nagai, J. and Bloch, K. (1966). *J. Biol. Chem.*, **241,** 1925
81. Harris, R. V., Harris, P. and James, A. T. (1965). *Biochim. Biophys. Acta*, **106,** 465
82. Harris, R. V. Personal Communication

83. Nagai, J. and Bloch, K. (1968). *J. Biol. Chem.*, **243,** 4626
84. Nagai, J. and Bloch, K. (1967). *J. Biol. Chem.*, **242,** 357
85. Vijay, I. K. and Stumpf, P. K. (1971). *J. Biol. Chem.*, **246,** 2910
86. Erwin, J. and Bloch, K. (1963). *Biochem. Z.*, **338,** 496
87. Korn, E. D. (1964). *J. Lipid Res.*, **5,** 352
88. Rosenberg, A. (1963). *Biochemistry*, **2,** 1148
89. Rosenberg, A., Pecker, M. and Moschides, E. (1965). *Biochemistry*, **4,** 680
90. Gurr, M. I. and Bloch, K. Unpublished results
91. Nichols, B. W., James, A. T. and Breuer, J. (1967). *Biochem. J.*, **104,** 486
92. Nichols, B. W. (1968). *Lipids*, **3,** 354
93. Nichols, B. W. and Moorhouse, R. (1969). *Lipids*, **4,** 311
94. Safford, R. and Nichols, B. W. (1970). *Biochim. Biophys. Acta*, **210,** 57
95. Appleby, R. S., Safford, R. and Nichols, B. W. (1971). *Biochim. Biophys. Acta*, **248,** 205
96. Roughan, P. G. (1970). *Biochem. J.*, **117,** 1
97. Baker, N. and Lynen, F. (1971). *Eur. J. Biochem.*, **19,** 200
98. Gurr, M. I. and Robinson, M. P. (1970). *Eur. J. Biochem.*, **15,** 335
99. Fukuba, H. and Mitsuzawa, T. (1957). *Nippon Nogei Kagaku Kaishi*, **31,** 151 (*Chem. Abstr.* **52,** 12 001g (1958))
100. Fukuba, H., Yamazawa, N. and Mitsuzawa, T. (1957). *Nippon Nogei Kagaku Kaishi*, **31,** 132 (*Chem. Abstr.* **53,** 22 133i, (1959))
101. Talamo, B., Chang, N. and Block, K. (1973). *J. Biol. Chem.*, **248,** 2738
102. Kates, M. and Paradis, M. (1973). *Can. J. Biol.*, **51,** 184
103. Barron, E. J. and Mooney, L. A. (1968). *Analyt. Chem.*, **40,** 1742
104. Stumpf, P. K. and James, A. T. (1963). *Biochim. Biophys. Acta*, **70,** 20
105. Fulco, A. J. (1965). *Biochim. Biophys. Acta*, **106,** 211
106. Barron, E. J. and Stumpf, P. K. (1962). *J. Biol. Chem.*, **237,** PC 613
107. Mazliak, P. (1965). *Compt. Rend.*, **261,** 2716
108a. Canvin, D. T. (1965). *Can. J. Bot.*, **43,** 71
108b. Drennan, C. H. and Canvin, D. T. (1969). *Biochim. Biophys. Acta*, **187,** 193
109. Zilversmit, D. B., Entenman, C. and Fishler, M. C. (1943). *J. Gen. Physiol.*, **26,** 325
110. Inkpen, J. A. and Quackenbush, F. W. (1969). *Lipids*, **4,** 539
111. Gurr, M. I., Hammond, E., W. Appleby, R. S., Nichols, B. W. and Morris, L. J. Unpublished results
112. Chesterton, C. J., Butterworth, P. H. W. and Porter, J. W. (1968). *Arch. Biochem. Biophys.*, **126,** 864
113. Willecke, K., Ritter, E. and Lynen, F. (1969). *Eur. J. Biochem.*, **8,** 503
114. Harwood, J. L., Sodja, A. and Stumpf, P. K. (1972). *Biochem. J.* **130,** 1013
115. Harwood, J. L. and Stumpf, P. K. (1971). *Arch. Biochem. Biophys.*, **142,** 281
116. Schultz, J. and Lynen, F. (1971). *Eur. J. Biochem.*, **21,** 48
117. Brenner, R. R. and Peluffo, R. O. (1966). *J. Biol. Chem.*, **241,** 5213
118. Kinsella, J. E. (1972). *Lipids*, **7,** 349
119. Bickerstaffe, R. and Annison, E. F. (1970). *Comp. Biochem. Physiol.*, **35,** 653
120. Pande, S. V. and Mead, J. F. (1970). *J. Biol. Chem.*, **245,** 1856
121. Oshino, N., Imai, Y. and Sato, R. (1966). *Biochim. Biophys. Acta*, **128,** 13
122. Jones, P. D., Holloway, P. W., Peluffo, R. O. and Wakil, S. J. (1969). *J. Biol. Chem.*, **244,** 744
123. Gurr, M. I., Davey, K. W. and James, A. T. (1968). *FEBS Lett.*, **1,** 320
124. Kannangara, C. G. and Stumpf, P. K. (1972). *Arch. Biochem. Biophys.*, **148,** 414
125. Raju, P. K. and Reiser, R. (1967). *J. Biol. Chem.*, **242,** 379
126. Marcel, Y. L., Christiansen, K. and Holman, R. T. (1968). *Biochim. Biophys. Acta*, **164,** 25
127. Mohrhauer, H. and Holman, R. T. (1963). *J. Nutr.*, **81,** 67
128. Brenner, R. R., de Tomas, M. E. and Peluffo, R. O. (1965). *Biochim. Biophys. Acta*, **106,** 640
129. Brenner, R. R. and Peluffo, R. O. (1967). *Biochim. Biophys. Acta*, **137,** 184
130. Brenner, R. R. (1969). *Lipids*, **4,** 621
131. Brenner, R. R., Peluffo, R. O., Nervi, A. M. and de Tomas, M. E. (1969). *Biochim. Biophys. Acta*, **176,** 420
132. Brenner, R. R. and Peluffo, R. O. (1969). *Biochim. Biophys. Acta*, **176,** 471
133. Dato, S. M. A. and Brenner, R. R. (1970). *Lipids*, **5,** 1013

134. Nervi, A. M., Brenner, R. R. and Peluffo, R. O. (1968). *Biochim. Biophys. Acta*, **152**, 539
135. Uchiyama, M., Nakagawa, M. and Okui, S. (1967). *J. Biochem (Tokyo)*, **62**, 1
136. Mohrhauer, H., Christiansen, K., Gan, M. V., Deubig, M. and Holman, R. T. (1967). *J. Biol. Chem.*, **242**, 4507
137. Christiansen, K., Marcel, Y., Gan, M. V., Mohrhauer, H. and Holman, R. T. (1968). *J. Biol. Chem.*, **243**, 2969
138. Budney, J. and Sprecher, H. (1971). *Biochim. Biophys. Acta*, **239**, 190
139. Stoffel, W., Ecker, W., Assad, H. and Sprecher, H. (1970). *Hoppe-Seyler's Z. Physiol. Chem.*, **351**, 1545
140. Nugteren, D. H. (1965). *Biochim. Biophys. Acta*, **106**, 280
141. Wit-Peeters, E. M. (1969). *Biochim. Biophys. Acta*, **176**, 453
142. Harris, P. and James, A. T. (1969). *Biochem. J.*, **112**, 325
143. Harris, P. and James, A. T. (1969). *Biochim. Biophys. Acta*, **187**, 13
144. Canvin, D. T. (1965). *Can. J. Bot.*, **43**, 63
145. Meyer, F. and Bloch, K. (1963). *Biochim. Biophys. Acta*, **77**, 671
146. Brown, C. M. and Rose, A. H. (1969). *J. Bacteriol.*, **99**, 371
147. Baranska, J. and Wlodawer, P. (1969). *Comp. Biochem. Physiol.*, **28**, 553
148. Fulco, A. J. (1972). *J. Biol. Chem.*, **247**, 3511
149. Holloway, P. W., Peluffo, R. and Wakil, S. J. (1963). *Biochem. Biophys. Res. Commun.*, **12**, 300
150. Fulco, A. J. and Bloch, K. (1964). *J. Biol. Chem.*, **239**, 993
151. Ginter, E., Ondreicka, R., Bobek, P. and Simko, V. (1969). *J. Nutr.*, **99**, 261
152. Wakil, S. J. (1964). Metabolism and Physiological Significance of Lipids, 3, (R. M. C. Dawson and D. N. Rhodes, editors) (London: John Wiley)
153. Oshino, N. (1972). *Arch. Biochem. Biophys.*, **149**, 378
154. Phelps, R. A., Shenstone, F. S., Kemmerer, A. R. and Evans, R. J. (1965). *Poultry Sci.*, **44**, 358
155. Johnson, A. R., Pearson, J. A., Shenstone, F. S. and Fogerty, A. C. (1967). *Nature (London)*, **214**, 1244
156. Bickerstaffe, R. and Johnson, A. R. (1972). *Br. J. Nutr.*, **27**, 561
157. Raju, P. K. and Reiser, R. (1969). *Biochim. Biophys. Acta*, **176**, 48
158. Raju, P. K. and Reiser, R. (1972). *J. Biol. Chem.*, **247**, 3700
159. Allen, E., Johnson, A. R., Fogerty, A. C., Pearson, J. A. and Shenstone, F. S. (1967). *Lipids*, **2**, 419
160. Johnson, A. R., Fogerty, A. C., Pearson, J. A., Shenstone, F. S. and Bersten, A. M. (1969). *Lipids*, **4**, 265
161. Fogerty, A. C., Johnson, A. R. and Pearson, J. A. (1972). *Lipids*, **7**, 335
162. Donaldson, W. E. (1967). *Biochem. Biophys. Res. Commun.*, **27**, 681
163. James, A. T., Harris, P. and Bezard, J. (1968). *Eur. J. Biochem.*, **3**, 318
164. Nordby, H. E., Heywang, B. W., Kircher, H. W. and Kemmerer, A. R. (1962). *J. Amer. Oil Chem. Soc.*, **39**, 183
165. Dupuis, R. G. and Favarger, P. (1965). *Helv. Physiol. Pharmacol. Acta*, **21**, 300
166. Pearson, J. A., Fogerty, A. C., Johnson, A. R. and Shenstone, F. S. (1970). *J. Amer. Oil Chem. Soc.*, **47**, Abstr. 65
167. Gurr, M. I., Robinson, M. P., James, A. T., Morris, L. J. and Howling, D. (1972). *Biochim. Biophys. Acta*, **280**, 419
168. Brett, D., Howling, D., Morris, L. J., and James, A. T. (1971). *Arch. Biochem. Biophys.*, **143**, 535
169. Oshino, N., Imai, Y., and Sato, R. (1971). *J. Biochem. (Tokyo)*, **69**, 155
170. Talamo, B. R. and Bloch, K. (1969). *Anal. Biochem.*, **29**, 300
171. Johnson, A. R. and Gurr, M. I. (1971). *Lipids*, **6**, 78
172. Gurr, M. I. and Robinson, M. P. (1972). *Anal. Biochem.*, **47**, 146
173. Cater, B. R., Poulter, J. and Hallinan, T. (1970). *FEBS Lett.*, **10**, 346
174. Hallinan, T. and Gurr, M. I. Unpublished results.
175. Scholan, N. A. and Boyd, G. S. (1968). *Biochem. J.*, **108**, 27P
176. Widnell, C. C. and Unkeless, J. C. (1968). *Proc. Nat. Acad. Sci. USA*, **61**, 1050
177. Holloway, P. W. and Wakil, S. J. (1970). *J. Biol. Chem.*, **245**, 1862
178. Holloway, P. W. (1971). *Biochemistry*, **10**, 1556
179. Lu, A. Y. H. and Coon, M. J. (1968). *J. Biol. Chem.*, **243**, 1331

180a. Gaylor, J. L., Moir, N. J., Seifried, H. E. and Jeffcoate, C. R. E. (1970). *J. Biol. Chem.*, **245,** 5511
180b. Shimakata, T., Mihara, K. and Sato, R. (1972). *J. Biochem. (Tokyo)*, **72,** 1163
181. Shimakata, T., Mihara, K. and Sato, R. (1971). *Biochem. Biophys. Res. Commun.*, **44,** 533
182. Oshino, N. and Sato, R. (1971). *J. Biochem (Tokyo)*, **69,** 169
183. Oshino, N. and Sato, R. (1972), *Arch. Biochem. Biophys.*, **149,** 369
184. Gompertz, D. and Greenbaum, A. L. (1966). *Biochim. Biophys. Acta*, **116,** 441
185. Mercuri, O., Peluffo, R. O. and Brenner, R. R. (1966). *Biochim. Biophys. Acta*, **116,** 409
186. Mercuri, O., Peluffo, R. O. and Brenner, R. R. (1967.) *Lipids*, **2,** 284
187. Inkpen, C. A., Harris, R. A. and Quackenbush, F. W. (1969). *J. Lipid Res.*, **10,** 277
188. de Gomez Dumm, I. N. T., de Alaniz, M. J. T. and Brenner, R. R. (1970). *J. Lipid Res.*, **11,** 96
189. Lee, C. J. and Sprecher, H. (1971). *Biochim. Biophys. Acta.*, **248,** 180
190. Gellhorn, A. and Benjamin, W. (1966). *Biochim. Biophys. Acta*, **116,** 460
191. Morris, L. J. (1970). *Biochem. J.*, **118,** 681
192. Morris, L. J. (1967). *Biochem. Biophys. Res. Commun.*, **29,** 311
193. James, A. T., Webb, J. P. W. and Kellock, T. D. (1961). *Biochem. J.*, **78,** 333
194. Lennarz, W. J. and Bloch, K. (1960). *J. Biol. Chem.*, **235,** PC 26
195. Light, R. J., Lennarz, W. J. and Bloch, K. (1962). *J. Biol. Chem.*, **237,** 1793
196. Davidoff, F. F. (1964). *Biochim. Biophys. Acta*, **90,** 414
197. Elovson, J. (1964). *Biochim. Biophys. Acta*, **84,** 275
198. Gurr, M. I. and Bloch, K. (1966). *Biochem. J.*, **99,** 16C
199. Howling, D., Morris, L. J. and James, A. T. (1968). *Biochim. Biophys. Acta*, **152,** 224
200. Howling, D., Morris, L. J., Gurr, M. I. and James, A. T. (1972). *Biochim. Biophys. Acta*, **260,** 10
201. Paulsrud, J. R., Stewart, S. E., Graft, G. and Holman, R. T. (1970). *Lipids*, **5,** 611
202. Raju, P. K. and Reiser, R. (1970). *Lipids*, **5,** 487
203. Nakagawa, M. and Uchiyama, M. (1969). *J. Biochem (Tokyo)*, 66, 95
204. Alfin-Slater, R. B. and Aftergood, L. A. (1968). *Physiol. Rev.*, **48,** 758
205. Fulco, A. J. and Mead, J. F. (1960). *J. Biol. Chem.*, **235,** 3379
206. Schlenk, H., Sand, D. M. and Sen, N. (1964). *Biochim. Biophys. Acta*, **84,** 361

6
The Prostaglandins

E. W. HORTON
University of Edinburgh

6.1	PROSTAGLANDIN NOMENCLATURE	239
	6.1.1 *Origin of the term*	239
	6.1.2 *Related compounds*	239
	6.1.3 *Prostanoic acid*	239
	6.1.4 *Trivial names and parent prostaglandins*	240
6.2	EXTRACTION AND SEPARATION	241
	6.2.1 *Solvent extraction*	241
	6.2.2 *Chromatographic methods*	242
6.3	ESTIMATION OF PROSTAGLANDINS	243
	6.3.1 *Biological assay*	243
	6.3.2 *Radioimmunoassay*	244
	6.3.3 *Enzymic assay*	244
	6.3.4 *Gas chromatography*	244
	6.3.5 *Gas chromatography–mass spectrometry*	245
6.4	IDENTIFICATION OF PROSTAGLANDINS IN ORGANS AND BODY FLUIDS	246
	6.4.1 *Distribution in tissues*	246
	6.4.2 *Output from organs and tissues*	246
6.5	BIOSYNTHESIS	248
	6.5.1 *Prostaglandin synthetase*	248
	6.5.2 *Co-factors*	249
	6.5.3 *Non-enzymic synthesis*	249
	6.5.4 *Other products*	249
	6.5.5 *Substrate specificity*	249
	6.5.6 *Inhibitors of prostaglandin synthetase*	249
	6.5.7 *Origin of substrate*	250
	6.5.8 *Mechanism*	250
	6.5.9 *Biosynthesis of prostaglandin A*	251
	6.5.10 *Prostaglandin A isomerase*	251
	6.5.11 *Prostaglandin C*	252

6.6	METABOLISM AND FATE	252
	6.6.1 *Distribution of injected prostaglandins*	252
	6.6.2 *Prostaglandin dehydrogenase*	253
	6.6.3 *Prostaglandin 13,14-reductase*	254
	6.6.4 *β-Oxidation*	254
	6.6.5 *ω-Oxidation*	254
	6.6.6 *Metabolism in man*	255
6.7	PHARMACOLOGICAL ACTIONS ON FEMALE REPRODUCTIVE TRACT SMOOTH MUSCLE	255
	6.7.1 *Parturition*	255
	6.7.2 *Abortion*	256
	6.7.3 *Human myometrial strips* in vitro	256
	6.7.4 *Human Fallopian tubes*	256
	6.7.5 *Uterus and oviduct of other species*	257
6.8	LUTEOLYSIN	257
	6.8.1 *Evidence that guinea-pig luteolysin is* $PGF_{2\alpha}$	257
	6.8.2 *Evidence that sheep luteolysin is* $PGF_{2\alpha}$	258
6.9	ROLE AT ADRENERGIC NERVE TERMINALS	258
	6.9.1 *Release of prostaglandins at adrenergic synapses*	258
	6.9.2 *Inhibition of noradrenaline release*	259
	6.9.3 *Inhibition of prostaglandin biosynthesis*	259
	6.9.4 *Interaction between prostaglandins and sympathomiametic amines on effector cells*	260
	6.9.5 *Concluding remarks*	260
6.10	PROSTAGLANDINS AND THE CENTRAL NERVOUS SYSTEM	260
	6.10.1 *Central nervous actions*	260
	6.10.2 *Distribution in the central nervous system*	260
	6.10.3 *Release of prostaglandins from the central nervous system*	262
	6.10.4 *Concluding remarks*	262
6.11	FEVER, INFLAMMATION AND PAIN	262
	6.11.1 *Fever*	263
	6.11.2 *Inflammation*	263
	6.11.3 *Pain*	263
6.12	ADENYL CYCLASE AND ADENOSINE 3′5′-MONOPHOSPHATE (CYCLIC AMP)	264
	6.12.1 *Adipose tissue*	264
	6.12.2 *Permeability to water*	264
	6.12.3 *Gastric secretion*	265
	6.12.4 *Systems in which PGE compounds mimic cyclic AMP*	265
	6.12.5 *Conclusions*	265
6.13	CONCLUDING REMARKS	265

6.1 PROSTAGLANDIN NOMENCLATURE

6.1.1 Origin of the term

The word 'prostaglandin' was coined by U.S. von Euler[1] in 1934 to describe an acidic lipid which he discovered in human semen, characterised by its stimulant action on isolated smooth muscle and its depressor action in the intact animal and believed to originate from the prostate gland. von Euler showed that prostaglandin belonged to a new class of naturally occurring substances and demonstrated convincingly that it differed from all other known substances with similar biological activity. Its chemical identification, however, had to await the development of appropriate analytical methods.

Professor Sune Bergström took up the problem of prostaglandin purification and in 1960, he and his colleagues announced the isolation of two prostaglandins from extracts of sheep vesicular glands[2-4]. The two prostaglandins could be separated by solvent partition between ether and a phosphate buffer, the one more soluble in ether being called prostaglandin E and the one more soluble in the buffer, prostaglandin F. The two compounds were assigned the empirical formulae $C_{20}H_{34}O_5$ and $C_{20}H_{36}O_5$ respectively. The full chemical structures of these and related naturally occurring prostaglandins were announced in 1962 and 1963[5]. The absolute configuration was finally established in 1966[6].

6.1.2 Related compounds

Many other unidentified substances related to prostaglandins have been described independently, for example, darmstoff originates from the frog intestine[7], irin from the rabbit iris[8,9], the menstrual stimulants from human endometrium[10], smooth muscle stimulating lipids in the brain[11-14], vesiglandin[15], medullin[16] and the slow reacting substances of normal guinea-pig[17] and pig[18] lung. It seems probable from recent work that the presence of one or more prostaglandins may account for most of the biological activity in these extracts.

6.1.3 Prostanoic acid

The basic C_{20} skeleton of the prostaglandins has been named prostane and the corresponding monocarboxylic acid, prostanoic acid (1), from which the

(1)

chemical names of all prostaglandins can be derived. Such chemical names tend to be too long for everyday use and so for the major prostaglandins, trivial names are commonly used.

6.1.4 Trivial names and parent prostaglandins

Five major series of natural prostaglandins have so far been described corresponding to differences in the cyclopentane ring (2)–(6). These parent prostaglandins, designated by the letters E, F, A, C and B are all hydroxylated

F_α (2)   E (3)   A (4)   C (5)   B (6)

in the 15-position and contain a 13,14 *trans* double bond. Within each series there are compounds differing in the degree of unsaturation. This is indicated in the trivial names by a subscript numeral after the letter. Thus, prostaglandins E_1, F_1, A_1, C_1 and B_1 have only the *trans* double bond, whereas E_2, F_2, A_2, C_2 and B_2 have an additional *cis* double bond in the 5,6-position. A further series with an additional *cis* double bond in the 17,18-position has been described but little investigated.

A system of abbreviations has been introduced by Andersen[19] to take into account the stereochemical configurations at C-9, C-11 and C-15. This so far, has found little application and will not be discussed further. There is much to be said for a system based upon the use of trivial names of the parent prostaglandins as this proves useful for rapid understanding both in written and oral communication. It is easier to use and can be quite accurate, provided certain rules are followed. Structures of six parent prostaglandins are shown ((7)–(11)).

(7) E_1 (8) $F_{1\alpha}$
(9) E_2 (10) $F_{2\alpha}$
(11) E_3 (12) $F_{3\alpha}$

The prefix *epi* denotes an inversion of the normal configuration of a particular substituent, for example, 15-epiprostaglandin A_2. The prefix *iso* has generally been used similarly for inversion at one of the side chains, for

Table 6.1   Abbreviations and chemical names of prostaglandins

Abbreviations used in this chapter*	Chemical name
PGE_1	(15S)-11a,15-dihydroxy-9-oxoprost-*trans*-13-enoic acid
PGE_2	(15S)-11a,15-dihydroxy-9-oxoprosta-*cis*-5,*trans*-13-dienoic acid
PGE_3	(15S)-11a,15-dihydroxy-9-oxoprosta-*cis*-5,*trans*-13,*cis*-17-trienoic acid
PGF_{1a}	(15S)-9a,11a,15-trihydroxyprost-*trans*-13-enoic acid
PGF_{2a}	(15S)-9a,11a,15-trihydroxyprost-*cis*-5,*trans*-13-dienoic acid
PGF_{3a}	(15S)-9a,11a,15-trihydroxyprost-*cis*-5,*trans*-13-*cis*-17-trienoic acid
PGA_1	(15S)-15-hydroxy-9-oxoprosta-10,*trans*-13-dienoic acid
PGA_2	(15S)-15-hydroxy-9-oxoprosta-*cis*-5,10,*trans*-13-trienoic acid
PGC_1	(15S)-15-hydroxy-9-oxoprosta-11,*trans*-13-dienoic acid
PGC_2	(15S)-15-hydroxy-9-oxoprosta-*cis*-5,11,*trans*-13-trienoic acid
PGB_1	(15S)-15-hydroxy-9-oxoprosta-8(12),*trans*-13-dienoic acid
PGB_2	(15S)-15-hydroxy-9-oxoprosta-*cis*-5,8(12),*trans*-13-trienoic acid
ent-PGE_1	(15R)-11β,15-dihydroxy-9-oxo-*ent*-prost-*trans*-13-enoic acid
15-epi-PGA_2	(15R)-15-hydroxy-9-oxoprosta-*cis*-5,10-*trans*-13-trienoic acid
8-iso-PGE_2	(15S)-11a,15-dihydroxy-9-oxo-8-iso-prost-*cis*-5,*trans*-13-enoic acid
19-hydroxy-PGA_1	(15S,19R)-15,19-dihydroxy-9-oxoprosta-10,*trans*-13-dienoic acid
13,14-dihydro PGE_2	(15S)-11a,15 dihydroxy-9-oxoprost-*cis*-5-enoic acid
5-*trans*-PGE_2	(15S)-11a,15 dihydroxy-*trans*-5,*trans*-13-dienoic acid
18-oxo-PGE_2	(15S)-11a,15 dihydroxy-9,18-dioxoprosta-*cis*-5,*trans*-13-dienoic acid
15-oxo-PGF_{2a}	9a,11a-dihydroxy-15-oxoprost-*cis*-5,*trans*-13-dienoic acid
tetranoic PGE_2	(15S)-11a,15 dihydroxy-9-oxo-2,3,4,5-tetranoic prost-*cis*-5,*trans*-13-dienoic acid

* To obtain the commonly accepted trivial name substitute the word prostaglandin for PG throughout

example, 8-isoprostaglandin E_2. When a carbon–carbon double bond is reduced, as for example following the action of 13,14-reductase, the product is referred to by the appropriate numbered prefix, for example, 13,14-dihydroprostaglandin E_2. Conversely, the introduction of a new double bond is referred to, for example, as 18,19-*cis*-didehydroprostaglandin E_1. Shortening of the side chains by the loss of one or more methylene groups is indicated by the prefixes 'nor', 'dinor', 'trinor' etc. Thus, β-oxidation of the prostaglandins results in dinor and tetranor derivatives. Conversely, lengthening of the side chains by one methylene group is identified by the use of the term 'homo', 'dihomo', trihomo' etc. Such compounds are not natural but are formed enzymically from analogues of the prostaglandin precursors. The full chemical names of the prostaglandins referred to in this chapter are shown in Table 6.1.

6.2 EXTRACTION AND SEPARATION

6.2.1 Solvent extraction

Prostaglandins can be extracted from an organ by homogenising it in four volumes of 96% ethanol, centrifuging and extracting the residue again with a further four volumes of ethanol[20]. Other methods of lipid extraction may be used. Excesses of pH and temperature should be avoided, the former should be kept within the range 4–8 and the temperature should not exceed 45°.

Contact with oxygen should be minimised and in some instances work should be carried out under nitrogen. Small quantities of prostaglandins can be formed non-enzymically during extraction, particularly when they are being extracted from tissues known to be rich in precursor acids[21].

A PGA may be formed as an artefact by dehydration of the corresponding PGE during the extraction process and the unstable PGC may be isomerised to PGB. 8-Iso-PGE may also appear as an artefact.

After evaporation of the alcohol to dryness, the extract is partitioned twice between water at pH 4.5 (using hydrochloric or citric acid) and an equal volume of ethyl acetate or diethyl ether. The separated organic phases are pooled, washed with a small volume of water to remove excess acid and concentrated to a smaller volume under reduced pressure. The concentrate is then extracted twice with an equal volume of pH 8.0 phosphate buffer. The combined aqueous phases are then adjusted to pH 4.5 and extracted twice with ethyl acetate. The combined ethyl acetate phases are washed and evaporated to dryness. The residue is partitioned between 67% aqueous ethanol and petroleum spirit. The ethanol is finally evaporated to dryness in preparation for chromatography.

Prostaglandins added to blood or plasma are not extracted well by the ethanol method described above[21], possibly due to adsorption to the precipitated plasma proteins. Better recoveries are obtained when heparinised blood is cooled and centrifuged at 4 °C. The plasma is separated and the cells washed with saline. The diluted plasma is adjusted to pH 4.5 and extracted as described above taking care to avoid emulsification with the ethyl acetate.

With this method, recoveries of PGE_1 and $PGE_{2\alpha}$ (added to blood to give concentrations of 10–100 ng ml^{-1}) are between 40 and 55%. For extraction of PGA compounds partition between ethyl acetate and pH 8 buffer should be avoided since these less polar prostaglandins are not extracted efficiently into the aqueous phase. Even then, recoveries of PGA_1 with this method tend to be low (*ca.* 10%).

An alternative method of extracting prostaglandins from plasma has been described[23]. Plasma from heparinised blood is diluted with an equal volume of saline (used to wash the blood cells). The diluted plasma is mixed with an equal volume of ethanol. Neutral fat and non-polar fatty acids are removed by partition with petroleum. Formic acid is then added sufficient to make a 1% v./v. concentration and the acidified aqueous ethanol extract is shaken with chloroform which extracts the prostaglandins. The chloroform layer is evaporated until all traces of formic acid have been removed.

Whether prostaglandins are extracted from tissues, blood or other body fluids, chromatographic purification is essential before any quantitative estimations can be made. Attempts to estimate the prostaglandin levels in such crude extracts by either gas chromatography or biological assay will yield results of very doubtful value.

6.2.2 Chromatographic methods

Chromatography on a column of silicic acid is an effective method of separating prostaglandins into several broad groups. The column is developed with a mixture of ethyl acetate and either benzene or toluene; the ethyl acetate concentration is increased either stepwise or as a gradient. Prostaglandins A and B are eluted first (with the 30% ethyl acetate), then PGE (60%) and finally PGF (80%). Each of these fractions may of course contain other substances including prostaglandin metabolites. Thus the 19-hydroxy-PGA derivatives are eluted with the PGF fraction.

Owing to variations in silicic acid from one batch to another, the behaviour of authentic prostaglandins must be checked by adding isotopically-labelled prostaglandins as internal standards. With some batches the addition of a small percentage of methanol to the eluant may be required for efficient elution and separation of the prostaglandins.

Silica gel (acid washed) suspended in chloroform and developed with increasing concentrations of methanol in chloroform is an alternative system to silicic acid.

Änggård and Bergkvist[24] have described a method for the separation of the methyl esters of prostaglandins A (and B), E and F on Sephadex LH-20. When blood pigments are present in an extract, these are eluted before the PGA fraction—a separation which is difficult to achieve on silicic acid. Unfortunately, this system has not yet been applied to the free acids and so its use is limited.

Reversed phase chromatography is an important method for the purification of prostaglandins and their metabolites. By selecting the appropriate solvent system, compounds differing only slightly in their physico-chemical characteristics can be separated. For example, this technique can be used to separate prostaglandins A, B and C[71].

Amberlite XAD-2 can be used for the concentrations of prostaglandins or their metabolites in urine or in large volumes of dialysate. The aqueous solution is allowed to percolate through the column at 2 ml cm^{-2} min^{-1}. After washing with water, the organic molecules including the prostaglandins are eluted with ethanol.

Numerous solvent systems have been described for the thin layer chromatographic separation of prostaglandins and their metabolites. Thin layer chromatography has the advantages of simplicity and speed. It is useful for small samples and plates can be scanned for radioactivity. Trace amounts of radioactive prostaglandins are added to the unknown mixture to be applied to the plate—the areas containing the radioactivity are then scraped off and eluted. The addition of $AgNO_3$ to the silica gel enhances the separation of prostaglandins which differ only in their degree of unsaturation.

6.3 ESTIMATION OF PROSTAGLANDINS

6.3.1 Biological assay

Biological preparations have been used extensively for the estimation of prostaglandins because of their great sensitivity. Furthermore, evidence of identification can be obtained using quantitative parallel assays. Thus a PGE can be distinguished from a PGF by estimating the unknown in terms of a compound of both series on rabbit jejunum and guinea-pig ileum[23, 24]. A combination of the cat blood pressure and rat fundus easily distinguishes PGA from both PGE and PGF[25].

The cascade of superfused tissues developed by Vane[26] may obviate the need for preliminary extraction. Blood or perfusion fluid from an animal or an organ is allowed to drip over the surface of smooth muscle preparations. By recording the contractions or relaxations of several tissues arranged in series it is possible to measure changes in levels of circulating prostaglandins or their output from organs. Greater specificity is obtained by the use of specific blocking drugs. Concentrations are estimated by comparing responses with those produced by authentic prostaglandins infused into the bathing fluid. One obvious advantage of the method is that information is obtained very rapidly. Unfortunately, the evidence of identification of the substance being estimated is inconclusive. Moreover, if more than one prostaglandin is present, accurate estimation is very difficult.

6.3.2 Radioimmunoassay

There have been rapid developments in the radioimmunoassay field. Methods have been developed in numerous laboratories and are being used to estimate PGF, PGE and PGB in sub-nanogram amounts. It is likely that in view of its sensitivity, specificity and relative speed, radioimmunoassay will form the basis for routine estimations of prostaglandin levels in body fluids in the future.

6.3.3 Enzymic assay

A very sensitive method which is specific for prostaglandins hydroxylated in the 15-position has been developed using a 15-hydroxyprostaglandin dehydrogenase from swine lung[27]. It seems unlikely that this method will now find wide application since, in spite of its specificity and sensitivity, it does not distinguish between different prostaglandins which have therefore to be separated before estimations can be made. It may be useful for distinguishing qualitatively between prostaglandins as a group and other substances which stimulate smooth muscle.

6.3.4 Gas chromatography

Much of the early work on the gas chromatographic behaviour of the prostaglandins has been summarised in the excellent review by Ramwell *et al.*, 1968[28].

Using a flame-ionisation detector and trimethylsilyl ether derivatives, Bygdeman and Samuelsson[29] could measure PGF in amounts down to 50–100 ng. Improved sensitivity has been achieved using an electron capture detector[30]. PGB compounds esterified and silylated can be detected in amounts down to 1 ng. Internal standards of ω-homo-E_1 and ω-nor-E_2 are added at the beginning of the extraction procedure and these together with any PGE in the mixture are converted to the corresponding PGB by treatment with methanolic KOH. This method can be applied to any prostaglandin which can be converted to PGB (i.e. PGE, PGA and PGC). The method of Albro and Fishbein[31] can be used for several different prostaglandins.

Radio gas chromatography has played an important role in the separation and identification of prostaglandin metabolites.

6.3.5 Gas chromatography–mass spectrometry

Bergström, Samuelsson and their co-workers used combined gas chromatography–mass spectrometry in the identification and estimation of prostaglandins and their metabolites. This technique is applicable to the nanogram range[32, 33].

PGA_1, PGA_2, PGB_1 and PGB_2 which co-chromatograph on silicic acid and Sephadex LH20 can be separated from each other as their methyl ester–trimethyl silyl ethers by gas chromatography and their identity established by a mass spectrum of the column effluent taken at the peak of the gas chromatographic trace.

PGE_1 and PGE_2 can be chromatographed as their methoximetrimethylsilyl ether–methyl ester derivatives (each compound showing two peaks corresponding to the two epimers formed). The trifluoroacetates of PGE_1 and PGE_2 (which are dehydrated to the corresponding PGA compound) can also be separated.

$PGF_{1\alpha}$ and $PGF_{2\alpha}$ together with the 19-hydroxy derivatives of PGA and PGB can be separated by gas chromatography either as the trimethyl ether–methyl esters or the trifluoroacetates.

If all the major peaks of a particular prostaglandin derivative are to be measured, then the lower limit of sensitivity is likely to be *ca.* 50 ng, though with partially purified extracts of biological origin, additional peaks contributing to background 'noise' may raise this to 200 ng. If the occurrence of one or two of the most abundant *m/e* peaks in the spectrum of the effluent occurring at the retention time corresponding to the carbon value of the derivative being studied is accepted as sufficient evidence of identification, greater sensitivity can be achieved. The techniques of mass fragmentography and multiple ion detection increase sensitivity to the very low nanogram, or even picogram, range at the price of having less conclusive identification.

There are numerous substances which may interfere with good gas chromatography–mass spectrometry. Compounds used in cleaning glassware, traces of silicone grease, impurities from the reagents used in the extraction procedure, blood pigments and other polar lipids must all be kept to a minimum (preferably excluded) when working in the low nanogram range.

Since losses on the gas chromatograph column and in the separator are variable and sometimes large, quantitation at these low levels is difficult. Samuelsson, Hamberg and Sweeley[33] introduced a reversed isotope-dilution method using a deuterated methoxime of methyl PGE_1 as a carrier and internal standard. Since the natural and deuterated compounds co-chromatograph, the excess of the deuterated derivative acts as a carrier for the natural prostaglandin both through the column and the separator, and by scanning for abundant m/e values and the corresponding $m/e + x$, in the case of the deuterated material (where x is the number of deuterium atoms), the ratio of unknown to deuterated prostaglandin can be determined. Since the amount of added prostaglandin is known, the amount of unknown can be estimated. This method has now been applied to prostaglandins which contain deuterium in the parent molecule. The carrier must, of course, contain a very high percentage of deuterium.

6.4 IDENTIFICATION OF PROSTAGLANDINS IN ORGANS AND BODY FLUIDS

6.4.1 Distribution in tissues

Prostaglandins have been isolated from various human and animal organs and fully characterised by rigorous chemical methods. From the data in Table 6.2, it is evident that these compounds are widely distributed in the body. The concentrations are generally small so that extensive chemical analysis may not be possible if only limited amounts of tissue are available.

Table 6.2 Mammalian sources from which prostaglandins have been isolated and identified chemically

Semen	Human, Sheep
Menstrual fluid	Human
Endometrium	Human
Lung	Human, monkey, ox, pig, sheep, guinea-pig
Brain	Ox
Thymus	Calf
Iris	Sheep
Pancreas	Ox
Kidney	Pig, rabbit

Some of the reports of identification, although not based upon conclusive evidence, contain such detailed evidence of various kinds that the identification of the particular prostaglandin is hardly in doubt. In other instances the evidence only justifies the term 'prostaglandin-like' or, possibly, for example, 'PGE-like'. In spite of the relative paucity of conclusive evidence, it is apparent that substances of the prostaglandin type occur in almost every mammalian tissue[34, 35]. They are present in all regions of the brain[36, 37] They have been detected in several subcellular fractions; for example, in whole rabbit brain homogenates $PGF_{2\alpha}$ is predominantly in the cytoplasmic fraction[38], but in rat cortical tissue, a high proportion of the prostaglandins is associated with the mitochondrial and synaptosomal fractions[39].

6.4.2 Output from organs and tissues

Release of prostaglandins or prostaglandin-like substances from numerous tissues and organs has been reported (Table 6.3). In some instances gas

Table 6.3 Release of prostaglandins

Site	Stimulus
Cat superfused somatosensory cortex	Spontaneous Sensory nerve stimulation Contralateral cortical stimulation
Cat superfused cerebellar cortex	Spontaneous
Cat perfused cerebral ventricles	Spontaneous, pyrogen
Dog perfused cerebral ventricles	Spontaneous 5-Hydroxytryptamine
Frog superfused spinal cord	Spontaneous Sensory nerve stimulation
Rat phrenic nerve-diaphragm *in vitro*	Electrical stimulation Noradrenaline
Rat epididymal fat pad *in vitro*	Nerve stimulation Noradrenaline
Rat gastro-intestinal tract	Pentagastrin Histamine Vagal stimulation Transmural stimulation Carbachol 5-Hydroxytryptamine
Dog blood-perfused spleen	Nerve stimulation Adrenaline Colloidal particles
Cat Ringer-perfused adrenals	Acetylcholine
Guinea-pig Ringer-perfused lungs	Phospholipase A
Dog Ringer-perfused lungs	Stretch
Rabbit eye	Mechanical stimulation
Frog intestine *in vitro*	Spontaneous
Human medullary carcinoma of the thyroid *in vivo*	Spontaneous
Rat carrageenin pouch	Inflammatory response
Dog kidney	Ischaemia
Guinea-pig uterus	Distension, oestrogen End of oestrous cycle
Sheep uterus	End of oestrous cycle Parturition
Rabbit kidney	Spontaneous Nerve stimulation

chromatographic–mass spectrometric identification of the prostaglandin has been achieved. Similar conclusive evidence is needed with respect to prostaglandin output from other sources before any quantitation can be regarded as reliable.

6.5 BIOSYNTHESIS

6.5.1 Prostaglandin synthetase

The enzymic conversion of arachidonic acid to PGE_2 was discovered independently by D. A. van Dorp[40] and S. Bergström[41]. $PGF_{2\alpha}$ is also formed from arachidonic acid[42] whilst PGE_1 and $PGE_{1\alpha}$ are similarly formed from dihomo-γ-linolenic acid, the immediate precursor of arachidonic acid[41-44]. The conversion of all-cis-5,8,11,14,17-eicosapentaenoic acid to PGE_3 has also been demonstrated[41].

The enzyme system which catalyses the conversion of polyunsaturated fatty acids to prostaglandins has been named prostaglandin synthetase. It is found in many tissues (Table 6.4) but a particularly rich source is the sheep vesicular gland, homogenates of which were formerly used for the large-scale production of prostaglandins.

Prostaglandin synthetase is associated with the microsomal fraction, but so

Table 6.4 Organs and tissues which contain prostaglandin synthetase[45, 45a]

Prostate	Human
Seminal vesicle	Sheep, ox, human, fowl
Lung	Guinea-pig, sheep, cow, pig, rabbit, fowl, duck, frog, toad
Intestine	Sheep, guinea-pig, rabbit, hamster, toad, duck, carp
Uterus	Human (endometrium), sheep, guinea-pig
Thymus	Sheep
Heart	Sheep
Liver	Sheep, guinea-pig, rat
Kidney	Sheep, guinea-pig, rat, fowl, duck, pig, cow, rabbit, hamster, carp
Pancreas	Sheep
Brain	Guinea-pig
Stomach	Rat, cow
Iris	Pig, rabbit
Spleen	Dog, sheep, rabbit, fowl, carp
Urinary bladder	Frog, toad
Adrenals	Cow, duck
Red bone marrow	Guinea-pig
Testis	Fowl
Skin	Fowl, frog
Ovary	Fowl
Gills	Mussel, lobster, carp, tench

far no reports have been published of preparations with high purity. An enzyme-rich particulate fraction has been obtained by centrifugation of sheep vesicular gland homogenates for 1 h at 100 000 g after removal of other cell particles at lower speeds. The freeze-dried sediment retains its activity for several months when stored at $-20°C$.

6.5.2 Co-factors

The addition of glutathione greatly enhances the yield of PGE compounds though this is partly at the expense of the corresponding PGF. Other SH-containing compounds (cysteine, homo-cysteine, thiophenol, thioglycollic acid) are far less effective than glutathione. Hydroquinone also increased the yield moderately but neither ATP nor NADH had any effect. Although less effective than glutathione, ascorbic acid and various phenols all increase the yield of prostaglandin[45-47]. The enzyme is partly inhibited by divalent cations such as copper, zinc and cadmium[47].

6.5.3 Non-enzymic synthesis

The formation of prostaglandins in low yield (0.1 %) from precursor acids in the presence of oxygen can occur in the absence of an enzyme[21].

6.5.4 Other products

On incubation of all-*cis*-8,11,15-eicosatrienoic acid with sheep vesicular gland homogenates, at least five products[47] have been reported in addition to PGE_1 and $PGF_{1\alpha}$ including 11-didehydro-$F_{1\alpha}$[48].

6.5.5 Substrate specificity

The structural requirements for the biosynthesis of prostaglandin-type compounds from unsaturated straight chain acids have been studied using a variety of synthetic precursors. Of the all-*cis*-tetraenoic acids tested, the C_{20} and C_{19} homologues gave high yields of PGE_2 (71 %) and *a*-nor-PGE_2 (41 %) respectively, the C_{21} and C_{22} gave somewhat lower yields (*ca.* 25 %). Neither the esters nor the alcohols (all-*cis*-8,11,14-eicosatrienol and all-*cis*-5,6,11,14-eicosatetraenol) are substrates for the enzyme[49].

6.5.6 Inhibitors of prostaglandin synthetase

Certain analogues of the essential fatty acids act as inhibitors of prostaglandin synthetase. In particular, 8-*cis*-12-*trans*-14-*cis*-eicosatrienoic acid and 5-*cis*-8-*cis* 12-*trans*-14-*cis*-eicosatetraenoic acid are effective blocking agents *in vitro*. The former of these has been fed to rats at 70 mg day^{-1}; growth rate was reduced by 20 % but little of the abnormal acid was incorporated into the lipids of the liver[50,51].

Linoleic (18:2ω6) and α-linolenic (18:3ω3) acids can inhibit the action of prostaglandin synthetase on arachidonic acid[52]. Oleic acid is a less active inhibitor.

The acetylenic analogue of arachidonic acid, 5,8,11,14-eicosatetraynoic acid inhibits prostaglandin biosynthesis[53] and has proved a useful tool in biological work.

One of the most significant recent advances in this field has been the discovery that prostaglandin formation both *in vivo* and *in vitro* is blocked by the aspirin group of anti-inflammatory drugs[54-56]. One of the most potent compounds is indomethacin which has already been widely used as a research tool to study biological systems in which the biosynthesis and release of prostaglandins may be implicated. The site of action of these compounds is not yet established.

6.5.7 Origin of substrate

There is little free arachidonic acid and dihomo-γ-linolenic acid in cells. These acids are incorporated into membrane phospholipids from which they must be hydrolysed before prostaglandin synthetase can act. It has not been possible to detect any prostaglandin in the phospholipid fraction and none of the newly synthesised prostaglandin is incorporated into the phospholipid[57,58]. Bartels, Vogt and Wille[59] have shown that phospholipase A injection will increase prostaglandin output, presumably by making the substrate available for prostaglandin synthetase. Eliasson[60] made similar observations *in vitro*. It is probable that activation of phospholipase A may be the rate-limiting step in prostaglandin synthesis[61].

6.5.8 Mechanism

By using 8,11,14-eicosatrienoic acid tritiated at either the 8,11- or 12-position, Klenberg and Samuelsson[62] demonstrated that the hydrogen atoms at these positions in the precursor acids are retained in the same positions in the PGE_1 molecule following bond formation between the carbon atoms at the 8 and 12 positions.

Using $^{18}O_2$ gas in the biosynthesis of PGE_1, it has been conclusively proved by mass spectrometric analysis that the oxygen atoms at the 11- and 15- positions in PGE_1 are derived from molecular oxygen[63,64] and that the oxygen at C-9 is derived from the same molecule of oxygen as the oxygen atom in the C-11 hydroxyl[47]. Hamberg and Samuelsson[66] showed that the *pro*-S hydrogen is lost from C-13 during the biosynthesis of PGE_1. Furthermore, using precursor acid tritiated in the 9 position, it has been proved that $PGF_{1\alpha}$ is not formed via PGE_1.

Further evidence in favour of this mechanism comes from the observation that oxygenation of 8,11,14-eicosatrienoic acid in the presence of the sheep vesicular gland enzyme yields 12-hydroxy-8,10-heptadecadienoic acid. By means of 8,11,14-eicosatrienoic acids tritiated at either 9,10,11,13(D- or L-) or 15 positions, it is shown that carbon atoms 8, 10 and 11 of the precursor

THE PROSTAGLANDINS 251

are eliminated as malonaldehyde and that the hydrogen removal from C-13 is stereochemically identical with that occurring during the biosynthesis of PGE_1. It thus seems very probable that both prostaglandins and 12-hydroxy-8,10-heptadecadienoic acid originate from the same cyclic peroxide of 8,11,14-eicosatrienoic acid[66, 67].

6.5.9 Biosynthesis of prostaglandin A

The biosynthetic pathway for PGA_1 and PGA_2 has not yet been elucidated although the formation of PGA_1 has been reported during incubation of 8,11,14-eicosatrienoic acid with sheep seminal vesicles[68]. The facility with which these compounds can be derived chemically from PGE_1 and PGE_2 suggests the possibility that PGA compounds are formed by an enzymatically-catalysed dehydration of PGE compounds.

6.5.10 Prostaglandin A isomerase

An enzyme which converts PGA_1 and PGA_2 to PGC_1 and PGC_2 respectively (Figure 6.1) was discovered in cat plasma by R. L. Jones[69-72].

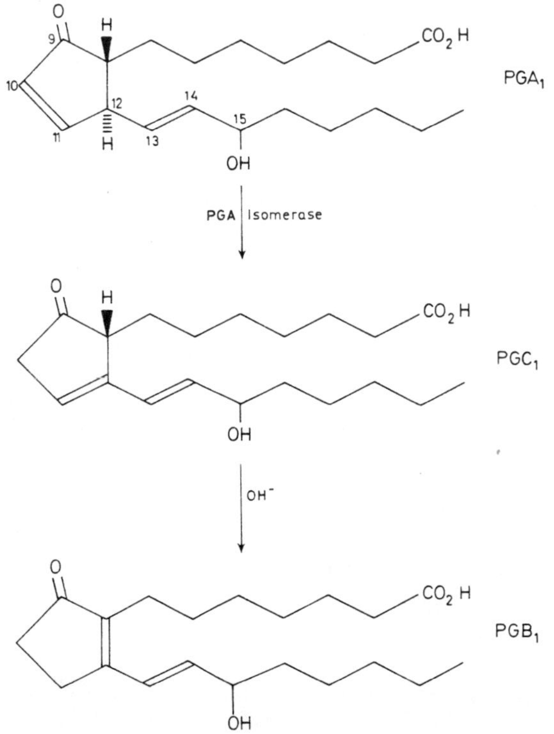

Figure 6.1 Conversion of PGA_1 to PGC_1 by PGA isomerase and of PGC_1 to PGB_1 under alkaline conditions

The enzyme, prostaglandin A isomerase, catalyses the shift of the 10,11-double bond to the 11,12-position. The mechanism of action at the molecular level has not yet been investigated. The isomerase has been detected in the plasma of cat, rabbit, rat, pig and dog but not in the plasma of ox, sheep, guinea-pig or man.

The Km value for the isomerase using PGA_1 as substrate at pH 7.0 and 25°C was calculated to be 2.5×10^{-5} M[72]. The enzyme is completely denatured by incubation at 62°C for 10 min. The optimum pH lies between 8 and 9. At pH 8.5, however, the product PGC is unstable and undergoes a base-catalysed isomerisation to PGB. At pH 7.0 negligible conversion to PGB occurs.

PGA_1, PGA_2, 19-OH-PGA_1 and 15-epi-PGA_2 are all substrates for the enzyme, although the latter two are poorer substrates than PGA_1 and PGA_2. On the other hand, 13,14-dihydro-PGA_1 is not a substrate but inhibits the enzyme. PGE_1 and $PGF_2\beta$ are inhibitors at concentrations 10 to 15-fold greater than 13,14 dihydro PGA_1.

6.5.11 Prostaglandin C

The structure of the product of the isomerase, PGC_1, has been established by a combination of physico-chemical techniques[71] and confirmed by chemical synthesis (unpublished data).

6.6 METABOLISM AND FATE

6.6.1 Distribution of injected prostaglandins

Autoradiographic studies in mice, using $PGF_{2\alpha}$ (tritiated at the C-17,18 positions) and PGE_1 (tritiated at the C-5,6 positions) showed that 15 min after the injection, radioactivity was concentrated in the liver, kidneys and subcutaneous connective tissue. No significant amounts could be found in the heart, blood vessels, brain, adipose tissue, endocrine glands, spleen, thymus or lymph nodes[73, 74]. One hour after the injection appreciable excretion of radioactivity occurred in the bile and urine; the labelled material is then concentrated in the liver, gall bladder, intestinal lumen, kidneys and thoracic duct (indicating some absorption from the intestinal lumen).

The rapid uptake of radioactivity by liver and kidney and its excretion via bile and urine agrees with findings in the rat[75]. In this animal, following a 20 min infusion of labelled PGE_1 intravenously, approximately two-thirds of the administered radioactivity was recovered in the urine (85%) and faeces (15%). In rats with biliary cannulae there was negligible radioactivity in the faeces and approximately one-fifth of the total radioactivity recovered was present in the bile.

Levels of radioactivity were measured in different organs following subcutaneous injections of PGE_1 in female rats. The levels in the kidney rose rapidly, reaching a peak 15 min after injection and always exceeded the levels found in any other organ. Radioactivity accumulated in the liver more slowly, reaching a peak after 20 min. Concentration in these two organs accounted for most of the injected material. The lungs contained only moderate amounts with a maximum at 10–15 min after injection. Small amounts accumulated in the heart, pituitary and adrenals. Some was found in the uterus. Brain, adipose tissue and skeletal muscle contained barely detectable amounts at any time during the experiment. After 60 min, most of the radioactivity not yet excreted was concentrated in the kidney and liver.

Two labelled metabolites of PGE_1 were isolated from rat plasma during these experiments and identified respectively as 11α-15-dihydroxy-9-oxoprostanoic acid and 11α-hydroxy-9,15-di-oxoprostanoic acid. Neither of these metabolites could be detected in the liver. The major part of the radioactivity in both liver and bile was attributable to the presence of compounds more polar than PGE_1. Similarly, there were more polar derivatives in the kidney and urine. Numerous urinary metabolites of PGE_2 have been identified in the rat[76, 79]. The major urinary metabolite of $PGF_{1\alpha}$ in the rat is α-dinor-$PGF_{1\alpha}$[77, 78]. This metabolite, and others, including α-tetranor-PGF, are also excreted following $PGF_{2\alpha}$ administration.

In the guinea-pig the principal urinary metabolites of PGE_2 and $PGF_{2\alpha}$ are $5\beta,7\alpha$-dihydroxy-11-oxo-tetranorprostanoic acid and $5\alpha,7\alpha$-dihydroxy-11-oxo-tetranorprostanoic acid respectively[76, 80, 81]. Oxidation of the secondary alcohol at the C-15 position and reduction of the 13,14-*trans* double bond occur readily in the lungs[42, 82, 83]. β-Oxidation of the carboxyl side chain probably occurs in the liver. Reduction of the 9-oxo group has been encountered so far only in the guinea-pig. This configuration is opposite to that found in the parent prostaglandins of the F series.

Ferreira and Vane[84] established using the blood-bathed organ technique that in the cat, dog and rabbit, PGE_1, PGE_2 and $PGF_{2\alpha}$, although stable in blood, are 95% removed on one circulation through the lungs, and on infusion into the portal vein are removed up to 80% by the liver. Vane[26] concludes that the body has a very efficient mechanism for preventing prostaglandins from reaching the arterial circulation.

Independently, Horton and Jones[85] and McGiff, Terragno, Strand, Lee, Lonigro and Ng[86] demonstrated that PGA_1 and PGA_2, unlike PGE and PGF compounds are not biologically inactivated on passage through the lungs of the cat and dog, although they are metabolised by the Krebs-perfused guinea-pig lung[87]. PGA_1 and PGA_2 are fairly effectively removed on circulation through the liver (55–90%)[85]. Since PGA is rapidly isomerised to PGC in plasma[72] it seems likely that it is the PGC which is not removed by the lungs. PGA released from a tissue into the venous effluent will be converted to PGC and reach target organs before substantial inactivation occurs (in those species which have plasma prostaglandin A isomerase).

6.6.2 Prostaglandin dehydrogenase

Homogenates of pig lung contain an enzyme system, 15-hydroxyprostaglandin dehydrogenase, which converts PGE_1 to its 15-oxo-derivative[88]. The enzyme has been purified from high-speed supernatant fractions by ammonium sulphate fractionation and by chromatography on TEAE–cellulose, DEAE-Sephadex and Sephadex G-100. An 11-fold purification with a 30% yield has been obtained.

The purified enzyme is NAD^+-dependent and is highly specific for the C-15 secondary alcohol substituent of the prostaglandins. The enzyme is denatured by temperatures above 55°C and the reaction rate is slow at 30°C. Most assays have been performed at 44°C and within the pH range 6–8. PGE, PGF, PGA, α-nor-PGE and ω-homo-PGE compounds are all substrates for the enzyme. The enzyme is also stereospecific with regard to the configuration at C-15[89,90]. Both PGB compounds and 15-epi-PGE_1 are non-competitive inhibitors of the enzyme. 13-14-dihydro-PGE_1 is not a good substrate for the enzyme and so it is probable that *in vivo* oxidation of the secondary alcohol at C-15 precedes reduction of the C-13,14 double bond[75].

Prostaglandin dehydrogenase has also been identified in the lungs of several species; it is also widely distributed in other organs, for example, the gastrointestinal tract, spleen, kidney and liver[75].

A histochemical method for the localisation of this enzyme in tissue sections has been described by Nissen and Andersen[91]. This method has been used to demonstrate the high concentration of the enzyme in the thick ascending limb of the loop of Henle and in the distal tubule of the rat, and also in the Purkinje cell layer of the cerebellar cortex[92].

In view of its high specificity, its widespread distribution and the relative inactivity of its end-products, prostaglandin dehydrogenase may prove to be the most important of the enzymes which metabolise prostaglandins E and F produced locally in tissues.

6.6.3 Prostaglandin 13,14-reductase

This enzyme reduces the C-13,14 double bond and was first reported in the guinea-pig lung[81]. Organs in the pig rich in this enzyme included adipose tissue, spleen, kidney, liver, adrenals and small intestine[76]. The enzyme is mainly located in the particle-free fraction of cells. Its substrate specificity has not yet been determined.

6.6.4 β-Oxidation

Many prostaglandins are substrates for the β-oxidising enzyme system in liver mitochondria, the carboxyl side chain being shortened by either two or four carbons. Although PGE_1 is converted to the C-18 metabolite, 13,14-dihydro-PGE_1 is degraded to the C_{16} derivative[93].

6.6.5 ω-Oxidation

19-Hydroxy-PGA_1 and 20-hydroxy-PGA_1 are formed when PGA_1 is incubated with the microsomal fraction of guinea-pig or human liver[76,94]. Furthermore, 19-hydroxy prostaglandins are natural constituents of human seminal plasma[95]. PGE_1 itself is not a substrate for this microsomal ω-oxidising enzyme, but ω-carboxylated metabolites of PGE and PGF have been found in the urine of rat, guinea-pig and man. Recent evidence indicates that shortening of the ω-side chain is also a metabolic pathway for the prostaglandins[96].

PGE_1 is not inactivated by blood[22]. In contrast PGA_1 and PGA_2 lose *ca.* 50% of their activity in 30 min on incubation with cat plasma[70]. A high proportion of the PGA compounds is converted (isomerised) to the corresponding PGC by the plasma enzyme, prostaglandin isomerase, and the unstable PGC slowly isomerises non-enzymically to the inactive PGB, thus accounting for the loss of activity on incubation of PGA with blood over a period of 0.5–1.5 h[72].

6.6.6 Metabolism in man

When labelled PGE_2 is injected intravenously in man, *ca.* 4% of the original PGE_2 is present unchanged in the circulation 1 min later and 40% is present in the blood as the metabolite, 11a-hydroxy-9,15-dioxo-5-*cis*-prostenoic acid. Approximately 50% of the radioactivity is recovered in the urine within 5 h. One urinary metabolite of PGE_2 in man has been identified as 7a-hydroxy 5,11-dioxo-tetranorprosta-1,16-dioic acid[97].

The reaction sequence may be as follows: oxidation at C-15 (particularly in the lungs), reduction of the C-13,14 double bond, two steps of β-oxidation and finally ω-oxidation. The order of the last two steps is not established, but the first two reactions probably precede β-oxidation, since the tetranorprostaglandins are poor substrates for prostaglandin dehydrogenase.

When labelled $PGF_{2\alpha}$ is injected intravenously in man, over 90% of the radioactivity is recovered from the urine within about 5 h. The main metabolite is 5a,7a-dihydroxy-11-oxotetranorprosta-1,16-dioic acid[98]. Additional urinary metabolites of $PGF_{2\alpha}$ in man have been identified.

PGE_1 and $PGF_{1\alpha}$ give rise to the same human urinary metabolites as PGE_2 and $PGF_{2\alpha}$ respectively. Thus the estimation of these compounds in urine will reflect the total production of the four prostaglandins of the E and F series.

6.7 PHARMACOLOGICAL ACTIONS ON FEMALE REPRODUCTIVE TRACT SMOOTH MUSCLE

6.7.1 Parturition

The pregnant human myometrium is very sensitive to prostaglandins. An increase in amplitude and frequency of contractions in women at mid-pregnancy can be induced by intravenous infusions of PGE_1 (0.6–9.0 µg min^{-1} [99].

Prostaglandins occur in umbilical cord vessels, in amniotic fluid and possibly in the circulation of women during labour[100-102]. This led to the suggestion that $PGF_{2\alpha}$ might be implicated physiologically in the uterine contractions of parturition[103]. Moreover, both this compound and PGE_2 can be used to induce labour[104, 105]. These prostaglandins now have an established place in the induction of parturition following successful controlled clinical trials. Investigations into the physiological role of prostaglandins in parturition continue.

6.7.2 Abortion

$PGF_{2\alpha}$ (50 µg min^{-1}) infused intravenously between the 9th and 22nd week of pregnancy induces abortion[106]. Considerable stimulation of uterine activity is observed. Side effects include pain, vomiting, nausea and diarrhoea.

PGE_2 and PGE_1 have also been used to terminate pregnancy. The total dose of PGE_2 varies between 1 and 10 mg infused over 6–12 h. The dose of $PGF_{2\alpha}$ is higher (10–100 mg) infused over a similar period.

Other routes, including extra- and intra-amniotic administration are also successful. Bygdeman and Wiqvist[107] have administered PGE_2 and $PGF_{2\alpha}$ into the uterine cavity between the foetal membranes and the uterine wall. The prostaglandin is administered intermittently via an in-dwelling catheter, producing sustained and intensive contractions. The total dose required was approximately one-tenth of that needed by the intravenous route. Neither nausea nor diarrhoea was experienced.

The suggestion that prostaglandins may be implicated in spontaneous abortion needs further study.

6.7.3 Human myometrial strips *in vitro*

Many observations have been made on the response of human myometrial strips *in vitro* to prostaglandins. In general, both amplitude and frequency of spontaneous contractions are *reduced* by PGE compounds. The threshold concentrations for PGE_1 and PGE_2 is *ca.* 10–100 ng ml^{-1}. PGA compounds and their 19-hydroxy derivatives, all natural constituents of human semen, are also inhibitory but 10–30 times less potent than PGE_1[108-110]. $PGF_{1\alpha}$ and $PGF_{2\alpha}$ stimulate contractions of the isolated human myometrium. Preparations obtained late in the menstrual cycle, or during pregnancy, are very sensitive to this stimulant action of $PGF_{2\alpha}$ suggesting that hormonal status affects the sensitivity of these isolated strips. This is confirmed by the observations that the uterus is some five times more sensitive to the inhibitory action of PGE_1 at the time of ovulation. Moreover, the myometrial strip from a pregnant woman is often contracted by PGE_1, though higher doses can cause inhibition[111].

6.7.4 Human Fallopian tubes

PGE_1 and PGE_2 contract the most proximal (uterine) end of the human Fallopian tube *in vitro* but relax the distal three-quarters[112, 113]. The difference is most marked in tissue removed during the secretory phase of the menstrual cycle. All parts of the tube are relaxed by PGE_2 and contracted by $PGF_{1\alpha}$ and $PGF_{2\alpha}$[113, 114]. All these prostaglandins occur in human semen and their action may result in the retention of the ovum within the tube and so increase the chance of fertilisation occurring following coitus.

6.7.5 Uterus and oviduct of other species

The response of female reproductive tract smooth muscle of several laboratory animals to PGE_1 has been investigated both *in vivo* and *in vitro*. Rat, guinea-pig and rabbit uterus *in vitro* contracts, guinea-pig uterus *in vivo* contracts, but both rabbit uterus and oviduct *in vivo* are inhibited by PGE_1[23a, 24a, 109, 163, 207-211]. Isolated rat uterus has been used as a bioassay preparation. This tissue is more sensitive to $PGF_{1\alpha}$ and $PGF_{2\alpha}$ than PGE_1 and PGE_2[23a, 24a]. Marked tachyphylaxis has been observed on this tissue with several prostaglandins[115, 116]. Guinea-pig uterus *in vitro* is not only very sensitive to PGE compounds but also shows the phenomenon of enhancement to the action of other oxytocic substances and electrical stimulation[117]. Interactions between prostaglandins and oxytocin on the human myometrium have been observed clinically.

6.8 LUTEOLYSIN

There is evidence in several species for the existence of a hormone, luteolysin, which is released from the uterus and, acting locally, causes regression of the ipsilateral corpora lutea. Recent work in the guinea-pig and the sheep strongly supports the conclusion that, in these two species, prostaglandin $F_{2\alpha}$ has a physiological role and is identical with the uterine hormone, luteolysin.

6.8.1 Evidence that guinea-pig luteolysin is $PGF_{2\alpha}$

All the evidence suggests that under the influence of progesterone, the endometrium releases a hormone which reaches the adjacent ovary by a local vascular route and there causes the corpus luteum to regress. Premature luteolysis in the guinea-pig can also be induced by the presence of foreign bodies in the lumen of the uterus or by treatment with oestrogen on the 6th day of the cycle. The following evidence seems to implicate $PGF_{2\alpha}$ in each of these mechanisms.

The presence of a foreign body in the lumen of the uterine horn *in vitro* stimulates the release of $PGF_{2\alpha}$[118]. Moreover, oestrogen treatment of guinea-pigs on day 6 of the oestrous cycle elevated $PGF_{2\alpha}$ levels in the utero-ovarian venous blood, in none of the control animals could $PGF_{2\alpha}$ be detected[119, 120]. Utero-ovarian venous blood was collected from guinea-pigs at different times of the cycle. At the end of the cycle greatly elevated levels of $PGF_{2\alpha}$ and PGE_2 were detected[120]. This corresponded to the time of fall in progesterone level. A parallel study on prostaglandin formation by homogenates of the uterus taken from guinea-pigs on different days in the cycle showed that $PGF_{2\alpha}$ production was far greater at the end of the cycle[121]. In all these experiments $PGF_{2\alpha}$ was identified by gas chromatography–mass spectrometry.

Since prostaglandin biosynthesis in uterine homogenates can be inhibited by indomethacin[121], another approach was offered for the study of luteolysin. If the uterine luteolytic hormone is $PGF_{2\alpha}$, then inhibition of its synthesis should prevent normal luteal regression and thus lengthen the oestrous cycle. Implants of paraffin wax impregnated with indomethacin were placed in the uterine horns of guinea-pigs. In six animals the oestrous cycle was in excess of 47 days, whereas six control animals with paraffin wax implants alone had a first cycle of mean length 19.5 ± 1.0 days and subsequently three cycles of 14.8 ± 0.2 days[122].

Since $PGF_{2\alpha}$ is luteolytic in the guinea-pig[123] and blockade of its synthesis prevents luteal regression, and since it is released in response to stimuli known to cause luteolysis, the cumulative evidence very strongly supports the conclusion that in the guinea-pig luteolysin is $PGF_{2\alpha}$.

6.8.2 Evidence that sheep luteolysin is $PGF_{2\alpha}$

Transplantation of the uterus or ovary, or both uterus and ovary, to the neck of the sheep has facilitated the investigation of the possible luteolytic role of $PGF_{2\alpha}$ in this species[124, 125]. The technique has firmly established that a substance is released from the uterus into the uterine venous blood and that this substance causes luteal regression as indicated by a fall in progesterone level and a subsequent rise in oestrogen, corresponding to the onset of oestrus. Moreover, infusion of $PGF_{2\alpha}$ at 25 µg hr^{-1} into the arterial supply to the transplanted ovary mimics these effects. Thorburn and Nicol[126] have shown that infusion of $PGF_{2\alpha}$ into either the ovarian artery or uterine vein in the *intact* sheep also cause rapid luteal regression as indicated by the fall in peripheral plasma progesterone level. The levels of $PGF_{2\alpha}$ required were within the range found in uterine venous blood of ewes at the end of the oestrous cycle[127]. The rise in uterine venous plasma levels of $PGF_{2\alpha}$ at this time has been confirmed[125]. In both investigations the identity of $PGF_{2\alpha}$ in uterine venous blood was established by mass spectrometry.

Some evidence in the sheep also indicates the pathway by which $PGF_{2\alpha}$ released into the uterine vein may reach the ipsilateral ovary without entering the systemic circulation[124, 125]. The ovarian artery runs a highly convuluted course, deeply embedded in the utero-ovarian vein. Infusion of labelled $PGF_{2\alpha}$ into the uterine vein resulted in levels of radioactivity in the ovarian artery which exceeded those in the iliac artery. It is suggested, therefore, that a counter-current mechanism operates by which some of the $PGF_{2\alpha}$ in the utero-ovarian vein is taken up into the ovarian artery and so reaches the target organ, the corpus luteum. Further work is needed to establish unequivocally that this is indeed the route by which uterine $PGF_{2\alpha}$ reaches the ovary.

6.9 ROLE AT ADRENERGIC NERVE TERMINALS

6.9.1 Release of prostaglandins at adrenergic synapses

Shaw and Ramwell[128], in 1968, reported that noradrenaline stimulates the release of a prostaglandin E-like substance, or substances, from rat epididymal fat pads *in vitro*. It was of interest to know whether this was a general phenomenon associated with adrenergic nerve stimulation. Davies, Horton and Withrington[129], in 1968, examined the venous effluent from a spleen perfused with blood from a donor dog for the presence of prostaglandins. Blood samples collected during and immediately after splenic nerve stimulation contained several prostaglandin-like substances. One of these, on the basis of its behaviour in three chromatographic systems, and on quantitative parallel biological assay on four tissues, was identified as PGE_2. This was subsequently confirmed by mass spectrometry. Blood collected before nerve stimulation contains no detectable prostaglandin. Using a cascade of superfused tissues to detect prostaglandin release, Vane and his co-workers[130, 131] detected the release of both PGE_2 and $PGF_{2\alpha}$. They established that the prostaglandin was probably released in association with the capsular, rather than the vascular smooth muscle. Prostaglandin-like substances are also released in response to adrenergic nerve stimulation or catecholamines from frog skin[132], rat diaphragm[133, 134] and cat spleen[135]. The evidence of identification in these reports is insufficiently conclusive to form the basis of a hypothesis for the physiological role of prostaglandins at adrenergic nerve endings.

6.9.2 Inhibition of noradrenaline release

When the splenic nerve is stimulated, the saline perfused cat spleen responds by vasoconstriction and capsular contraction. These responses are reduced if PGE_2 or PGE_1 (2×10^{-6}M) are infused intra-arterially[136, 137]. This is accompanied by decreased output of the transmitter, noradrenaline. Hedqvist concludes that PGE_2 reacts presynaptically, reducing noradrenaline output in response to continued nerve stimulation. Similar observations have been made by this group on the rabbit heart[138], guinea-pig *vas deferens*[139] and cat hind limb[140]. All the evidence is compatible with the view that prostaglandins of the E series released in response to adrenergic nerve stimulation can modulate effects of further stimulation by controlling the output of transmitters from presynaptic sites. The role of prostaglandin $F_{2\alpha}$ also released on nerve stimulation is not accounted for by this hypothesis.

6.9.3 Inhibition of prostaglandin biosynthesis

If endogenous prostaglandins are implicated in the events at adrenergic nerve synapses, blockade of their biosynthesis would be expected to modify responses to adrenergic nerve stimulation. Eicosatetraynoic acid, an inhibitor of prostaglandin biosynthesis, increases the pressor responses to nerve stimulation, increases transmitter output and abolishes the output of previously detectable prostaglandin-like activity in the effluent[135]. Similarly, indomethacin treatment increases perfusion pressure in the cat spleen accompanied by a fall in prostaglandin output as detected by a cascade of superfused tissues[141]. Both splenic contraction and rise in perfusion pressure in response to nerve stimulation were enhanced by indomethacin. Furthermore, these increases were reduced by infusion of PGE_2.

From the rabbit kidney *in vivo* PGE_2, $PGF_{2\alpha}$ and a PGA-like compound are released into the venous blood, the levels are increased on stimulating the renal nerve. Indomethacin greatly reduced the amount of prostaglandins released and simultaneously raised the arterial blood pressure[142]. Indomethacin also increases the urinary noradrenaline levels in the rat[143]. This may be attributable to hypersecretion of noradrenaline on nerve stimulation. There is increased release of noradrenaline and decreased release of prostaglandins from the isolated rabbit heart after treatment with eicosatetraynoic acid[144]. Also, eicosatetraynoic acid enhances responses of the guinea-pig *vas deferens* to nerve stimulation[145]. Moreover, the inhibitor abolishes the output of prostaglandin-like substances into the incubation fluid.

All these results with inhibitors of prostaglandin biosynthesis are compatible with the hypothesis that endogenous prostaglandins reduce transmitter output at adrenergic synapses.

6.9.4 Interaction between prostaglandins and sympathomiameticamines on effector cells

There is substantial evidence that prostaglandins of the E series not only act presynaptically to inhibit noradrenaline release but also post-synaptically. Such interactions are complex and vary greatly from one tissue to another[146,147]. It has been suggested that at some sites such actions of prostaglandins may be exerted on the adenyl cyclase system, for example, in adipose tissue and in the Purkinje cells of the cerebellum.

6.9.5 Concluding remarks

The weight of evidence is in favour of Hedqvist's hypothesis that release of PGE_2 on adrenergic nerve stimulation exerts a local inhibitory action on further release of noradrenaline in response to continued nerve stimulation. However, this may be an over-simplified picture since the hypothesis does nothing to account for the actions of $PGF_{2\alpha}$ nor the possible post-synaptic actions of, PGE_2 in physiological concentrations.

6.10 PROSTAGLANDINS AND THE CENTRAL NERVOUS SYSTEM

Three lines of evidence suggest that the prostaglandins may have a physiological role in the central nervous system. Prostaglandins have numerous central nervous actions, they have been identified in the central nervous tissues and they are released from the brain and spinal cord in response to nerve and chemical stimulation.

6.10.1 Central nervous actions

Direct evidence that prostaglandins act upon central neurones has been obtained by microiontophoretic administration to single cells in the brain stem of the cat[148]. The spontaneous firing of some of these neurones was increased by PGE_1, PGE_2 and $PGF_{2\alpha}$, a smaller proportion being inhibited by PGE_1 and $PGF_{2\alpha}$. Tachyphylaxis quickly develops to these responses although there is no cross-tachyphylaxis between different prostaglandins. PGE_1 and PGE_2 also excite Purkinje cells of the cerebellum[149]. Moreover, at this site the inhibitory action of noradrenaline is reversed, whereas the similar inhibitory action of cyclic AMP is not reversed by PGE. It is concluded therefore that the reversal of the effect of noradrenaline by PGE_1 and PGE_2 may be attributable to an action on the enzyme, adenyl cyclase, through which the action of noradrenaline appears to be mediated.

PGE_1 and $PGE_{2\alpha}$ have actions on neurones in the spinal cord[150-152]. When PGE_1 is injected intravenously in a cat with cervical spinal trans-section, there is a powerful contraction of the gastrocnemius muscle. The effect cannot be mimicked by close arterial injection of PGE_1 to the gastrocnemius muscle, although it can be mimicked by direct application to the exposed spinal cord. The muscular response is abolished by denervating the muscle but not by section of the ipsilateral dorsal roots. These observations show that the contraction is due to an action of PGE_1 on the spinal cord, not on the muscle and not on the neuromuscular junction. It is concluded that PGE_1 either directly, or indirectly, increases the excitability of α-motoneurones. This action of PGE_1 may also account for the potentiation of crossed extensor reflexes in the spinal cat following intravenous administration[151]. $PGF_{2\alpha}$ produces a similar contraction but its site of action has not been located. In the spinal chick, however, $PGF_{2\alpha}$ appears to act on motor pathways in the spinal cord[151].

Close arterial injection of PGE_1 to the spinal cord of the chloralosed anaesthetised cat, inhibits the evoked monosynaptic potentials recorded in the ventral roots in response to electrical stimulation of the dorsal roots[150]. The time course and duration of these responses are different from those described above. Usually the effect begins after a period of 15–20 min and progresses over a 2–3 h period. A similar, slow onset and long-lasting effect in the cat was observed when PGE_1 was injected in the cerebral ventricles of unaesthetised animals[153]. Such an injection is followed by the slow onset of stupor and catatonia, effects which last for many hours. In contrast, $PGF_{2\alpha}$ injected into the cerebral ventricles, in the same dose range, produced no detectable changes whatsoever[154]. Details of other pharmacological actions on the central nervous system are described elsewhere[155]. Two further points deserve mention: first, studies with radio-isotopes indicate that very small proportions of injected prostaglandins, even via the intra-ventricular route, reach the central nervous system or are present at the time when the pharmacological effects are maximal[73-75, 156]. Recently evidence has been presented that prostaglandins injected intravascularly have potent and long-lasting effects on cerebral blood vessels[157-159].

6.10.2 Distribution in the central nervous system

Prostaglandins have been identified as natural constituents of the brain and spinal cord of several species[37, 160, 161]. In the dog, four prostaglandins were found and they were distributed widely throughout all regions of the brain and spinal cord. The concentrations did not vary widely from one region to another[37].

Subcellular fractionation of homogenates of rat cerebral cortex indicated that *ca.* 40% of the PGE fraction was associated with 'nerve ending fraction'. Central nervous prostaglandins are, however, not confined to this fraction[163] and there is no evidence that prostaglandins at nerve terminals are present in the synaptic vesicles. Prostaglandins are synthesised by cerebral tissue and the biosynthesis is inhibited by anti-inflammatory drugs[164].

6.10.3 Release of prostaglandins from the central nervous system

Prostaglandin-like substances are released from the surface of the brain and spinal cord and the release is increased in response to certain chemicals and nerve stimulation. Stimulation of afferent nerves increases the output of prostaglandins from the superfused spinal cord of the frog[165] and from the somatosensory cortex of the anaesthetised cat[166]. The cortical output in the cat is also increased by stimulating the contralateral cortex, an effect which is abolished by section of the corpus callosum. Prostaglandin-like substances are also released spontaneously from the cerebellar cortex of the cat[167] and from the ventricular system of cats and dogs[168,169]. The output from the cerebral ventricles of the dog of prostaglandin E-like material is increased by perfusing 5-hydroxytryptamine through the ventricles but the output of prostaglandins is unaffected by stimulating the foot pad, infusion of catecholamines or intraperitoneal injections of amphetamine, tranylcypromine or chlorpromazine in doses sufficient to produce central effects[169].

6.10.4 Concluding remarks

It is apparent from the evidence presented that a role for prostaglandins in the central nervous system is far from established. It is known that the brain contains high concentrations of cyclic AMP. The results of Siggins *et al.*[149] may therefore be significant in relation to a possible function of prostaglandins E at certain sites where adenyl cyclase is implicated in neuronal mechanisms. On the other hand, this does nothing to explain the pharmacological effects of $PGF_{2\alpha}$. There is no good evidence to support the hypothesis that prostaglandins are central transmitters, though such a possibility has not been excluded. On balance it is likely that prostaglandins in the brain and spinal cord have a modulator role, possibly analogous to that suggested for peripheral adrenergic terminals.

6.11 FEVER, INFLAMMATION AND PAIN

The discovery by Vane and his colleagues[54] that aspirin and related drugs inhibit the biosynthesis of prostaglandins (see Biosynthesis) provides a very plausible explanation for the mechanism of antipyretic, anti-inflammatory and analgesic actions of these drugs[170]. Such an explanation implies that prostaglandins are concerned in the production of fever, inflammation and pain. Some evidence has accumulated in favour of such involvement.

6.11.1 Fever

Milton and Wendlandt[171] first reported that PGE_1 injected in submicrogram doses into the third ventricle of unanaesthetised cats causes a rise in body temperature. Since this prostaglandin occurs in cat brain[36, 160] and has been identified in the hypothalamus of the dog[37] and since a PG-like substance is released spontaneously into the third ventricle of the cat the possible involvement of PGE compounds in temperature regulation was suggested. It has now been shown that during pyrogen fever in the cat, a PGE-like substance is released into the third ventricle at three times the spontaneous rate. Furthermore, if the fever is abolished with an antipyretic drug of the aspirin type, PGE output is reduced[173]. This is convincing evidence that PGE compounds are involved in the mediation of pyrogen fever.

6.11.2 Inflammation

The role of prostaglandins in inflammation is a more controversial issue. PGE compounds are potent arteriolar vasodilators and can increase capillary permeability[163, 174]. These effects can be observed on intradermal injection[175]. There have been reports that prostaglandins are released in experimentally-induced inflammation in the rat[176] and from fluid perfusing the skin of patients with eczema. Prostaglandins are also released following burn injury[177, 178] but this must be considered as a very special kind of inflammation. Evidence that the output of prostaglandins in these conditions is reduced by the blockade of prostaglandin biosynthesis with aspirin is so far lacking.

The possible role of prostaglandins in inflammatory conditions is made even more complex by reports that PGE_1 can supress adjuvant arthritis in the rat[179]. It is conceivable, however, that these apparently contradictory results can be attributed to differences in prostaglandin concentration.

6.11.3 Pain

A role for prostaglandins in the neural responses to nociceptive stimuli which lead to the sensation of pain, is suggested by the observations that prostaglandins on injection into the peritoneal cavity of mice cause a writhing response.[180] When this response is elicited by other substances such as bradykinin or acetylcholine, it is blocked by analgesics of the aspirin-type. It has yet to be shown that the activity of these pain-producing substances depends upon stimulation of prostaglandin synthesis. The demonstration that prostaglandins are released in these circumstances would be an important piece of evidence in favour of the hypothesis that prostaglandins are involved in peripheral mechanisms subserving pain.

6.12 ADENYL CYCLASE AND ADENOSINE 3'5'-MONO PHOSPHATE (CYCLIC AMP)

6.12.1 Adipose tissue

PGE_1 is a potent inhibitor of lipolysis, both *in vivo* and *in vitro*. Using the rat isolated epididymal fat pad preparation, lipolysis can be induced by catecholamine, corticotrophin, glucagon, thyroid-stimulating hormone, arginine vasopressin, sympathetic nerve stimulation and cold stress. The lipolytic response to each of these is reduced in the presence of PGE_1[181-813]. Similar inhibition, *in vitro*, with PGE_1 has been observed using adipose tissue from other species including man[184, 185]. It is known that hormonally-induced lipolysis involves the second messenger, cyclic AMP[186, 187]. Since the antilipolytic action of PGE_1 is exerted against all hormones, it seems likely that it either inhibits the enzyme system, adenyl cyclase, or prevents cyclic AMP from activating lipase. Experiments with the dibutyryl derivative of cyclic AMP (which has the same pharmacological activity but is able to penetrate cell membranes more readily) have shown that PGE_1 probably does not inhibit the lipolytic action of cyclic AMP. There has been some argument as to the possible additional action of PGE_1 on the phosphodiesterase which inactivates cyclic AMP but, on balance, the present evidence is in favour of an inhibitory action on adenyl cyclase.

Since prostaglandins are released from adipose tissue *in vitro*, in response to nerve stimulation, or to catecholamines[128] and since the amounts released are sufficient to have an inhibitory action on the lipolytic effect of catecholamines, it has been proposed that hormones which act via cyclic AMP on adipose tissue also release prostaglandins which then have an inhibitory action on adenyl cyclase, thus tending to limit the action of the hormone by a local negative feed-back mechanism. According to this hypothesis, adrenergic nerve stimulation to adipose tissue, acting via the transmitter, noradrenaline, would increase cyclic AMP formation and so stimulate lipolysis. Prostaglandin release accompanying this would inhibit the action of noradrenaline on adenyl cyclase so that the lipolytic response to noradrenaline released from continued nerve stimulation would be limited. This mechanism, which was proposed at the Nobel Symposium on Prostaglandins in 1966, may well apply to the action of other hormones on other target organs.

6.12.2 Permeability to water

A possible interaction between PGE_1 and adenyl cyclase could explain some of its actions on other tissues. For example, vasopressin increases the water permeability of the isolated toad bladder and isolated rabbit renal tubule, effects that can be mimicked by cyclic AMP or theophylline[188]. There is good evidence that vasopressin activates adenyl cyclase in these systems, thus increasing cyclic AMP formation which in turn leads to an increase in permeability to water. PGE_1 inhibits vasopressin on these systems, possibly by an action on adenyl cyclase[189, 190].

6.12.3 Gastric secretion

Gastric secretion in response to a variety of secretagogues is mediated via cyclic AMP formation. PGE_1 both in the rat[191] and in the dog inhibit the secretion of acid in response to both histamine and pentagastrin[192, 193]. It is suggested that this action of PGE_1 can be attributed to its inhibitory action on adenyl cyclase, thus reducing the amount of cyclic AMP formed in response to the secretagogues. Since prostaglandins are released from gastric mucosa[194] this may be another example of a local feed-back mechanism.

6.12.4 Systems in which PGE compounds mimic cyclic AMP

On many tissues PGE_1 and PGE_2 mimic the action of cyclic AMP. Thus PGE_2, like corticotrophin, stimulates steroidogenesis in the adrenal cortex[195]. Similarly, PGE_1 mimics the action of thyroid stimulating hormone on the thyroid, both effects being mediated via cyclic AMP[196]. Platelet aggregation is associated with a decrease in cyclic AMP formation. Inhibition of aggregation by PGE_1[197, 198] may be due to stimulation of platelet adenyl cyclase[199].

6.12.5 Conclusions

There is a growing literature on the interaction between prostaglandins and adenyl cyclase. It seems unlikely that all the pharmacological actions of all prostaglandins can be accounted for by such interaction, although many of the actions of PGE compounds may be attributable to such a mechanism. The postulated inter-relationship of adenyl cyclase, prostaglandins and ionic concentrations of calcium and sodium[133] deserves further investigation. It has often been suggested that there is some, as yet, ill-defined connection between membrane calcium and the mode of action of prostaglandins[200-202, 212, 213].

6.13 CONCLUDING REMARKS

The prostaglandin literature is so vast and diversified that it is difficult for a single author to do justice to all aspects of the subject. Moreover, papers are appearing so rapidly that any review is likely to be quickly out-dated. It is hoped that this account has presented a balanced picture of the current position in the prostaglandin field. Notable omissions include any reference to chemical synthesis, prostaglandin antagonists and analogues and their cardiovascular and respiratory actions.

Readers who desire more specialised information should consult one of the numerous reviews or monographs[155, 203-206]. Finally, the bibliography prepared by the Upjohn Company, the newsletter 'Research in Prostaglandins' published by the Worcester Foundation, and the journal 'Prostaglandins' are valuable sources of information and of references to recent papers.

References

1. Euler, U. S. von (1934). *Arch. Exp. Pathol. Pharmak.*, **175**, 78
2. Bergström, S. and Sjövall, J. (1960). *Acta Chem. Scand.*, **14**, 1693
3. Bergström, S. and Sjövall, J. (1960). *Acta Chem. Scand.*, **14**, 1710
4. Bergström, S., Krabisch, L. and Sjövall, J. (1960). *Acta Chem. Scand.*, **14**, 1706
5. Bergström, S., Ryhage, R., Samuelsson, B. and Sjövall, J. (1963). *J. Biol. Chem.*, **238**, 3555
6. Nugteren, D. H., van Dorp, D. A., Bergström, S., Hamberg, M. and Samuelsson, B. (1966). *Nature (London)*, **212**, 38
7. Vogt, W. (1949). *Arch Exp. Path. Pharmak.*, **206**, 1
8. Ambache, N. (1957). *J. Physiol.*, **135**, 114
9. Ambache, N. (1959). *J. Physiol.*, **146**, 255
10. Pickles, V. R. (1957). *Nature (London)*, **180**, 1198
11. Ambache, N. and Reynolds, M. (1960). *J. Physiol.*, **154**, 40P
12. Ambache, N. and Reynolds, M. (1961). *J. Physiol.*, **159**, 63P
13. Kirschner, H. and Vogt. W. (1961). *Biochem. Pharmacol.*, **8**, 224
14. Toh, C. C. (1963). *J. Physiol.*, **165**, 47
15. Euler, U. S. von (1935). *J. Physiol.*, **84**, 21P
16. Lee, J. B., Covino, B. G., Takman, B. H. and Smith, E. R. (1965). *Circulat. Res.*, **17**, 57
17. Babilli, S. and Vogt, W., (1965). *J. Physiol. (London)*, **177**, 31P
18. Linn, B. O., Shunk, C. H., Folkers, K., Ganley, O. and Robinson, H. J. (1961). *Biochem. Pharmacol.*, **8**, 339
19. Andersen, N. H. (1969). *J. Lipid Res.*, **10**, 316
20. Samuelsson. B. (1963). *J. Biol. Chem.*, **238**, 3229
21. Nugteren, D. H., Vonkeman, H. and Dorp, D. A. van (1967). *Recl. Trav. Chim. Pays-Bas Belg.*, **86**, 1237
22. Holmes, S. W., Horton, E. W. and Stewart, M. J. (1968). *Life Sci.*, **7**, 349
23. Unger, W. G., Stamford, I. F. and Bennett, A. (1971). *Nature (London)*, **233**, 336
23a. Horton, E. W. and Main, I. H. M. (1963). *Brit. J. Pharmac. Chemother.*, **21**, 182
24. Änggård, E. and Bergkvist, T. H. (1970). *J. Chromatogr.*, **48**, 542
24a. Horton, E. W. and Main, I. H. M. (1965). *Brit. J. Pharmac. Chemother.*, **24**, 470
25. Horton, E. W. and Jones, R. L. (1969). *J. Physiol. (London)*, **200**, 56P
26. Vane, J. R. (1969). *Brit. J. Pharmac.*, **35**, 209
27. Änggård, E., Matschinsky, F., and Samuelsson, B. (1968). *Brit. J. Pharmac.*, **34**, 190P
28. Ramwell, P. W., Shaw, J. E., Clarke, G. B., Grostic, M. F., Kaiser, D. G. and Pike, J. E. (1968). *Progress in the Chemistry of Fats and other Lipids*, Vol. 9, 231. (R. T. Holman, editor) (Oxford: Pergamon)
29. Bygdeman, M. and Samuelsson, B. (1966). *Clinica Chim. Acta*, **13**, 465
30. Jouvenaz, G. H., Nugteren, D. H., Beerthius, R. K. and Dorp, D. A. van (1970). *Biochim. Biophys. Acta*, **202**, 231
31. Albro, P. W. and Fishbein, L. (1969). *J. Chromatogr.*, **44**, 443
32. Thompson, C. J., Los, M. and Horton, E. W. (1970). *Life Sci.*, **9**, 983
33. Samuelsson, B., Hamberg, M. and Sweeley, C. C. (1970). *Anal. Biochem.*, **38**, 301
34. Karim, S. M. M., Hillier, K. and Devlin, J. (1968). *J. Pharm. Pharmac.*, **20**, 749
35. Karim, S. M. M., Sandler, M. and Williams, E. D. (1967). *Brit. J. Pharmac. Chemother.*, **31**, 340
36. Horton, E. W. and Main, I. H. M. (1967). *Nobel Symposium 2 Prostaglandins*, 253. (S. Bergström, B. Samuelsson, editors) (Stockholm: Almqvist and Wiksell)
37. Holmes, S. W. and Horton, E. W. (1968). *J. Physiol. (London)*, **195**, 731
38. Hopkin, J. M., Horton, E. W. and Whittaker, V. P. (1968). *Nature (London)*, **217**, 71
39. Kataoka, K., Ramwell, P. W. and Jessup, S. (1967). *Science*, **157**, 1187
40. Dorp, D. A. van, Beerthuis, R. K., Nugteren, D. H. and Vonkeman, H. (1964). *Nature (London)*, **203**, 839
41. Bergström, S., Danielsson, H. and Samuelsson, B. (1964). *Biochim. Biophys. Acta*, **90**, 207
42. Änggård, E. and Samuelsson, B. (1965). *J. Biol. Chem.*, **240**, 3518
43. Dorp. D. A. van, Beerthuis, R. K. and Vonkeman, H. (1964). *Biochim. Biophys. Acta*, **90**, 204

44. Kupiecki, F. P. (1965). *Life Sci.*, **4**, 1811
45. Dorp, D. A. van (1966). *Mem. Soc. Endocr.*, **14**, 39
45a Christ, E. J., and Dorp, D. A. van (1972). *Biochim. Biophys. Acta*, **270**, 537
46. Dorp, D. A. van (1967). *Prog. Biochem. Pharmac.*, **3**, 71
47. Nugteren, D. H., Beerthuis, R. K. and Dorp, D. A. van (1966). *Rec. Trav. Chim. Pay-Bas. Belg.*, **85**, 405
48. Granström, E., Lands, W. E. M. and Samuelsson, B. (1968). *J. Biol. Chem.*, **243**, 4104
49. Struijk, C. B., Beerthuis, R. K. and Dorp, D. A. van (1967). *Nobel Symposium 2. Prostaglandins*, 51. (S. Bergström, B. Samuelsson, editors) (Stockholm: Almqvist and Wiksell)
50. Dorp, D. A. van (1971). *Ann. N.Y. Acad. Sci.*, **180**, 181
51. Nugteren, D. H. (1970). *Biochim. Biophys. Acta*, **210**, 171
52. Pace-Asciak, C. and Wolfe, L. S. (1968). *Biochim. Biophys. Acta*, **152**, 784
53. Ahern, D. G. and Downing, D. T. (1970). *Biochim. Biophys. Acta*, **210**, 456
54. Vane, J. R. (1971). *Nature New Biol.*, **231**, 232
55. Smith, J. B. and Willis, A. L. (1971). *J. Clin. Invest.*, **50**, 432
56. Ferreira, S. H., Moncada, S. and Vane, J. R. (1971). *Nature New Biol.*, **231**, 237
57. Lands, W. E. M., and Samuelsson, B. (1968). *Biochim. Biophys. Acta*, **164**, 426
58. Vonkeman, H. and Dorp, D. A. van (1968). *Biochim. Biophys. Acta*, **164**, 430
59. Bartels, J., Vogt, W. and Willie, G. (1968). *Nauyn-Schmiedebergs Arch. Pharmak. Exp. Path.*, **259**, 153
60. Eliasson, R. (1958). *Nature (London)*, **182**, 256
61. Kunze, H. and Bohn, R. (1969). *Nauyn-Schmiedebergs Arch. Pharmak.*, **264**, 263
62. Klenberg, D. and Samuelsson, B. (1965). *Acta Chem. Scand.*, **19**, 534
63. Nugteren, D. H. and Dorp, D. A. van (1965). *Biochim. Biophys. Acta*, **98**, 654
64. Ryhage, R. and Samuelsson, B. (1965). *Biochem. Biophys. Res. Commun.*, **19**, 279
65. Samuelsson, B. (1965). *J. Amer. Chem. Soc.*, **87**, 3011
66. Hamberg, M. and Samuelsson, B. (1966). *J. Amer. Chem., Soc.*, **88**, 2349
67. Hamberg, M. and Samuelsson, B. (1967). *J. Biol. Chem.*, **242**, 5233
68. Daniels, E. G., Hinman, J. W., Johnson, B. A., Kupiecki, F. P., Nelson, J. W. and Pike, J. E. (1965). *Biochem. Biophys. Res. Commun.*, **21**, 413
69. Jones, R. L. (1970). *Biochem. J.*, **119**, 64P
70. Horton, E. W., Jones, R. L., Thompson, C. J. and Poyser, N. L. (1971). *Ann. N.Y. Acad. Sci.*, **180**, 351
71. Jones, R. L. (1972). *J. Lipid Res.*, **13**, 511
72. Jones, R. L., Cammock, S. and Horton, E. W. (1972). *Biochim. Biophys. Acta*, **280**, 588
73. Gréen, K., Hansson, E. and Samuelsson, B. (1967). *Prog. Biochem. Pharmac.* **3**, 86
74. Hansson, E. and Samuelsson, B. (1965). *Biochim. Biophys. Acta*, **106**, 379
75. Samulesson, B. (1964). *J. Biol. Chem.* **239**, 4091
76. Samuelsson, B., Granström, E., Gréen, K. and Hamberg, M. (1971). *Ann. N.Y. Acad. Sci.*, **180**, 138
77. Granström, E., Inger, U. and Samuelsson, B. (1965). *J. Biol. Chem.*, **240**, 457
78. Gréen, K. and Samuelsson, B. (1968). *Prostaglandin Symposium of the Worcester Foundation for Exp. Biol.* p. 389 (P. W. Ramwell and J. E. Shaw, editors) (New York: Interscience)
79. Gréen, K. (1969). *Acta Chem. Scand.*, **23**, 1453
80. Granström, E. and Samuelsson, B. (1969). *Europ. J. Biochem.*, **10**, 411
81. Hamberg, M. and Samuelsson, B. (1969). *Biochem. Biophys. Res. Commun.*, **34**, 22
82. Änggård, E. and Samuelsson, B. (1964). *J. Biol. Chem.*, **239**, 4097
83. Änggård, E. and Samuelsson, B. (1965). *Biochemistry, N.Y.*, **4**, 1864
84. Ferreira, S. H. and Vane, J. R. (1967). *Nature (London)*, **216**, 868
85. Horton, E. W. and Jones, R. L. (1969). *Brit. J. Pharmac.*, **37**, 705
86. McGiff, J. C., Terragno, N. A., Strand, J. C., Lee, J. B., Lonigro, A. J. and Ng, K. K. F. (1969). *Nature (London)*, **223**, 742
87. Piper, P. J., Vane, J. R. and Wyllie, J. H. (1970). *Nature (London)*, **225**, 600
88. Änggård, E. and Samuelsson, B. (1966). *Ark. Kemi.*, **25**, 293
89. Nakano, J. R., Änggård, E. and Samuelsson, B. (1969). *Europ. J. Biochem.*, **11**, 386
90. Shio, H., Andersen, N. H., Corey, E. J., and Ramwell, P. W. (1969). *Abstracts 4th Int. Congr. Pharmac., Basle.* 100
91. Nissen, H. M. and Andersen, H. (1968). *Histochemie*, **14**, 189

92. Siggins, G., Hoffer, B. and Bloom, F. (1971). *Ann. N.Y. Acad. Sci.*, **180**, 302
93. Hamberg, M. (1968). *Europ. J. Biochem.*, **6**, 135
94. Israelsson, U., Hamberg, M. and Samuelsson, B. (1969). *Europ. J. Biochem.*, **11**, 390
95. Hamberg, M. and Samuelsson, B. (1965). *Biochim. Biophys. Acta*, **106**, 215
96. Granström, E. (1973). *Prostaglandins 1972* (S. Bergström and S. Bernhard, editors) Oxford: Pergamon Press
97. Hamberg, M. and Samuelsson, B. (1969). *J. Amer. Chem. Soc.*, **91**, 2177
98. Granström, E. and Samuelsson, B. (1969). *J. Amer. Chem., Soc.*, **91**, 3398
99. Bygdeman, M., Kwon, S. and Wiqvist, N. (1967). *Nobel Symposium 2 Prostaglandins* 77 (S. Bergström, B. Samuelsson, editors) (Stockholm: Almqvist and Wiksell)
100. Karim, S. M. M. (1967). *Brit. J. Pharmac.*, **29**, 230
101. Karim, S. M. M. and Devlin, J. (1967). *J. Obstet. Gynaec. Brit. Commonw.*, **74**, 230
102. Karim, S. M. M. (1968). *Brit. Med. J.*, **4**, 618
103. Karim, S. M. M. (1969). *Prostaglandins, Peptides and Amines* 65 (P. Mantegazza and E. W. Horton, editors) (London: Academic Press)
104. Karim, S. M. M., Trussell, R. R., Patel, R. C. and Hillier, K. (1968). *Brit. Med. J.*, **4**, 621
105. Karim, S. M. M., Hillier, K., Trussell, R. R., Patel, R. C. and Tamusange, S. (1970). *J. Obstet. Gynaec. Brit. Commonw.*, **77**, 200
106. Karim, S. M. M. and Filshie, C. M. (1970). *Lancet*, **1**, 157
107. Bygdeman, M. and Wiqvist, N. (1971). *Ann. N.Y. Acad. Sci.*, **180**, 473
108. Bygdeman, M., Hamberg, M. and Samuelsson, B. (1966). *Mem. Soc. Endocr.*, **14**, 49
109. Sullivan, T. J. (1966), *Brit. J. Pharmac.*, **26**, 678
110. Bygdeman, M. and Hamberg, M. (1967). *Acta Physiol. Scand.*, **69**, 320
111. Embrey, M. P. and Morrison, D. L. (1968). *J. Obstet. Gynaec. Brit. Commonw.*, **75**, 829
112. Sandberg, F., Ingleman-Sundberg, A. and Rydén, G. (1963). *Acta Obstet. Gynec. Scand.*, **42**, 269
113. Sandberg, F., Ingleman-Sundberg, A. and Rydén, G. (1964). *Acta Obstet. Gynec. Scand.*, **43**, 95
114. Sandberg, F., Ingleman-Sundberg, A. and Rydén, G. (1965). *Acta Obstet. Gynec. Scand.*, **44**, 585
115. Adamson, U., Eliasson, R. and Wiklund, B. (1967). *Acta Physiol., Scand.* **70**, 451
116. Eliasson, R., Brzdekiewicz, Z. and Wiklund, B. (1969). *Prostaglandins, Peptides and Amines*, 57 (P. Mantegazza and E. W. Horton, editors) (London: Academic Press)
117. Clegg, P. C., Hall, W. J. and Pickles, V. R. (1966). *J. Physiol. (London)*, **183**, 123
118. Poyser, N. L., Horton, E. W., Thompson, C. J. and Los, M. (1971). *Nature (London)*, **230**, 526
119. Blatchley, F. R., Donovan, B. T., Poyser, N. L., Horton, E. W., Thompson, C. J. and Los, M. (1971)., *Nature (London)*, **230**, 243
120. Blatchley, F. R., Donovan, B. T., Horton, E. W. and Poyser, N. L. (1972). *J. Physiol. (London)*, **223**, 69
121. Poyser, N. L. (1972). *J. Endocr.*, **54**, 147
122. Horton, E. W. and Poyser, N. L. (1973). *Brit. J. Pharmac.* (in the press)
123. Blatchley, F. R. and Donovan, B. T. (1969). *Nature (London)*, **221**, 1065
124. McCracken, J. A. (1971). *Ann. N.Y. Acad. Sci.*, **180**, 456
125. McCracken, J. A., Carlson, J. C., Glew, M. E., Goding, J. R., Baird, D. T., Gréen, K. and Samuelsson, B. (1972). *Nature New Biol.*, **238**, 129
126. Thorburn, G. D. and Nicol, D. H. (1971). *J. Endocr.*, **51**, 785
127. Bland, K. P., Horton, E. W. and Poyser, N. L. (1971). *Life Sci.*, **10**, 509
128. Shaw, J. E. and Ramwell, P. W. (1968). *J. Biol. Chem.*, **243**, 1498
129. Davies, B. N., Horton, E. W. and Withrington, P. G. (1968). *Brit. J. Pharmac. Chemother.*, **32**, 127
130. Ferreira, S. H. and Vane, J. R. (1967). *Nature (London)*, **216**, 868
131. Gilmore, N., Vane, J. R. and Wyllie, J. H. (1968). *Nature (London)*, **218**, 1135
132. Ramwell, P. W. and Shaw, J. E. (1970). *Rec. Prog. Hormone Res.*, **26**, 139
133. Ramwell, P. W., Shaw, J. E. and Kucharski, J. (1965). *Science*, **149**, 1390
134. Laity, J. L. H. (1969). *Brit. J. Pharmac.*, **37**, 698
135. Hedqvist, P., Stjarne, L. and Wennmalm, A. (1971). *Acta Physiol. Scand.*, **83**, 430
136. Hedqvist, P. (1969). *Acta Physiol. Scand.*, **75**, 511
137. Hedqvist, P. and Brundin, J. (1969). *Life Sci.*, **8**, 389
138. Hedqvist, P., Stjarne, L. and Wennmalm, A. (1970). *Acta Physiol. Scand.*, **79**, 139

139. Hedqvist, P. and Wennmalm, A. (1970). *Acta Physiol. Scand.* **79,** 19A
140. Hedqvist, P. (1971). *Ann. N.Y. Acad. Sci.,* **180,** 410
141. Ferreira, S. H. and Moncada, S. (1971). *Brit. J. Pharmac.,* **43,** 419P
142. Davis H. A. and Horton, E. W. (1972). *Brit. J. Pharmac.,* **46,** 658
143. Stjarne, L. (1971). *Acta Physiol. Scand.,* **83,** 574
144. Samuelsson, B. and Wennmalm, A. (1971). *Acta Physiol. Scand.,* **83,** 163
145. Hedqvist, P. and Euler, U. S. von (1972). *Neuropharmacology,* **11,** 177
146. Horton, E. W. (1973). *Brit. Med. Bull.,* **29,** 148
147. Clegg, P. C. (1966). *Mem. Soc. Endocr.,* **14,** 119
148. Avanzino, G. L., Bradley, P. B. and Wolstencroft, J. H. (1966). *Brit. J. Pharmac.,* **27,** 157
149. Siggins, G. R., Hoffer, B. and Bloom, F. E. (1969). *Science,* **165,** 1018
150. Duda, P., Horton, E. W. and McPherson, A. (1968). *J. Physiol. (London),* **196,** 151
151. Horton, E. W. and Main, I. H. M. (1967). *Brit. J. Pharmac.* **30,** 568
152. Phillis, J. W. and Tebecis, A. K. (1968). *Nature (London),* **217,** 1076
153. Horton, E. W. (1964). *Brit. J. Pharmac.,* **22,** 189
154. Horton, E. W. and Main, I. H. M. (1965). *Int. J. Neuropharmac.,* **4,** 65 (see erratum **4,** 359)
155. Horton, E. W. (1972). *Prostaglandins. Monographs on Endocrinology,* **7.** (Heidelberg: Springer-Verlag)
156. Holmes, S. W. and Horton, E. W. (1968). *Brit. J. Pharmac.,* **34,** 32
157. Yamamoto, L., Feinderl, W., Wolfe, L., Katoh, H. and Hodge, C. (1972). *J. Neurosurg,* **37,** 385
158. Steiner, L., Forster, D. M. C., Bergvall, U. and Carlson, L. A. (1972). *Neuroradiology,* **4,** 20
159. Denton, I. C., Jnr., White, R. P. and Robertson, J. R. (1972). *J. Neurosurg.,* **36,** 34
160. Horton, E. W. and Main, I. H. M. (1967). *Brit. J. Pharmac.,* **30,** 582
161. Samuelsson, B. (1964). *Biochim. Biophys. Acta,* **84,** 218
162. Kataoka, K., Ramwell, P. W. and Jessup, S. (1967). *Science,* **157,** 1187
163. Horton, E. W. (1963). *Nature (London),* **200,** 892
164. Flower, R. J. and Vane, J. R. (1972). *Nature (London),* **240,** 410
165. Ramwell, P. W., Shaw, J. E. and Jessup, R. (1966). *Amer. J. Physiol.,* **211,** 998
166. Ramwell, P. W. and Shaw, J. E. (1966). *Amer. J. Physiol.,* **211,** 125
167. Coceani, F. and Wolfe, L. S. (1965). *Can. J. Physiol. Pharmac.,* **43,** 445
168. Feldberg, W. and Myers, R. D. (1966). *J. Physiol. (London),* **184,** 837
169. Holmes, S. W. (1970). *Brit. J. Pharmac.,* **38,** 653
170. Vane, J. R. (1972). *Hosp. Pract.,* **7** (3) 61
171. Milton, A. S. and Wendlandt, S. (1970). *J. Physiol. (London),* **207,** 76P
172. Feldberg, W. and Myers, R. D. (1966). *J. Physiol. (London),* **184,** 837
173. Feldberg, W. and Gupta, K. P. (1973). *J. Physiol. (London),* **228,** 41
174. Kaley, G. and Weiner, R. (1968). *Prostaglandin Symposium of the Worcester Foundation for Exp. Biol.,* 321. (P. W. Ramwell and J. E. Shaw, editors) (New York: Interscience)
175. Crunkhorn, P. and Willis, A. L. (1969). *Brit. J. Pharmac.,* **36,** 216P
176. Willis, A. L. (1969). *J. Pharm. Pharmac.,* **21,** 126
177. Jonsson, C-E. (1972). *Dissertation,* Uppsala University Uppsala
178. Jonsson, C-E. and Änggård, E. (1972). *Scand. J. Clin. Invest.,* **29,** 289
179. Aspinall, R. L. and Cammarata, P. S. (1969). *Nature (London),* **224,** 1320
180. Collier, H. O. J. and Schneider, C. (1972). *Nature (London), New Biol.,* **234,** 141
181. Steinberg, D., Vaughan, M., Nestel, P. J. and Bergström, S. (1963). *Biochem. Pharmacol.,* **12,** 764
182. Steinberg, D., Vaughan, M., Nestel, P. J., Strand, O. and Bergström, S. (1964). *J. Clin. Invest.,* **43,** 1533
183. Steinberg, D. and Vaughan, M. (1967). Proceedings *Nobel Symposium 2. Prostaglandins,* 109. (S. Bergström, B. Samuelsson, editors) (Stockholm: Almqvist and Wiksell)
184. Fain, J. N. (1967). *Ann. N.Y. Acad. Sci.,* **139,** 879
185. Mandel, L., Humes, J. L. and Kuehl, F. A., Jr. (1968). *Prostaglandin Symposium of the Worcester Foundation for Exp. Biol.,* p. 79. (P. W. Ramwell and J. E. Shaw, editors) (New York: Interscience)

186. Steinberg, D. (1966). *Pharmacol. Rev.*, **18**, 217
187. Vaughan, M. (1966). *Pharmacol. Rev.*, **18**, 215
188. Orloff, J. and Handler, J. (1967). *Amer. J. Med.*, **42**, 757
189. Orloff, J. and Grantham, J. (1967). *Nobel Symposium 2. Prostaglandins.* 143. (S. Bergström and B. Samuelsson, editors) (Stockholm: Almqvist and Wiksell)
190. Orloff, J., Handler, J. S. and Bergström, S. (1965). *Nature (London)*, **205**, 397
191. Shaw, J. E. and Ramwell, P. W. (1968). *Symposium of the Worcester Foundation for Exp. Biol.* 55. (P. W. Ramwell and J. E. Shaw, editors) (New York: Interscience)
192. Robert, A. (1968). *Symposium of the Worcester Foundation for Exp. Biol.* 47 (P. W. Ramwell and J. E. Shaw, editors) (New York: Interscience)
193. Robert, A., Nezamis, J. E. and Phillips, J. P. (1967). *Amer. J. Dig. Dis.*, **12**, 1073
194. Bennett, A., Friedmann, C. A. and Vane, J. R. (1967). *Nature (London)*, **216**, 873
195. Flack, J. D., Jessup, R. and Ramwell, P. W. (1969). *Science*, **163**, 691
196. Kaneko, T., Zor, U. and Field, J. B. (1969). *Science*, **163**, 1062
197. Kloeze, J. (1967). *Nobel Symposium 2. Prostaglandins.* 241. (S. Bergström and B. Samuelsson, editors) (Stockholm: Almqvist and Wiksell)
198. Kloeze, J. (1969). *Biochim. Biophys. Acta*, **187**, 285
199. Marquis, N. R., Vigdahl, R. L. and Tavormina, P. A. (1969). *Biochem. Biophys. Res. Commun.*, **36**, 965
200. Pickles, V. R., Hall, W. J., Clegg, P. C. and Sullivan, T. J. (1966). *Mem. Soc. Endocr.*, **14**, 89
201. Coceani, F. and Wolfe, L. S. (1966). *Can. J. Physiol. Pharmac.*, **44**, 933
202. Coceani, F., Dreifuss, J. J., Puglisi, L. and Wolfe, L. S. (1969). *Prostaglandins, Peptides and Amines.* 73. (P. Mantegazza, E. W. Horton, editors) (London: Academic Press)
203. Karim, S. M. M. (1972). *The Prostaglandins* (Lancaster: M. T. P.)
204. Ramwell, P. W. and Pharriss, B. B. (1972). *Prostaglandins in Cellular Biology* (New York: The Plenum Press)
205. *Prostaglandins, 1972.* Proceedings of the Vienna Conference on Prostaglandins, September, 1972. (S. Bergström and S. Bernhard, editors) Oxford: Pergamon Press
206. *Annals New York Academy of Science* **180** (R. W. Ramwell and J. E. Shaw, editors)
207. Bergström, S., Eliasson, R., Euler, U. S. von and Sjövall, J. (1959). *Acta Physiol. Scand.* **45**, 133
208. Berti, F. and Naimzada, M. (1965). *Boll. Soc. Ital. Biol. Sper.*, **41**, 1324
209. Horton, E. W. and Main, I. H. M. (1966). *Mem. Soc. Endocr.*, **14**, 29
210. Horton, E. W., Main, I. H. M. and Thompson, C. J. (1965). *J. Physiol. (London)*, **180**, 514
211. Hawkins, R. A., Jessup, R. and Ramwell, P. W. (1968). *Symposium of the Worcester Foundation for Exp. Biol.* 11. (P. W. Ramwell and J. E. Shaw, editors) (New York: Interscience)
212. Eagling, E. M., Lovell, H. G. and Pickles, V. R. (1972). *Brit. J. Pharmac.*, **44**, 510
213. Kirtland, S. J. and Baum, H. (1972). *Nature New Biol.*, **236**, 47

7
The Halogenated Sulphatides

T. H. HAINES
City College of New York

7.1	INTRODUCTION	271
7.2	NOMENCLATURE	275
7.3	THE CHLOROSULPHATIDES OF *Ochromonas danica*	275
	7.3.1 *Isolation*	275
	7.3.2 *Structure*	276
	7.3.3 *Analysis*	278
	7.3.4 *Biosynthesis*	278
	7.3.5 *Metabolism*	280
7.4	BROMOSULPHATIDES	283
7.5	THE OCCURRENCE OF HALOSULPHATIDES IN MEMBRANES	283
	ACKNOWLEDGEMENTS	285

7.1 INTRODUCTION

The occurrence of natural products containing halogen is not new. Early discovery of such compounds in the biosphere was often reported where the counter-ion of an alkaloid was chloride. However, organic chloride is now well known to occur in fungi, actinomycetes, and other microbes.

A few of the more well known halogenated natural products that have been reported include griseofulvin, chloramphenicol, and chlortetracycline[1]. It is not a surprise that these compounds are antimicrobial. In fact, the biogenesis of halogenated natural products for many years was thought to be primarily antimicrobial. Indeed, the relatively small quantities of these materials made by the organisms and their affect on other microbes would seem to support the notion that halogenated substances were biosynthesised primarily for defence purposes.

At the present time, approximately 200 such natural halogenated compounds have been isolated[2].

Not including marine organisms, virtually all these compounds are chloro compounds. A few fluoro and iodo substances are also found in the non-marine world. The marine organisms that have yielded halogenated compounds have yielded nearly exclusively bromo compounds, with occasional exceptions. It is possible to obtain a bromo analogue of a natural chloro compound in almost all cases. Such analogues are obtained by culturing the microorganism in a chloride-free medium, and adding bromide in appropriate concentration. Thus, most organisms are capable of producing the halogenated

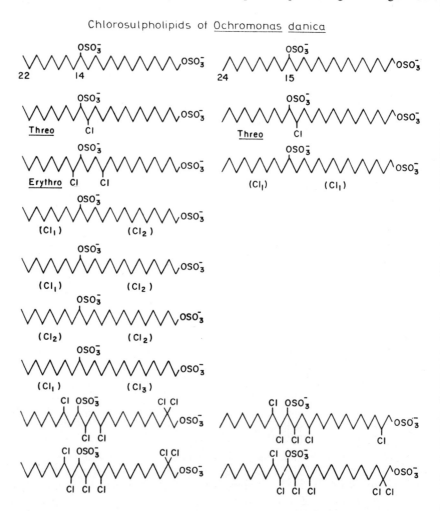

Figure 7.1 Structures of the chlorinated alkyldiol disulphates characterised to date

compound in the bromo form. It has not been possible to repeat this in any known case with fluoride or with iodide.

The discovery of chloride in lipid extracts was therefore somewhat of a surprise[3]. The lipids in question were the sulphatides of *Ochromonas danica*. These compounds had earlier been described by Mayers and Haines[4] and by Mayers, Pousada and Haines[5]. Alkyl sulphates, themselves, are unknown in the biosphere, whereas the chlorination of such alkyl sulphates is even more surprising.

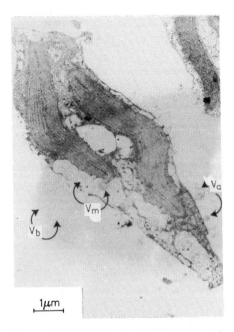

Figure 7.2 Electron micrograph of *Ochromonas danica* showing excreted vesicles of membrane ($\times$ 6900). [V_a: membrane extruding; V_b: extruded vesicles; V_m: myelinic vesicles]

This chapter will be devoted almost exclusively to the discussion of the chlorinated alkyl sulphates of *Ochromonas danica*.

These curious lipids, chlorinated alkyl sulphates, are the only known chlorolipids to date. They are rather well characterised chemically by the two research groups indicated above, and have only been reported in phytoflagellates. In addition to unusual chemical structures, these substances are also characterised by an unusual physiology. For example, the sulphatides are not metabolised by the microbe, since the sulphate ester is not cleaved[6]. Furthermore, the concentration of chloride ion in the medium determines the extent of chloride fixation[3,7].

A very significant aspect of these unusual lipids is the structure of the alkyl disulphates (Figure 7.1). The occurrence of the second sulphate ester on the

14 or 15 carbon of an alkyl chain implies that the lipid molecule has polar groups at *both* ends of the chain. This is unique in lipid chemistry. The fact that all other lipids whose structures are known have polar groups at only one end of the aliphatic chain is a foundation stone of the current theories of membrane structure. This unique structure has broad ramifications as a membrane component. Figures 7.2 and 7.3 show electron micrographs of *O. danica*. The organism appears to extrude cell membrane[8].

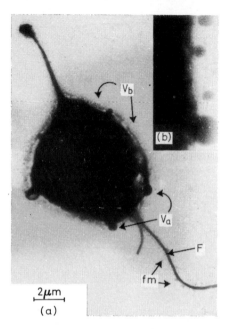

Figure 7.3 Negative stained electron micrograph of *O. danica*. (a) Micrograph shows pseudopod, flagella (F) with mastigonemes (fm) attached and membrane vesicles [membrane extruding: Va; extruding vesicles: Vb] associated with the cell. ($\times$ 6380). (b) High magnification of one area of the cell membrane showing extruding vesicles $\times$ 19 220) (reduced $\frac{6}{10}$ ths on reproduction)

These electron micrographs and others obtained by Gibbs show that *Ochromonas danica* has all of the usual organelles including chloroplast, mitochondria, Golgi apparatus and nucleus, in addition to an extremely thin membrane (*ca.* 80Å). There are many significant implications of the structures of the halosulphatides in terms of the physiology and the cellular locations of these substances. It is these topics which will be discussed in this article.

Unique findings in scientific investigations often provide impetus, and at the same time methods or approaches, for investigation in seemingly unrelated areas. This is not an unlikely development in this case. This will become clearer in the development of the article.

7.2 NOMENCLATURE

Sulphur-containing lipids are designated as follows:
(a) a *sulpholipid* is any lipid containing sulphur; (b) a *sulphatide* is a sulpholipid in which the sulphur occurs at a sulphate ester; (c) a *sulphonolipid* is a sulpholipid in which the sulphur occurs as sulphonic acid; and (d) a *halosulphatide* is a sulphatide which contains halogen atoms. It should be noted that this nomenclature is essentially that expressed by Haines[9, 10].

7.3 THE CHLOROSULPHATIDES OF *OCHROMONAS DANICA*

The chlorosulphatides of *Ochromonas danica* represent a complex family of compounds. As *aliphatic sulphate esters* with polar groups at *both ends* of the lipid molecule, which contain up to six *halogen* atoms replacing hydrogen atoms on the otherwise saturated chain, they are triply unique. Consisting of 10% of the total lipids of the organism—including both the typical chloroplast lipids—and the typical mitochondrial lipids, which together dominate the lipids of the cell, they represent 3% of the dry weight of the cell, and over 50% of the total sulphur in the cell[3, 6].

The chlorosulphatides are not restricted to *O. danica*. The same group of compounds have been demonstrated in *O. malhamensis* by labelling cells with ^{36}Cl. Attempts to confirm by ^{36}Cl-labelling an earlier report[6] that sulpholipids were present in *Tetrahymena pyriformis* were not successful[11]. Definitive evidence for the occurrence of these substances in a variety of other microbes[6] is still forthcoming.

7.3.1 Isolation

Haines and Block[12] observed a group of metabolically inert sulphate esters in the phytoflagellate, *Ochromonas danica*. The substances were originally identified as four spots on paper chromatograms of ^{35}S- labelled extracts of the cells. Two of the spots have since been shown to be the mixture of disulphates of docosane and tetracosane, with from 0 to 6 chlorine atoms replacing hydrogen on the chain. The third spot was a mixture of monosulphates, which resulted from hydrolysis of the disulphates, probably by the trichloroacetic acid used in the extracting solvents. The fourth spot appears to have been an artifact of the extraction procedure. Haines and Block[12] have cultured *O. danica* in the logarithmic growth phase in the presence of each of the compounds, and established that the sulphatides were incorporated into the cells, whereas the artifact was not. Incorporation of ^{35}S-sulphatide into the cells did not produce labelled cystine or methionine, whereas incubation with ^{35}S-labelled sulphate did. It appeared, therefore, that the biologically significant sulphur compounds were the disulphates or sulphatides.

Considerable difficulties were encountered during the first isolation of the sulphatides. The substances as detergents were difficult to purify as preparations contained small amounts of organic contaminants and proteins. As organic sulphate esters, the sulphatides are extremely hygroscopic. A cumbersome procedure for their isolation and purification was developed by Haines[6], and improved by Mayers and Haines[4]. This procedure, although it yielded a rather pure disulphate appropriate for spectroscopy and elemental analysis, was tedious and difficult. The preparation of an ultra-pure sulphatide, either in the mixture form or as a single species, remains a difficult and challenging problem. However, for most purposes, including the structure determination of the remaining disulphates, the preparation of purified sulphatide is not necessary. It is possible, for example, to prepare the crude sulphatides to remove the sulphate groups by either hydrolysis or solvolysis, and obtain the corresponding diols. These diols are much easier to separate and purify by either thin layer chromatography or gas chromatography.

A simple procedure for the preparation of crude sulphatide extracts was developed by Elovson and Vagelos[3] which was essentially the Folch extraction procedure[13]. The application of the Folch procedure to the sulphatides is useful, as most methods are in the lower phase of the Folch extract after the addition of water, whereas the sulphatides are in the upper phase. This procedure is rapid, can be conducted in a matter of hours after the crude extraction of the cells by chloroform: methanol (2:1). Approximately 95% of the sulphatides are found in the upper phase of the Folch extract.

Hydrolysis of the crude sulphatide preparation produces a mixture of diols, which are easier to separate and purify. Figure 7.1 illustrates the complexity of the diol mixture obtained by this procedure. In order to obtain diols in milligram quantities or more, hydrolysis can be effected in moist dioxane (solvolysis). This procedure is highly selective for sulphate esters and therefore produces a very clean preparation of diols, even from the crude sulpholipid preparation. The success of the procedure depends upon limiting the amount of water present during hydrolysis[9,14]. Hydrolysis of micro-amounts of sulphatide cannot be conducted by solvolysis, since it is very difficult to limit the amount of water, and excess water inhibits or prevents solvolysis.

The mechanism of the solvolysis reaction is important, and was established by Mayers and Haines[15]. The mechanism is such that the solvolysis occurs with retention of configuration around the CO bond of the sulphate ester as it becomes a hydroxyl. This is important in determining the configuration of the original suphate ester. Additionally, solvolysis with retention of configuration allows synthesis of a single natural disulphate after its purification as a diol. This is possible because the synthesis of sulphate esters is always conducted with retention of configuration. Although it is not selective, small amounts of sulphatide are hydrolysed to diols in aqueous acid (1 N HCl, 2 h, 100 °C). A disadvantage to the hydrolysis procedure is that the product does not occur with complete retention of configuration.

7.3.2 Structure

Isolation of a pure preparation of the sulphatides allowed a chemical characterisation of the alkyl disulphates. It was first established that the ratio of sulphur to carbon was 1:11 by elemental analysis. Likewise, an infra-red study of model sulphate esters showed that it was possible to distinguish between primary and secondary aliphatic sulphates[4]. This study permitted the recognition of both primary and secondary sulphate esters which were identified in the preparation of sulphatides. Furthermore, mass spectra identified the diols which resulted from hydrolysis of the sulphate ester to be a 22 carbon chain. Thus, the structure was necessarily a 22 carbon disulphate with both a primary and secondary sulphate ester.

Sulphatides having been identified were then hydrolysed as a group. The products were separated by column chromatography and by thin layer chromatography. The separation on 2-dimensional chromatography is shown in Figure 7.4. The diols in the lower right-hand corner (C_{22} and C_{24}) are easily separated from the halogenated diols in the remainder of the chromatogram. The non-halogenated diols may be separated from the others by simple crystalisation from hexane. The diols are insoluble in hexane, whereas all the halogenated diols are soluble. The crystallisation procedure allowed early identification of the diols as 1,14-docosane diol and 1,15-tetracosane diol by mass spectrometry. These diols may be separated from each other only by gas chromatography. On thin layer chromatography and on column chromatography, they appear in a mixture as shown in Figure 7.4. The configuration of the secondary sulphate ester for the C_{22} compound was established by solvolysis and rotation of the resulting diol[5]. The structure was found to be 1-[S]-14-docosane disulphate. Synthesis of the 1,14-docosane diol has been achieved by Mayers and Haines[4]. The synthesised racemic material was compared with the natural diol by infra-red spectrophotometry and by mass spectrometry and the assignment of the structure verified.

The discovery of chlorine in the sulphatide preparation was first made by Elovson and Vagelos[3]. The identification of chloride was made possible by mass spectra of the diols after removal of the sulphate groups. The characteristic spectra of the isotopes of polychloro compounds is a classic example of the use of isotopic peaks in the interpretation of mass spectra[16]. The appearance of these clusters in the mass spectra made the identification unequivocal. With the identification of chlorine in the sulpholipid diol extracts, Haines et al.[17] were able to completely characterise the first of the chlorosulphatide series as *threo*-(R)-13-chloro-1-(R)-14-docosane disulphate. The characterisation was made by comparison of the natural material to corresponding *threo*- and *erythro*-synthetic compounds. Additionally, the natural compound was converted to the epoxide in base and chromatographic comparison was made to synthetic *trans*- and *cis*-epoxides. The rotation of the natural compound and comparison of the rotation to analogous compounds in the literature permitted a complete identification.

A dichloro 22-carbon disulphate was identified by Elovson and Vagelos[3] as 11,15-dichloro-1,14-docosanediol disulphate. Complete characterisation was not possible, since the identification was made with mass spectrometry; the orientation of the functional groups remains unknown. Unpublished data by M. Pousada and T. H. Haines suggests the chlorohydrin is *erythro* in configuration.

A second dichloro compound has been identified as 2,2-dichloro-1,14-docosane disulphate from mass spectra of column fractions[18]. Both groups have identified trichloro and tetrachloro compounds in the mixtures, but neither group has characterised these compounds, as they are small in quantity and difficult to separate at present.

In addition to the above compounds, there are two pentachloro and two hexachloro compounds found in the sulphatide family. The chlorinated sulphatide which dominates the mixture is the hexachloro docosane disulphate. This compound was obtained in sufficient quantity by Elovson and Vagelos[19] to allow them to locate the position of the chloro groups on the chain. This was accomplished by an elegant series of experiments including [36]Cl-labelled material combined with chemical degradation of the chain. A compound was identified as 2,2,11,13,15,16-hexachloro-1,14-docosane disulphate. This material constitutes approximately 34% of the total weight of the mixture. The structure assignment was in agreement with data obtained by Pousada et al.[18], who also characterised the C_{24} hexachloro compound as 2,2,12,14,16,17-hexachloro-1,15-tetracosane disulphate. The pentachloro compounds were analogous to the hexachloro compounds described above, with each missing a different chloro group. Their structures are shown in Figure 7.1.

7.3.3 Analysis

Perhaps the best method for identifying chlorosulpholipids is [36]Cl-labelling, followed by hydrolysis of the upper phase of the Folch extract of tissue. Thin layer chromatography of the diols would result in identification of chlorosulpholipids via autoradiography. The method, which will obviously not identify the structures of the sulphatides alone, will allow identification of halogenated disulphates and identification by co-chromatography is probably sufficient. Surely a pattern on a 2-dimensional chromatogram which is identical to that shown in Figure 7.4 would verify the identification of the chlorosulphatides described in this section. Such a procedure would take a few days of laboratory work, but would permit rapid screening of organisms for chlorosulphatides. Labelling with ^{35}S-sulphate, which was the original method of identification[12], is clearly not as specific for chlorosulphatides, but may be used as a preliminary screen, and would certainly subsequently identify the chlorinated diols as derived from the sulpholipids.

Since unknown sulpholipids may co-chromatograph with the halosulphatides in a given solvent system, it is advisable to solvolyse or hydrolyse the resulting diols for a positive identification. 1,12-Octadecane diol is commercially available as a standard for thin layer chromatography of 1,14-docosane diol. General methods for the analysis of sulphatides which apply in this case have been reviewed by Haines[9].

7.3.4 Biosynthesis

A general outline of the biosynthesis of the sulphatides has been described[20, 21]. The aliphatic chain is synthesised from acetate, presumably utilising the usual fatty acid synthesis system. The introduction of hydroxyl occurs on the saturated chain by way of the introduction and hydration of a *cis* double bond. It would appear[21] that oleic acid via erucic acid is the precursor that is hydrated. It is likely that elongation to the C_{24}-analogue precedes hydration for the C_{24} group of sulphatides. The evidence for the oleic acid intermediate is that

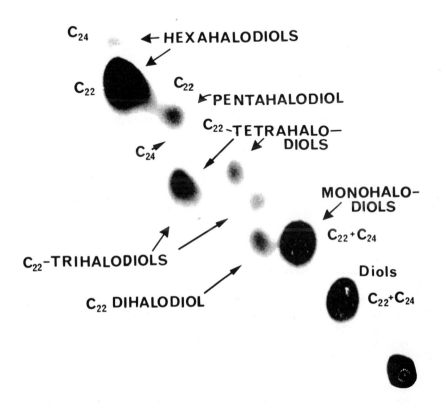

Figure 7.4 Autoradiogram of a two-dimensional chromatogram of halodiols after 1-^{14}-Claurate was incubated with *Ochromonas danica*

carboxyl-labelled ^{14}C-oleate and erucate were incorporated intact into the sulphatide group of compounds. In the case of hydration of the erucic acid double bond, the hydroxyl is placed on the carbon of the double bond distal to the carboxyl. The C_{24}-analogue, however, is hydrated so that the hydroxyl appears to be placed on the proximal carbon of the double bond.

Oleic acid

15-Hydroxy C_{24} series   14-Hydroxy C_{22} series

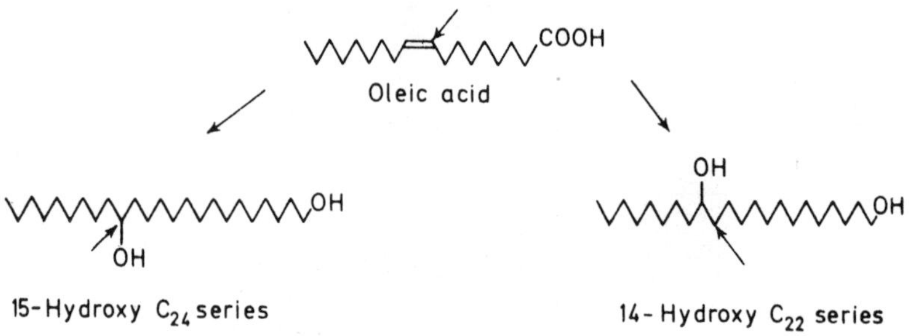

Figure 7.5

The incorporation of ^{14}C-labelled fatty acids directly into the chain was established by Mooney et al.[22], by degradation of the monochlorosulphatide. This was accomplished by solvolysis of the sulphatide mixture, separation of the diols on 2-dimensional thin layer chromatograms and degradation of the monohalodocosane diol via periodate oxidation. The resulting aldehyde fragments were separated and the activity was found exclusively in the expected fragment depending on whether the precursor was ^{14}C-carboxy- or ^{14}C-methyl-labelled fatty acid.

It is not clear how the carboxyl group of the C_{22} and C_{24} series has been reduced, but it could be assumed that the routes utilised for the formation of alcohol for waxes are also used in this case. It would also appear likely that the sulphation of the resulting diol is by way of the activated intermediate PAPS (3′-phosphoadenosine-5′phosphosulphate) (Figure 7.6). Mooney and Haines[21] have also shown that chlorination of the chain occurs after sulphation. This was established by incubating non-halogenated ^{14}C-labelled natural sulphatide with cells and identifying the usual pattern of chlorinated sulphatides. It had also been demonstrated that ^{35}S-labelled sulphatide was not degraded and re-utilised in the formation of these chlorinated sulphatides (Figure 7.7).

This implies a rather unusual mechanism for chlorination in view of the fact that other chlorinating enzymes generally chlorinate activated carbons[23]. It is clear that chlorination in this case must occur by a free-radical mechanism, since it occurs on a saturated hydrocarbon chain. The energy requirements for such a chlorination reaction are substantial.

7.3.5 Metabolism

In the very first paper which identified these unusual compounds, Haines and Block[12] noted that the sulphatides were metabolically inert. This inertness

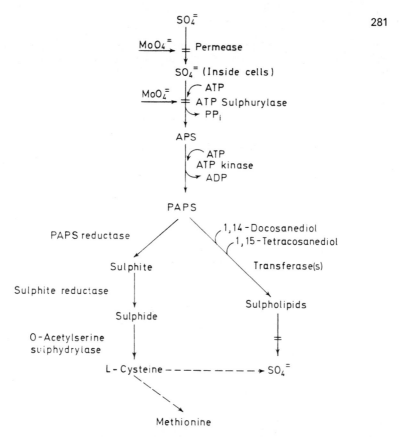

Figure 7.6 Pathway for the utilisation of sulphate by *Ochromonas danica*

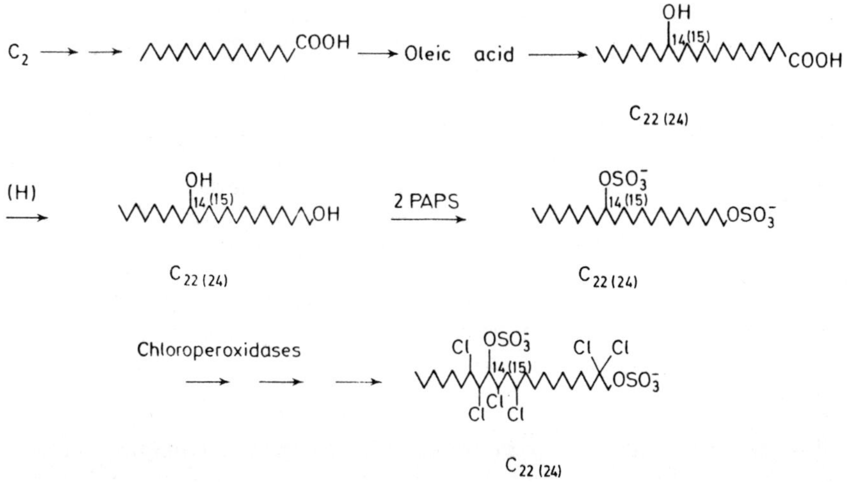

Figure 7.7 Pathway for the biosynthesis of the halogenated sulphatides (particularly the hexachlorodocosanediol disulphate) of *Ochromonas* as proposed by Mooney and Haines[20]

was identified by incubating growing cells with ^{35}S-labelled sulphatide and noting that the cystine and methionine of the cells remained unlabelled. This was significant because incubation of ^{35}S-sulphate labels the cystine and methionine immediately. In the same paper, it was noted that the polar sulphatides were incorporated by the cells. Thus, over 50% of the sulphatides were incorporated by the cells. Thus, over 50% of the sulphatides could be isolated from the cells after 1h incubation. Additionally, it was exceedingly difficult to obtain the sulphatide that had been incorporated by extraction with chloroform-methanol. It was only possible to extract completely these labelled sulpholipids by pronase digestion of the solvent-extracted residue. This is most unusual for a polar lipid, especially one in such concentration in cells.

The dominant halosulphatides in the sulphatide extract are the hexachlorosulphatide, the monochlorosulphatide, and the non-chlorinated sulphatide. The ratio of the hexachlorosulphatide to the chloro-free sulphatide shifts dramatically with a large increase in the concentration of chloride ion in the growth medium. Chloride concentrations above 0.375 saline are toxic to the fresh water phytoflagellate. The absence of chloride from the culture medium bleaches the phytoflagellate[24]. This effect is apparently due to a requirement of chloride for photosynthesis. This requirement was first discovered by Warburg[25], who noted that addition of chloride ion enhanced oxygen production in photosynthesis. Subsequent work by Arnon[26] identified the chloride requirement with Photosystem II and the Hill reaction. Besides bleaching the cell, it was also noted that *Ochromonas danica* cultured in the absence of chloride ion produced only the non-chlorinated docosane disulphate. The bleaching preceded the absence of chlorinated diols in the extract; that is, the cells had ceased to produce chloroplasts, while they still made monochloro and a few other halogenated chlorolipids. But after rigorous exclusion of chloride ion over many generations, the cells finally survived without chlorolipids at all[21]. Of interest here is the fact that cells grown in the absence of chloride but in the presence of bromide were normal in regard to chloroplast production, while they produced bromosulpholipids. This would suggest that bromide might replace the chloride requirement for photosynthesis. This observation is consistent with that of Bove et al.[26].

With regard to the cleavage of sulphate esters, although it appears that the *Ochromonas* species is not capable of cleaving the disulphates, there is a general gnome that anything produced in nature must be degraded by nature. One wonders where the enzyme is that degrades these relatives of commerical detergents. Payne and Painter[27] have identified sulphatases which cleave both primary and secondary sulphate esters. The enzymes were found in a *Pseudomonas* (strain $C_{12}B$). These enzymes hydrolysed effectively a sample of 1,12-octadecanediol disulphate synthesised by the Haines group. The ability of the organism to degrade this compound suggests that its enzymes may be a natural degradative system for the chlorosulphatides. There was no attempt in these experiments to explore the degradation of the chlorolipids.

The average of chloro groups from lipoid molecules is not unique, however. Maynard et al.[28] have shown that 4,6-dichloro-17α-hydroxypregna-4,6-diene, 3,20-dione is dechlorinated by enzyme systems. It should also be noted that DDT is degraded enzymically[29,30].

7.4 BROMOSULPHATIDES

In addition to the chloro compounds described above, a group of bromosulphatides have been identified in the cells, cultured in the absence of chloride ion and in the presence of bromide[31]. In addition to the hexabromo and monobromo compounds, both of which are identical to the hexachloro and monochloro compounds just described, the bromosulpholipids are in the same range of structures as was mentioned for the chlorosulphatides. This was established by the identification of corresponding spots on 2-dimensional chromatograms on thin layer chromatography of the bromo diols (see Figure 7.2, p. 273).

It should be noted that the relative amount of the bromo compounds is quite different from those of the chloro compounds. The Figure 7.2 chromatogram indicates a concentration of the hexachloro and the monochloro and the non-chlorinated diols, whereas lesser quantities of the other chlorinated lipids are indicated. The bromo analogues of these compounds are more dramatic in their differences. Only trace amounts of the intermediate compounds are noted in the 2-dimensional chromatograms of bromodiols. It should also be noted that the bromodiols are much less stable than the chlorinated analogues. In fact the bromodiols are unstable and have an odour of HBr immediately upon isolation. Thus these compounds cannot be stored, and their characterisation must be made quickly after they have been isolated. The characterisation work was always conducted within 6 or 7 days of isolation, the compounds were stored cold (5°C), and physical methods of characterisation were generally accomplished within a day or so. The compounds did appear to be relatively stable on silica columns during the isolation procedure.

Evidence for the occurrence of iodosulphatides was equivocal, since iodide streaked over the thin layer chromatogram in the region of the halodiols. Attempts to separate the iodosulphatides and iododiols from iodide were not successful in the half-life of the isotope. Fluoride was found to be toxic to the cultures at low concentrations.

7.5 THE OCCURRENCE OF HALOSULPHATIDES IN MEMBRANES

Although the bilayer theory of membrane structure has had its ups and downs since it was first proposed by Gorter and Grendel[32] and Danielli and Davson[33], it remains today as the dominant theory of membrane structure. Perhaps the most persuasive evidence for this theory is that obtained by freeze fracture electron micrographs[34]. These investigations have convincingly demonstrated a 'cleavage plane' down the centre of a variety of natural membranes. Likewise, the electron density profiles obtained by x-ray diffraction patterns obtained by Wilkins[35] and by others support the concept of a 'cleavage plane'. These and other physical data suggest that at least an important portion of natural membranes exist as a bimolecular leaflet. The stability of this model is dependent upon the structure of the polar lipids which are present in natural membranes. To date, all polar lipids consist of hydrocarbon chains which terminate as a methyl group at one end and a polar hydrophilic group at the other. The entire length of the lipid molecule from the methyl group to the hydrophilic end is hydrophobic. The lipids described in this article do not conform to this generalised structure of membrane lipids. The two sulphate groups on these substances are charged at all aqueous

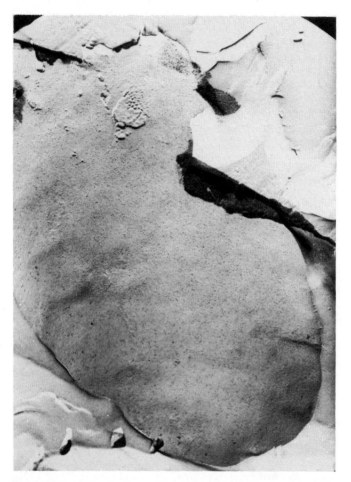

Figure 7.8 Freeze fracture of *O. danica* cell membrane prepared by James and Haines[38]

pHs and therefore the polar groups would be at one end of the molecule and close to the other end as well. Such lipids are not suitable for the formation of a monolayer and presumably a bilayer as well. In fact, these substances are most effective at developing monolayers as the compounds themselves are water-soluble and have a potent detergent action.

It is therefore of considerable value to establish whether or not these materials occur in natural membranes. The occurrence of a polar lipid, which is approximately one-third of the total lipids of the organism, strongly implies its presence in a cellular organelle. *O. danica* contains all the usual lipids which are found in chloroplasts and in mitochondria. Thus the occurrence of such a large quantity of chlorosulphatides in the lipid extracts is somewhat surprising, and suggests the compounds may be components of membrane. Chen and Haines[36] have obtained evidence which directly suggests the chlorosulphatides are concentrated in a membrane fraction of *Ochromonas danica*. The evidence is based upon its concentration in a band of a density gradient, which band has been identified with electron microscopy as membrane vesicles. Figure 7.3 shows an electron micrograph of *Ochromonas danica*. The micrograph shows the organelles previously mentioned, and was prepared as described by Aaronson *et al.*[37]. This electron micrograph illustrates the very thin cell membrane of *Ochromonas*, and provokes one to wonder about the strength of this membrane in a freshwater medium.

It should also be noted that *Ochromonas danica* is a phagotroph. It is not surprising, therefore, that the organism forms vesicles rather readily, and such vesicles are found in the external medium[8]. The subject of the latter work is the extrusion of membrane vesicles by *Ochromonas danica*. Figure 7.8 shows a negatively stained *Ochromonas* which was obtained as described by Orner *et al.*[38]. The electron micrograph illustrates the extrusion of membrane vesicles from the cell. It should be noted that vesicles are even extruded from the mastigonenes of the flagella. The occurrence of chlorosulphatides in a membrane pellet obtained by centrifuging the external medium is further evidence that the cell membrane is the membrane which contains the sulphatide. It should be noted that this suggestion has not been definitively established.

As was mentioned before, it is difficult to imagine how these halosulphatides would be components of a bilayer. Figure 7.8 is therefore of interest, as it shows a freeze fracture of *Ochromonas* prepared according to the procedure of Pinto de Silva and Branton[34]. This freeze fracture shows the interior of a cell membrane of *Ochromonas danica*. The freeze fracture, obtained by R. B. James and T. H. Haines[38], clearly shows the cell membrane is capable of fracturing down a 'cleavage plane'. Should the present evidence be confirmed that the chlorosulphatides are a component of cell membrane in this organism, their chemical relationship to each other and to the proteins will be of much significance in understanding membrane structure.

Acknowledgements

Acknowledgement is extended to the City University Research Foundation, the Office of Education, and the Petroleum Research Fund for their support of the author's work while this chapter was written. Appreciation is extended to Miss Susan Epstein for her help in preparing the manuscript. The author appreciates the use of the space and facilities of E. Lederer at Gif-sur-Yvette, France, and E. E. Snell at Berkeley, California, during some of the work described. Special acknowledgement is made to Dr. S. Aaronson and Dr. R. B. James for their expert aid in electron micrography (Figures 7.2, 7.3 and 7.4).

References

1. Haines, T. H. (1972). *Lipids and Biomembranes of Eukaryotic Microorganisms*, 1972 (J. Erwin, editor) (New York: Academic Press)
2. Siuda, J. F. (1973). *Lloydia*, **36**, 107
3. Elovson, J. and Vagelos, P. R. (1969). *Proc. Nat. Acad. Sci. (U.S.A.)*, **62**, 957
4. Mayers, G. L. and Haines, T. H. (1969). *Biochemistry*, **6**, 1965
5. Mayers, G. L., Pousada, M. and Haines, T. H. (1969). *Biochemistry*, **8**, 2981
6. Haines, T. H. (1965). *J. Protozool.*, **12**, 655
7. Haines, T. H. (1970). *Properties and Products of Algae*, 129 (J. E. Zajic, editor) (New York: Plenum Press)
8. Orner, R., Haines, T. H., Aaronson, S. and Behrens, U. (1973) (In preparation)
9. Haines, R. H. (1971). *Progress in the Chemistry of Fats and Other Lipids*, Vol. 11, 297 (R. T. Holman, editor) (Oxford: Pergamon)
10. Haines, T. H. (1973). *Annu. Rev. Microbiol.*, **27**, 1
11. Emanuel, D. L., Stern, A. and Haines, T. H. (1972). *J. Protozool.*, in the press
12. Haines, T. H. and Block, R. J. (1962). *J. Protozool.*, **9**, 33
13. Folch, J., Lees, M. and Sloane-Stanley, G. H. (1957). *J. Biol. Chem.*, **226**, 497
14. Grant, G. A. and Beall, D. (1960). *Rec. Progr. in Hormone Res.*, **5**, 307
15. Mayers, G. L., Pousada, M. and Haines, T. H. (1969). *Biochemistry*, **8**, 2981
16. McLafferty, F. W. (1966). *Interpretation of Mass Spectra* (NY: Benjamin)
17. Haines, T. H., Pousada, M., Stern, B. and Mayers, G. L. (1969). *Biochem. J.*, **113**, 585
18. Pousada, M., Das, B. P., Lederer, E. and Haines, T. H. (1973). (In preparation)
19. Elovson, J. and Vagelos, P. R. (1970). *Biochemistry*, **9**, 3110
20. Mooney, C. L. and Haines, T. H. (1973). *Biochemistry*, **12** (In the press)
21. Mooney, C. L. (1973). *Federation Proceedings*, **32**, 562
22. Mooney, C. L., Mahoney, E. M., Pousada, M. and Haines, T. H. (1972). *Biochemistry*, **11**, 4839
23. Hager, L. P., Thomas, J. A. and Morris, D. R. (1970). *Biochemistry of the Phagocytic Process* (J. Schultz, editor) (Amsterdam: North Holland)
24. Rosenbaum, W. and Haines, T. H. (Unpublished experiments)
25. Warburg, O. and Luttgens, W. (1946). *Biokhimiya*, **2**, 303
26. Bové, J. N., Bové, C., Whatley, F. R. and Arnon, D. I. (1963). *Z. Naturforsch.*, **18[b]**, 683
27. Payne, W. J. and Painter, B. G. (1971). *Microbios*, **3**, 199
28. Maynard, D. E. *et al.* (1971). *Biochemistry*, **10**, 355
29. Bunyan, P. J., Page, J. M. J. and Taylor, A. (1966). *Nature (London)*, **210**, 1040
30. Klein, A. K., Lang, E. P., Datta, P. R., Watts, J. O. and Chen, J. T. (1965). *J. Assoc. Off. Agr. Chem.*, **47**, 1129
31. Pousada, M., Bruckstein, A., Das, B. P. and Haines, T. H. (1973) (In preparation)
32. Gorter, E. and Grendel, F. (1925). *J. Experimental Medicine*, **41**, 439
33. Danielli, J. F. and Davson, H. (1935). *J. Cellular Physiology*, **5**, 495
34. Pinto de Silva, P. and Branton, D. (1970). *J. Cell Biol.*, **45**, 598
35. Wilkins, M. H. F. (1972). *Ann. N.Y. Acad. Sci.*, **195**, 346
36. Chen, L. L. and Haines, T. H. (1972). *J. Protozool.*, **19**, 36P
37. Aaronson, S., Behrens, U., Orner, R. and Haines, T. H. (1971). *J. Ultrastruct. Res.*, **35**, 418
38. James, R. B. and Haines, T. H. (Unpublished observations)

Index

Abortion, prostaglandins and, 256
Acetamide, iodo-, fatty acid desaturase inhibition by, 211
Acetobacter xylinum, cellulose biosynthesis in, 57
Acetylcholine, neural transmission and, 167
Acetyl-CoA carboxylase, 100–109
  of plants, structure, 104
Acetyl-CoA-ACP transacylase, 119
N-Acetyl glucosaminyl transfer in glycoprotein biosynthesis, 81, 82
a_1-Acid glycoprotein, 71
ACP hydrolase, 118
Acyl-carrier protein, 111–118
  fatty acid synthetase and, 109
Acyl-enzyme intermediates in desaturation of fatty acids, 204
Acyl-S-ACP, as substrate for unsaturated fatty acid biosynthesis, 194, 195
Acyl-S-CoA, as substrate for unsaturated fatty acid biosynthesis, 193, 194
Acyl transferase, fatty acid desaturation and, 202
Adenosine 3′,5′-cyclic phosphate, sterol biosynthesis and, 21
Adenosine 3′,5′-monophosphate, 264, 265
Adenyl cyclase, 264, 265
Adipose tissue, prostaglandins and adenyl cyclase in, 264
Adrenal cortex, prostaglandins and, 265
Adrenergic nerve terminals, prostaglandins and, 258–260
Adrenergic synapses, prostaglandin release at, 258, 259
Aerobic mechanism of unsaturated fatty acid biosynthesis, 190–231
Age, polyprenols in plants and, 51
Ageing of nervous system, lipids and, 170–172
Allosteric regulation
  of acetyl-CoA carboxylase, 105
  of fatty acid synthetases, 133, 134
Alternative pathway for unsaturated fatty acid biosynthesis, 203, 204

Amino acids, composition of acyl carrier protein, 111–114
Aminoacidurias, brain, 172, 173
Anaerobic pathway of monoenoic fatty acid biosynthesis, 187–190
Analysis of chlorosulphatides, 278
Animals
  acyl carrier protein, amino acid sequence in, 114
  polyprenols in, 51
Arachidonic acid, prostaglandin biosynthesis from, 248
Arsenites
  fatty acid elongation inhibition by, 204
  sterol biosynthesis and, 21
Arteriolar vasodilators, prostaglandins and, 263
Arthrobacter
  acyl-carrier protein from, 111
  amino acid sequence in, 114
Ascorbic acid
  fatty acid desaturation and, 211
  prostaglandin biosynthesis and, 249
Aspergillus niger
  mannosyl transfer in, 86, 87
  polyprenols, 46, 51
Aspergillus oryzae, glycoprotein biosynthesis in, 87
Aspirin
  prostaglandin biosynthesis inhibition by, 262
  prostaglandin synthetase inhibition by, 250
Avocado, acyl-carrier protein from, 111
AY-9944, sterol biosynthesis and, 21

Bacillus licheniformis, teichoic acids in, 67
Bacitracin
  'O'-antigen determinant biosynthesis, inhibition by, 63
  peptidoglycan biosynthesis inhibition by, 61
  sterol biosynthesis and, 21
  teichoic acid biosynthesis, inhibition by, 69

Bacteria
  polyprenols in, 51
  prenols, 43
  wall, glycan biosynthesis in, undecaprenol and, 55–70
  unsaturated fatty acids temperature and, 210
Benzoic acid, *p*-chloromercuri-, fatty acid desaturase inhibition by, 211
Benzoic acid, *p*-hydroxymercuri-, fatty acid desaturase inhibition by, 211
Betulaprenol-7, biosynthesis, 53
Betulaprenols, 43
Betulaprenol monophosphate in glycoprotein biosynthesis, 81
Bile salts
  control of rate of sterol biosynthesis by, 25
  sterol biosynthesis and, 22
Biological assay of prostaglandins, 243, 244
Biosynthesis
  bacterial wall glycan, undecaprenol and, 55–70
  of chlorosulphatides, 278–280
  of gangliosides in nervous tissue, 157–159
  glycans, in green plants, polyprenols and, 87–89
  lipids in, 39–97
    in yeasts and fungi, polyprenols and, 85–87
  mammalian glycan, polyprenols and, 70–85
  polyprenols, 52–55
  of prostaglandins, 248–252
    inhibition of, 259, 260
  saturated fatty acids, 99–140
  sterol, control of rate of, 24–28
    enzymes in, 1–37
    pathway, 5–19
    particle-bound enzymes in, 19–24
  of unsaturated fatty acids, 181–235
Biotin in palmitate biosynthesis, 100
Biotin carboxylase in palmitate biosynthesis, 102
Biotin carboxyl carrier protein in palmitate biosynthesis, 102
Blood-group substances, 72
Blood vessels, cerebral prostaglandins and, 261
Brain
  development, steryl esters in, 156
  disease, lipids and, 172–175
  fatty acids, metabolism, 150
  fatty acid synthetase in, 133
  lipids in, 143
    exchange of, 164–167
  prostaglandins and, 261
  steryl esters in, 156
  subcellular fractions, lipids in, 147
Brain cells metabolism, 162

Brain stem, sheep, lipids in, 144
Branch points in fatty acid synthesis, 187
Bromosulphatides, 283
Burn injury, prostaglandins and, 263

Callose in plant cells, 89
Candida lipolytica, dilinoleoyl-PC biosynthesis in, 198
Castaprenols, 43
Catabolism, fatty acids in brain, 151, 152
Cellulose, biosynthesis, 57
  in plant cells, 89
Cell walls, polyprenols in, 52
Central nervous system, prostaglandins and, 260–262
Ceramide biosynthesis, 73
Cerebellum, sheep, lipids in, 144
Cerebroside
  in brain, 161
  concentration in brain, 145
  in myelin, during development, 169
Cerebrum, sheep, lipids in, 144
Cetyl monophosphate in glycoprotein biosynthesis, 81
Chain elongation, fatty acid biosynthesis in brain by, 151
Chain termination
  in fatty acid synthetase in mammary glands, 130
  in liver fatty acid synthetase, 129
  in yeast fatty acid synthetase, 127
Chloramphenicol
  ACP-dependent synthetase inhibition by, 132
  teichoic acid biosynthesis, inhibition by, 68
Chlorella, $\Delta 12$ desaturases in, 199
Chloroplasts, polyprenols and, 52
Chlorosulphatides
  analysis, 278
  biosynthesis, 278–280
  isolation, 275, 276
  metabolism, 280–282
  of *Ochromonas danica*, 275–282
  structure, 276–278
Cholestan-$3\beta,5\alpha,6\beta$-triol, sterol biosynthesis and, 21
5α-Cholest-7-en-3β-ol, 4,4-dimethyl-,oxidative demethylation, 11
Cholesterol
  biosynthesis, 9, 16
    inhibition by bacitracin, 61
  in brain, 152–157
    exchange of, 164–166
  from lanosterol, 9
  multi-enzymic synthesis, other microsomal processes and, 28–33
Chondroitin structure, 72
Chromatography of prostaglandins, 242, 243

Citrate in fatty acid biosynthesis, 105
Clostridium butyricum, acyl-carrier protein from, 111
Codium fragile, mannolipid formation in, 88
Collagen
 biosynthesis, 76
 glucosylation, retinol in, 84
Cryptococcus laurentii, mannosyl transfer in, 87
Cyanides, fatty acid desaturase inhibition by, 211, 212
Cyclisation in sterol biosynthesis, 9–12
Cycloartenol, 9
Cycloproprene fatty acids, biosynthesis, 212–215
Cytochrome P-450 dependent oxidases, 31
Cytochrome b_5, microsomal, isolation, 23

Darmstoff, 239
Demyelinating diseases, 174
Deoxycholate
 in desaturase purification, 218
 in solubilisation of particle-bound enzymes, 22
Desaturases
 $\Delta 9$, 200
 assay, 215, 216
 classification, 230
 complex lipids as substrates for, 195–199
 complex, molecular organisation of, 216–221
 electron donors and, 210, 211
 purification, 217–219
 specificity, 225
Desaturation
 of fatty acids, 200–202
 control, 220, 221
 inhibitors of, 211–215
 products, 204–209
 inhibition by, 206
Desmosterol in myelin, 165
Detergent effects
 in fatty acid biosynthesis, 212
 in unsaturated fatty acid biosynthesis, 207
Detergents
 glucosyl transferase activity and, 84, 85
 in isolation of particle-bound enzymes, 22
Development of central nervous system, lipids and, 169
Diabetes
 fatty acid desaturation control and, 220
 fatty acid synthetase control and, 133
Diet, fatty acid synthetase and, 132
Dihomo-γ-linolenic acid, prostaglandin biosynthesis from, 248
Disease, brain, lipids and, 172–175
Diurnal rhythm of β-hydroxy-β-methylglutaryl coenzyme A reductase activity, 24

Dolichol monophosphate
 in *N*-acetylglucosaminyl transfer, in glycoprotein biosynthesis, 82
 glucosyl transfer to lipid and, 78
 in glycoprotein biosynthesis, 81
Dolichol monophosphate glucose, 76
Dolichol monophosphate mannose, 47
Dolichol phosphate in mammalian glycosyl transferase systems, 74
Dolichols, 43
 biosynthesis, 53
 radioactive, 50
Double bonds
 specificity of desaturases and, 225
 in unsaturated fatty acids, 185
Drugs
 fatty acid desaturase inhibition by, 212
 sterol biosynthesis and, 20

Eczema, prostaglandins and, 263
EDTA, fatty acid desaturase inhibition by, 211
Effector cells, prostaglandins and sympathomimetamines as, 260
Eicosatetraynoic acid
 prostaglandin biosynthesis inhibition by, 259
5,8,11,14-Eicosatetraynoic acid
 prostaglandin synthetase inhibition by, 250
Electrons, microsomal transport of, 5
Electron transport components of desaturases, methods for studying, 219, 220
Endoplasmic reticular system, 27
 cholesterol biosynthesis in, 29
Enduracidin, peptidoglycan biosynthesis inhibition by, 61
Enoyl-ACP reductase, 124
Enzymes
 membrane-bound, 11
 particle-bound in sterol biosynthesis, 19–24
 sterol biosynthesis and, 1–37
 similarities and specificities of, 15–19
Enzymic assay of prostaglandins, 244
Ergosterol biosynthesis, 16
Escherichia coli
 acyl carrier protein, amino acid sequence of, 115
 fatty acid synthetase, enzymes of, 118–124
 polyprenol monophosphate glucose in, 64
Esterification, sterol intermediates, 18
Ethanolamine plasmalogen biosynthesis, 149
Ethylamine, β-mercapto-, sterol biosynthesis, and, 21
Euglena gracilis, fatty acid synthetase in, 131, 132
Exopolysaccharides, capsular, 64–66

Extraction
 in isolation of particle-bound enzymes, 22
 prostaglandins, 241–243

Fallopian tubes, human, prostaglandins and, 256, 257
Farnesyl monophosphate in glycoprotein biosynthesis, 81
Farnesyl pyrophosphate, 8
 squalene from, 9
Fasting, acetyl-CoA carboxylase activity and, 107, 108
Fatty acids
 in brain, metabolism, 150
 desaturation, control of, 220, 221
 dienoic, biosynthesis, 228–230
 free, in brain lipids, 152
 monoenoic, biosynthesis, 225–228
 saturated, biosynthesis, 99–140
 unsaturated, biosynthesis, 181–235
Fatty acid synthetase, 109–135
 control, 132–135
Feeding trials, rats, acetyl-CoA carboxylase activity and, 107, 108
Female reproductive tract smooth muscle, prostaglandin effect on, 255, 257
Fever, prostaglandins and, 262, 263
Ficaprenol diphosphate N-acetyl glucosamine, 47
Ficaprenol diphosphate N-acetylmuramic acid, 47
Ficaprenol diphosphate galactose, 47
Ficaprenol monophosphate as glucose acceptor, 77
Ficaprenols, 43
Frogs, body lipid composition and temperature, 210
Fungi
 glycan biosynthesis in, polyprenols and, 85–87
 polyprenols in, 51

Galactosylceramide sulphate in brain, 161
Galactosyl glycerides in plants, 205
Galactosyl transfer retinol phosphate derivatives and, 84
Gangliosides
 biosynthesis, 72, 73, 76
 catabolism, 159–161
 in nervous tissue, 157–161
Gas chromatography in prostaglandin assay, 244
Gastric secretion, prostaglandins and, 265
Genetics, control of rate of sterol biosynthesis and, 28
Geranyl pyrophosphate, 8
Glucoproteins, mitochondrial, biosynthesis, 78

Glucosyl ceramide in ganglioside biosynthesis, 76
Glucosyl transfer
 in glycan biosynthesis in green plants, 89
 retinol phosphate derivatives and, 84
Glutathione
 prostaglandin biosynthesis and, 249
Glycan
 bacterial wall, biosynthesis and undecaprenol, 55
 biosynthesis, in green plants, polyprenols and, 87–89
 lipids in, 39–97
 mammalian, polyprenols and, 70–85
 in yeasts and fungi, polyprenols and, 85–97
Glycerol-3-phosphate, stearic acid desaturation stimulation by, 208
Glycogen biosynthesis, 77
Glycogen synthetase, 77
Glycoproteins
 biosynthesis, 70, 71
 retinol phosphate derivatives in biosynthesis of, 82
Glycosphingolipids in brain, 161
Glycosyl transfer in biosynthesis of mammalian glycan, 74–78
Gonadotrophic hormones, cholesterol formation and, 27
Grey matter, lipids in, 144
Guinea-pig, luteolysin, prostaglandin $F_{2\alpha}$ and, 257

Hansenula holstii, mannan biosynthesis in, 86
Hatching, acetyl-CoA carboxylase activity and, 108
Holo-ACP synthetase, 118
Hormones
 fatty acid desaturation control and, 220
 fatty acid synthetase control and, 133
Hydrocortisone, fatty acid synthetase control and, 133
Hydrogenolysis of polyprenols, 49
Hydrogen removal in fatty acid desaturation 223–225
Hydrolysis of polyprenols, 47
Hydroquinane, prostaglandin biosynthesis, and, 249
β-Hydroxyacyl-ACP dehydrase, 123, 124
3-Hydroxydecanoyl thiolester dehydratase in fatty acid synthesis, 189, 190
β-Hydroxy-β-methylglutaryl coenzyme A reductase, 7, 8
 in brain, 153
 in control of sterol biosynthesis, 24
 solubilisation, 22
15-Hydroxyprostaglandin dehydrogenase in prostaglandin assay, 244

INDEX

5α-Hydroxysterol dehydrase solubilisation, 23
Hypothyroid, fatty acid synthetase control and, 133

Immunoglobulins
  from mouse myeloma tumour, 80
  structure, 71
Indomethacin
  prostaglandin biosynthesis inhibition by, 259
  prostaglandin synthetase, inhibition by, 250
Inflammation, prostaglandins and, 263
Inhibition
  by desaturation products, 206
  of prostaglandin biosynthesis, 259, 260
  of prostaglandin synthetase, 249, 250
Iodosulphatides, 283
Irin, 239
Isoprene, biosynthetic units, in sterol biosynthesis, 8

β-Ketoacyl-ACP reductase, 123
β-Ketoacyl-ACP synthetase, 121–123
Klebsiella aerogenes, capsular exopolysaccharides in, 65
Krabbe's disease, 162

Lactobacillus casei, undecaprenol biosynthesis in, 55
Lanosterol
  cholesterol from, 9
  formation, 10
  reduction, 16
Leaves, polyprenols in, 52
Lewis blood-group determinants, synthesis, 72
Linoleic acid
  prostaglandin synthetase inhibition by, 250
α-Linolenic acid
  prostaglandin synthetase, inhibition by, 250
Lipidoses, brain, 172
Lipids
  in brain, 143
    exchange of, 164–167
  complex, brain, 151
  in desaturase complex, molecular organisation, 216, 217
  in glycan biosynthesis, 39–97
  glycosyl transferases and, 84
  metabolism in brain, 147–164
  in venous system, 141–179
  neural transmission in nervous system, 167–169
  peroxidation, in ageing brain, 171

Lipid transport, cholesterol esters and, 156
Lipofusin in aged brain, 171
Lipofusin particles in aged brain, 170
Lipolysis in isolation of particle-bound enzymes, 23
Liver
  fatty acid synthetase in, 128–129, 132
  pig, dolichols in, 51
  rat, biosynthesis of serum glycoproteins in, 74
    dolichol monophosphate in, 52
    mitochondria, glucoprotein biosynthesis in, 78
Luteolysin, 257, 258
Lysolecithin in multiple sclerosis, 174
Lysosomes in ageing brain, lipid peroxidation and, 171

Maleimide, *N*-ethyl, fatty acid desaturase inhibition by, 211
Malonyl-CoA ACP-transacylase, 120, 121
Malvalic acid biosynthesis, 212
Mammals, glycan biosynthesis, polyprenols and, 70–85
Mammary gland, fatty acid synthetase, 129–131
Man, prostaglandin metabolism in, 255
Mannosyl transfer
  in *Aspergillus niger*, 86
  in glycan biosynthesis in green plants, 87–89
  in yeasts and, 85
  in glycoprotein biosynthesis, 78–81
  retinol phosphate derivatives and, 83
Mass spectrometry—gas chromatography, in prostaglandin assay, 245, 246
Medullin, 239
Membranes
  enzymes in, 11
  halosulphatides in, 283–285
  reconstitution, 5
  undecaprenol bound to, 52
  working with, 4
Menstrual stimulants, 239
Metabolism
  of chlorosulphatides, 280–282
  connexion between different sequences, 202, 203
  of prostaglandins in man, 255
Metal chelates, fatty acid desaturase and, 211
Methyl sterol oxidase, 11, 20
Mevalonic acid,
  formation, 6, 7
  transformations, 7–9
Micobacterium phlei, fatty acid synthetase in, 131
Micrococcus lysodeikticus, peptidoglycan biosynthesis in, 58

Microsomes
  brain, phospholipid exchange in, 166
  enzymic processes of, 4
  processes, multi-enzymic synthesis of cholesterol and, 28–33
Mitochondria, enzymes from, 5
Mitochondrial membranes, glycoprotein biosynthesis in, 81
Molecular weight, polyprenols, 44
Monogalactosyl diglyceride in *Chlorella*, 196
Mouse myeloma tumour glycoprotein biosynthesis in, 80
Mucins, structures, 71, 72
Mucopolysaccharides
  retinol phosphates, in biosynthesis of, 82
  structure, 72
Multi-enzymic systems, 4, 5
  in fatty acid synthetases of yeast, 124–127
Multiple sclerosis
  lipids in brain in, 174
  steryl esters and, 156
Mycobacterium phlei, acyl-carrier protein from, 111
Myelin, lipid exchange in, 164
Myelination
  central nervous system, lipids and, 169
  lipid composition in brain and, 143
Myometrial strips
  human, *in vitro*, prostaglandins and, 256
Myometrium, human, prostaglandins and, 255

NAD(P)H, desaturase assay and, 216
Nerve cells, lipids in, 145–147
Nervous system
  lipids in, 141–179
  and neural transmission in, 167–169
Neural transmission of lipids in nervous system, 167–169
Neurofibrillary tangles in aged brain, 170, 172
Neurones
  brain lipids and, 163, 164
  prostaglandins and, 260, 261
Neurospora crassa, $\Delta 12$ desaturation of phospholipids by, 198
Nomenclature
  of halogenated sulphatides, 275
  of polyprenols, 42–44
  prostaglandins, 239–241
  unsaturated fatty acids, 183, 184
Noradrenaline inhibition of release by prostaglandins, 259
Nucleotides, reduced pyridine, in fatty acid desaturation, 210
Nutrition
  fatty acid desaturation control and, 220
  fatty acid synthetase and, 132

'O'-antigen determinents, 61
Obesity, acetyl-CoA carboxylase activity and, 108
Ochromonas danica, chlorosulphatides of, 275–282
Oleic acid
  biosynthesis, 203
  alternative pathway for, 214, 215
Oxidases, microsomal mixed-function, 29–31
Oxidation
  dimethyl sterols, 30
  fatty acid chain shortening by, 209
Oxidative demethylation of methyl sterols, 11
Oxygen
  aerobic desaturation of fatty acids and, 209
  in oleic acid biosynthesis, 203, 204
Oviducts, prostaglandins and, 257
Ovine submaxillary mucin, 71
Oxidation of prostaglandins, 254, 255

P-450, isolation, 23
Pain, prostaglandins and, 263
Palmitic acid
  biosynthesis, 100
  desaturation of, 199
Palmityl-CoA, lipogenic enzyme inhibition by, 107
Particles, enzymes bound to, in sterol biosynthesis, 19–24
Parturition, prostaglandins and, 255, 256
Pathways
  sterol biosynthesis. 3–19
    techniques for investigation, 13–15
Peptidoglycan, biosynthesis, 57–61
Peroxidation of lipids, in ageing brain, 171
Phaseolus aureus, glycan biosynthesis in, 88
o-Phenanthroline, fatty acid desaturase inhibition by, 211
Phenobarbital, microsomal enzyme induction by, 29
Phenols, prostaglandin biosynthesis and, 249
Phosphatases, prenol monophosphates and, 49
Phosphatidyl choline
  in *Chlorella*, 196
  in plants, 205
Phosphatidyl ethanolamine biosynthesis, 149
Phosphatidyl glycerol in *Chlorella*, 196
Phospholipase A, prostaglandin synthesis and, 250
Phospholipases, *p*-nitrophenol glucuronyl transferase activity and, 85
Phospholipids
  brain, biosynthesis, 147
  exchange of, 166, 167

4-Phosphopantotheine-peptides, primary sequence of, 115
Phosphorylation of polyprenols, 46
Pigs, liver, dolichols in, 51
Ping Pong Bi Bi kinetic scheme for malonyl transacylation, 120
Pisum sativum, mannolipid formation in, 88
Plants
  glycan biosynthesis in, polyprenols and, 87–89
  polyprenols in, 51
Platelet aggregation, prostaglandins and, 265
Polymannan biosynthesis, 56
Polyphosphoinositides in neural transmission, 168
Polyprenol diphosphate phosphohydrolase, 49
Polyprenol monophosphate mannose in polymannan biosynthesis, 56
Polyprenol phosphate glycosides, 46–49
Polyprenol phosphates, 46–49
Polyprenols
  chemistry, 41–50
  distribution, 50–52
  glycan biosynthesis, in green plants and, 87–89
  in yeasts and fungi and, 85–87
  mammalian glycan biosynthesis and, 70–85
  radioactive, 49, 50
  —hexahydro-, 43
  —*exo*-methylene-hexahydro-, 43, 46
Prenol phosphokinase, 53
Prephytoene pyrophosphate, 9
Presqualene pyrophosphate, 9
Prostaglandin A, biosynthesis, 251
Prostaglandin A isomerase, 251, 252
Prostaglandin C, 252
Prostaglandin dehydrogenase, 253, 254
Prostaglandin 13,14-reductase, 254
Prostaglandins, 237–270
  adrenergic nerve terminals and, 258–260
  biosynthesis, 248–252
    inhibition of, 259–260
    mechanism, 250–251
  central nervous system and, 260–262
  distribution, 246
  effect on female reproductive tract smooth muscle, 255–257
  estimation, 243–246
  extraction and separation, 241–243
  identification, 246–248
  metabolism and fate, 252–255
Prostaglandin synthetase, 248, 249
Prostanoic acid, 239
Proteins
  cyanide-sensitive, in fatty acid desaturation, 212
  glucosylation, 75, 76

Proteins *continued*
  non-catalytic, microsomal multi-enzymic system and, 32, 33
  sterol-carrier, 33
Proteolysis in isolation of particle-bound enzymes, 23
Prosthetic group turnover in acyl carrier protein, 117, 118
Psychosine in cerebroside biosynthesis in brain, 161

'Quaking' mouse mutant, sphingolipid deficiency and, 151

Radioimmunoassay of prostaglandins, 244
Rats, liver, dolichol monophosphate in, 52
Regulatory effects of fatty acids on liver microsomal desaturation, 206
Retinol phosphate derivatives, 83, 84
Retroconversion of unsaturated fatty acids, 209
Ristocetin, peptidoglycan biosynthesis, inhibition by, 61

Saccharomyces cerevisiae, glycan biosynthesis in, 85
Saturation, polyprenols, 45, 46
Senile plaques in ageing brain, 172
Separation, prostaglandins, 241–243
Sheep
  central nervous system, lipids in, 144
  luteolysin, prostaglandin $F_{2\alpha}$ and, 258
Shigella flexneri, polyprenol monophosphate glucose in, 64
Sitosterol biosynthesis, 16
Skin, sterol biosynthesis in, 18
Solanesol, 43
Solvent extraction of prostaglandins, 241, 242
Spadicol, 43
Spinach
  acyl-carrier protein from, 111
  amino acid sequence in, 114
Spinach leaf chloroplasts acetyl-CoA carboxylase, 104
Spinal cord
  prostaglandins and, 261
  sheep, lipids in, 144
SQ-10,591, sterol biosynthesis and, 21
Squalene
  biosynthesis, 9
  in brain, 153
  formation, 10
  reactions, 9
Squalene oxidase, 9
Squalene oxide, 9
  formation, 20

INDEX

Squalene-2,3-oxide formation, 10
Squalene oxide cyclases solubilisation, 22
Squalene-2,3-oxide cycloartenol cyclase, 9
Squalene-2,3-oxide sterol cyclase, 9
Squalene synthetase, 9, 22
Staphylococcus aureus
    peptidoglycan biosynthesis in, 57, 58
    undecaprenol in, 49, 51
Staphylococcus lactis, teichoic acids in, 66
Staphylococcus newington
    'O'-antigen determinant of, 61
    polyprenol phosphate biosynthesis in, 55
Staphylococcus typhimurium
    antigen lypopolysaccharide of, 61
    octasaccharide from, 63
Starch in plant cells, 89
Stearic acids, desaturation of, 199–201
Stearyl coenzyme A, 31
Sterculene, 1,2-dihydroxy, fatty acid synthesis inhibition by, 212
Sterculic acid biosynthesis, 212
Sterculyl alcohol, fatty acid synthesis by, 212
Stereochemistry
    in biosynthesis of polyprenols, 52
    in fatty acid desaturation, 221–223
    polyprenols, 44, 45
Steroids, fatty acid desaturase inhibition by, 212
Sterol carrier protein in brain, 153
Sterol esters in brain, 155–157
Sterols
    biosynthesis, control of rate of, 24–28
        enzymes of, 1–37
        particle-bound enzymes in, 19–24
        pathways, 5–19
    esterification of intermediates, 18
    protein carrier, 33
    in skin, 18
Subtilisin Carlsberg, proteolytic modification of BCCP by, 104
Sugars, phosphorylated, fatty acid synthetases and, 133
Sulphatides
    halogenated, 271–286
    in membranes, 283–285
Sympathomimetamines, interaction between prostaglandins and, effect on effector cells, 260
Synthetases
    in peptidoglycan biosynthesis, 57–60
    Staphylococcos newington, 'O'-antigen determinant of, 62
    for teichoic acids, 67–69

Teichoic acids, 66–70
Temperature regulation, prostaglandins and, 263

Terpenoids, biosynthesis, mevalonic acid in, 6
Testicular tissue, control of rate of sterol biosynthesis in, 26
Thiols, desaturases and, 211
Thyroid, prostaglandins and, 265
Thyroxine, fatty acid desaturation control and, 220
Tomatoes, roots, mannolipid biosynthesis in, 88
Torulopsis utilis
    desaturase activity at various temperatures, 210
    dilinoleoyl—PC biosynthesis in, 198
Transacylases, fatty acid desaturation and, 208
Transcarboxylase component in palmitate biosynthesis, 104
Tricarboxylic acids, as allosteric regulators for acetyl-CoA carboxylase, 105
Tri-iodothyronine, fatty acid synthetase control and, 133
Trimarinol, sterol biosynthesis and, 21
Triparanol, sterol biosynthesis and, 21
Tritium, in desaturase assay, 216

Undecaprenol
    bacterial wall glycan biosynthesis and, 55–70
    biosynthesis, 53
    in *Staphylococcus aureus*, 49
Uterus, prostaglandins and, 257

Vancomycin, peptidoglycan biosynthesis, inhibition by, 61
Vesiglandin, 239

Water, permeability, of tissues, prostaglandins and, 264
Weaning
    acetyl-CoA carboxylase activity and, 108
    fatty acid synthetase and, 132
Wheat germ enzyme, 104
White matter, lipids in, 144

Yeast
    acyl-carrier protein from, 111
    amino acid sequence in, 114
    glycan biosynthesis in, polyprenols and, 85–87
    multi-enzyme complexes, fatty acid synthetases in, 124–127
    polyprenols in, 51
Yeast squalene synthetase, solubilisation of, 22

DATE DUE